Labor Forces and Landscape Management

Hiroyuki Shimizu · Chika Takatori
Nobuko Kawaguchi
Editors

Labor Forces and Landscape Management

Japanese Case Studies

Springer

Editors
Hiroyuki Shimizu
Nagoya University
Nagoya, Aichi
Japan

Nobuko Kawaguchi
Nagoya University
Nagoya, Aichi
Japan

Chika Takatori
Nagoya University
Nagoya, Aichi
Japan

ISBN 978-981-10-2277-7 ISBN 978-981-10-2278-4 (eBook)
DOI 10.1007/978-981-10-2278-4

Library of Congress Control Number: 2016948263

Printed on acid-free paper

This Springer imprint is published by Springer Nature
The registered company is Springer Nature Singapore Pte Ltd.
The registered company address is: 152 Beach Road, #22-06/08 Gateway East, Singapore 189721, Singapore

The original version of the book was revised:
For detailed information please see erratum.
The erratum to this book is available at
10.1007/978-981-10-2278-4_26

Preface

The concept of "landscape management labor accounts" was developed through the on-site research experiences reported in *Basic and Clinical Environmental Approaches in Landscape Planning* (Shimizu et al. 2014). *Basic and Clinical Environmental Approaches in Landscape Planning* was published as a result of the Global Center of Excellence (GCOE) Program "From Earth System Science to Basic and Clinical Environmental Studies," funded by the Ministry of Education, Culture, Sports, Science and Technology starting in 2009 (Nagoya University Global Center for Excellence Program 2013). In the process of publishing, we were keenly aware of the importance of clinical environmental approaches in landscape planning, which is "diagnostic" by analyzing how the earth–life system interacts with human society and "treatment-based" by providing technological or regulatory solutions to environmental problems. Moreover, it became clear that local landscapes are faced with serious management challenges in the future, with a rapidly aging and shrinking population. Thus, gaining a holistic understanding of how much and what kind of management is needed across various landscapes today and into the future is critical. Landscape management labor accounts is one methodology for grasping them.

Contents of This Book

This book is divided into four parts.

Part I

In this part, the current state of landscape management in Japan is discussed and the basic concept of "landscape management labor accounts" is proposed. In Chap. 1, the background, concept, and perspective of this research are described. Recently, landscape management has been recognized as important to keeping landscapes sustainable. This is particularly true in mature countries such as Japan, which is grappling with a population that is both rapidly aging and shrinking, where it is becoming more and more difficult to manage landscapes. Some landscapes must be returned to nature while other landscapes must be maintained appropriately by human labor. Landscape management labor accounts are a new tool to evaluate the amount and density of labor forces needed to maintain different landscape types in an integrated manner, including urban gardens, farmlands, and forests. Chapter 2 shows typical landscape types extracted by principal component and cluster analysis, as well as their distribution patterns and characteristics. Seven basic landscape types are described, namely, the paddy field type, urban paddy field mixed type, paddy field Satoyama type, urban type, other field Satoyama type, nature type, and other field type. These landscape types have a strong relationship with Japanese geography. The basic landscape types are compared with land use and population changes from the past to the present. Finally, decreasing landscape management sustainability is described by comparing current population with projected population levels in 2050.

Part II

In this part, changing policies, research, and contemporary issues in landscape management are discussed, covering various landscapes including urban, urban periphery, agricultural, plantation forest, water, and coastal areas. In urban areas, the increase of sealed parking lots and unmanaged residential properties degrades the urban ecological environment, and inadequate landscape management budgets threaten the quality of parks and street trees. Unplanned urban sprawl in farmlands on the urban periphery causes chaotic landscapes that degrade land, as well as unmanaged secondary forests that lead to ecosystem disservices, such as loss of aesthetic value and damage from insects and animals. Land improvement projects in agricultural areas led by the national government bring production and efficiency but harm plant and animal habitat and reduce biodiversity. Additionally, many farmers are elderly and are expected to soon stop farming, leading to the rapid abandonment of farmland. In water landscapes, well-developed artificial structures, such as high levees and straightened rivers, can mitigate high-probability disasters but bring about habitat loss and reduced biodiversity. However, naturalized rivers, vegetated channels, and temporary ponds, as in alluvial riverscapes, need much labor management. In coastal landscapes, most parts of banks are covered by concrete, tidal flats are converted to artificial lands, and natural landscapes are mostly lost. Additionally, natural environments in the oceans, such as seaweed beds, are degrading rapidly. To recover from the degraded condition discussed above, a sufficient labor force is needed.

Part III

In this part, landscape labor accounts are calculated specifically for several case studies in the Chubu metropolitan area of Japan, the central part of the island of Honshu. Landscape elements, landscape units, and neighborhood landscape complex units comprise the calculation base for labor accounts and densities. Labor accounts and labor densities in various scales are also defined. As a case study in an urban landscape, labor accounts and densities in Nagoya City are considered, including public parks, roadside trees, private gardens, and kitchen gardens. Next, the labor accounts and densities of the secondary urban forest in Fujimaki Town in Nagoya City, where local residents are engaged in secondary forest management, are examined as an urban forest neighborhood landscape complex unit. Asami District in Mie Prefecture is examined as an example of a flatland farmland neighborhood landscape complex unit. Nyu District, which is managed by a farming cooperative, is studied as a paddy field Satoyama neighborhood landscape complex unit. Kayumi District, where tea plantations are developed, is examined as an example of another field Satoyama neighborhood landscape complex unit. Shima District, which is famous for its traditional women divers, is examined as an example of a coastal neighborhood landscape complex unit. Finally, the integration of landscape labor accounts, from the small scale of neighborhood landscape complexes to the large regional scale of the Chubu metropolitan area, is also considered.

Part IV

In Part IV, planning perspectives related to each landscape type are considered and discussed. Using the concept of landscape labor accounts and densities, problems related to the management of each landscape type may be seen more clearly and appropriate consideration with respect to sustainable management can be described.

In planning for existing urban areas, considering the fact that further development of new parks through land acquisition is financially difficult, it is important to maintain and increase green spaces within private properties and streets. This may be accomplished through the new integrated urban planning system, with a high priority on the ecological benefits of green spaces in existing urban areas. In urban periphery planning, urban–rural mixed land use and depopulation are serious challenges. Land-use realignment under the idea of "living in the garden" may be key in the future. In forest planning, improvements in the ecological, economic, and social aspects of unmanaged plantation forests are the most important issues, and the development of the system of "foresters" as forest managers is discussed. Landscape management labor accounts can contribute to recognizing unmanaged areas, estimating needed manpower, and anticipating costs. Regarding inland water systems, the process of ecosystem degeneration is reviewed and visions and policies to rehabilitate them are discussed. The most important issue in rural planning with an aging and shrinking society is the possibility of continued farming and survival of the local community. The development of corporate or cooperative farming and the maintenance of collaborative works of individual farmers or residents are discussed.

In coastal planning, the planning connectivity from forest areas to coastal areas and the development of planning methodologies based on river basins will be more important moving forward. Landscape management labor accounts may contribute to the establishment of such kinds of thinking. The relationship between landscape management and ecosystem services is also discussed, with a case study in Nagoya City. Finally, perspectives related to sustainable labor force management of landscapes are considered.

Nagoya, Japan

Hiroyuki Shimizu
Chika Takatori
Nobuko Kawaguchi

Acknowledgments

The Scientific Research of the Japan Society for the Promotion of Science supported this study. In addition, without scholarship donations by Sansei Technologies, Inc., KYB Corporation, Urban Formative Architect & Associates, and IIJIMA Structural Design Office, this research could not have been completed.

For academic support, firstly we would like to thank the Leibniz Institute of Ecological Urban and Regional Development in Dresden (IOER), Germany. The idea for landscape management labor accounts first came up during a stay at the IOER during a discussion on landscape planning. We would also like to thank Vivek Shandas of Portland State University and Claudia Cassatella of Polytechnic University of Turin, who provided us with important comments on our study. We would like to again thank all of our kind supporters

Contents

Part III Case Studies

Part IV Perspectives

Part I
Introduction

Chapter 1
Introduction

Hiroyuki Shimizu

Abstract Many mature countries, including Japan, are confronting population decline and aging societies. Such a radical decrease of the population could adversely affect the sustainability of landscape management, as the issue is exacerbated by the lack of skilled local labor forces in landscape management. It is important, therefore, to develop methods for providing or re-providing appropriate labor for the sustainable management of landscapes, and how to sustainably reduce or shrink landscape management areas according to the available labor. In this chapter, an original concept of landscape management labor accounts is proposed, and the background, fundamentals, and perspective is explained.

Keywords Landscape management · Land use · Labor force · Labor accounts · Sustainable · Landscape complex units

1.1 Introduction: Background, Concept and Perspective on the Research

Many mature countries are confronting population shrinkage and aging societies. Japan is one of such mature countries and the population will decrease from 126,597,000 in 2015 to 86,737,000 in 2060 (National Institute of Population and Social Security Research 2012). It will be a significant reduction of 31 % (Fig. 1.1). Such a radical decrease of the population might affect the sustainability of land or landscape management. Even today many farmland and forest abandonments have occurred because of lack of management labor in depopulated areas. This phenomenon might accelerate in the future. It is a concern that the degradation of landscape management may lead to the deterioration of ecosystem services and biodiversity.

H. Shimizu (✉)
Graduate School of Environmental Studies, Nagoya University,
Furo-cho, Chikusa-ku, Nagoya 464-8601, Aichi, Japan
e-mail: shimkankyou@msn.com

H. Shimizu et al. (eds.), *Labor Forces and Landscape Management*,
DOI 10.1007/978-981-10-2278-4_1

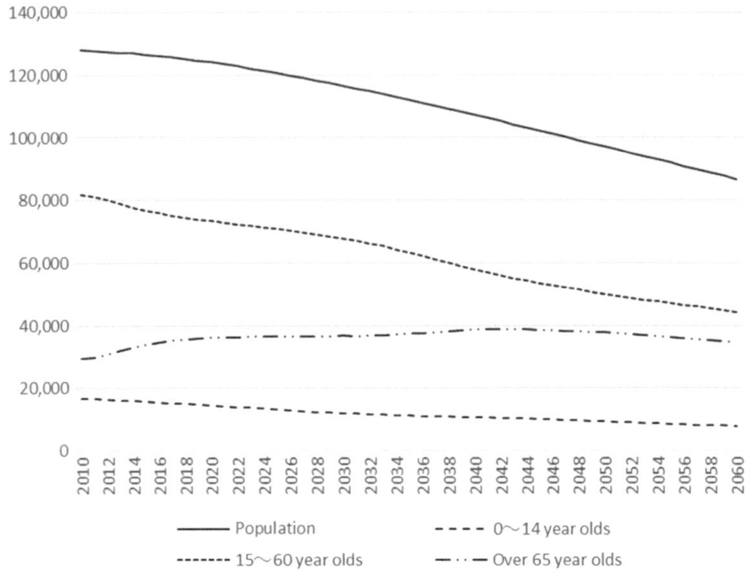

Fig. 1.1 Population estimation toward 2060 (National Institute of Population and Social Security Research 2012)

There are many reports on the facts that the expansion of human activities has led to deterioration of global environments. The *Fifth Global Environment Outlook* (*GEO-5*) (United Nations Environment Programme 2012) provides an overall view of these issues on land use change and management. The basic point of view of GEO-5 is that "changing climate patterns, economic globalization, population growth, increasing use of natural resources and rapid urbanization are putting pressure on terrestrial ecosystems as never before." There is another important report on land management, which is "Land Management (Hindmarch and Pienkowski 2000)." The report starts from the point of view that "Europe's countryside has undergone profound and damaging changes over recent decades, largely as a result of intensive, high-input land management practices." This report is very important, because it clarified the fact that there is a hidden cost behind intensive land use or intensive farming, which leads to land degradation such as widespread pollution and destruction of soil.

But there has been little attention paid to unsustainable landscape damage under a depopulating situation. Indeed there are strong pressures on terrestrial land caused by excessive human intervention. It is also important to focus on the fact that there are countries or regions where people worry about the lack of labor that can manage their land sustainably.

In Japan there are many beautiful countryside cultural landscapes named "Satoyama", which consist of paddy fields, croplands, secondary forests and small dwelling sites in sloped mountainous areas. These Satoyama landscapes have been maintained with sustainable collaboration of human beings and nature for over a thousand years. There is also another beautiful type of landscape in coastal areas with traditional half fishery and half agricultural lifestyles, which can be also named as one type of a Satoyama landscape in a broad sense. These traditional Satoyama landscapes are confronting a strong crisis of shrinking or disappearance. Depopulation and aging are in progress, because young people have been leaving to big cities in pursuit of contemporary culture and money. In the near future many beautiful Satoyama landscapes might disappear because of the lack of management labor for the sites.

Not only in the countryside but also in urban areas, there is also a phenomenon of land degradation because of shrinking. Depopulation and aging are also in progress at a rapid speed. In small and middle-housescale regional cities, there are many vacant houses with no appropriate care, which causes troubling landscape degradation.

There were originally few public green spaces such as parks in Japanese cities. The provision of green spaces depends much on private green sites such as trees and gardens around houses. But individual residential sites tend to be divided into smaller ones with fewer green spaces by inheritance, and the amounts of green spaces in urban areas have decreased consequently in total. Additionally, green management of private sites has become difficult because of aging, which accelerates the decrease of urban green spaces. Additionally public parks are confronting the difficulty of getting appropriate management costs, because of shrinking budgets of municipalities by aging, depopulation and moving out of industries.

These issues are basically caused by lack of local labor forces for appropriate landscape management. Therefore, in order to tackle these issues, it is very important to consider and develop methods on how to provide or re-provide appropriate labor for the sustainable management of landscapes or how to reduce or shrink landscape management areas sustainably according to the feasible labor.

For this purpose, it is necessary to count up labor forces needed in a targeted area, and distinguish places which must be preferentially managed and which could be abandoned or returned to nature for the appropriate distribution of available labor forces. For this purpose, a new concept of *landscape management labor accounts* will be introduced in this book. It is the number of necessary laborers, or manpower, for the ecological management of a certain type of local landscape.

We will examine all types of landscape management labor accounts, such as for paddy fields, other agricultural fields, forests, coasts, urban parks, roadside trees and private gardens in the case studies of Part 3. We will also define *landscape complex units* as a basic community for integrated local landscape management, in which landscape management labor accounts will be integrated into a complex one. Thus, we can learn the amount of excess, or deficiency of, landscape management labor and can value possibilities of sustainable landscape management.

It can easily be imagined that appropriate labor forces for appropriate landscape management are insufficient in shrinking districts. In these cases, it might need to be considered how to withdraw appropriately in the future. Or if there are some kinds of important man-nature collaborating landscapes, we might consider how appropriate labor should be imposed from outside professionally or by a voluntary basis.

And some kinds of economic support, such as direct payment systems, might be considered. For these purposes landscape management labor accounts can provide a foundation for consideration.

By using GIS technologies we can understand the spatial distribution patterns of landscape labor accounts. Fortunately the National Government of Japan provides much national spatial statistic information such as population, land uses or economics and so on. By reviewing these spatial statistics we can learn about the relations between populations shrinking or growing, land use changes, landscape type changes and landscape labor accounts. Especially it is very important to know the bias between the needs of landscape management labor and agglomeration of population. How much of a labor force can be supplied from core cities to their surrounding countrysides? Landscape labor accounts can help to answer this question.

This book consists of four main parts. The first part is concerned with forming concepts, definitions, and overviews. Issues on landscape management and land use and population changes are dealt with in the second part. The third part consists of case studies on landscape management labor accounts. Major landscape types chosen for case studies include urban, flatland farmlands, Satoyama, and coastal neighborhoods. In the last part of this section, integration methods to develop landscape management labor accounts on different scales are considered. The fourth part of the book is a detailed exposition of contemporary trial to solve the issues on the landscape management for the future in the field of urban, rural, forest, river and coastal planning. Also connection to ecosystem service studies and perspective of development of landscape management labor accounts to world landscape management researches are discussed.

1.2 Development of the Concept and Definitions

1.2.1 Landscape Management Labor Accounts

Activities by human beings on Earth make certain impacts on it, and causes land consumption and degradation. To prevent such damage, appropriate landscape management is needed. For the appropriate management, appropriate costs and labor must be considered. These appropriate costs and labor provide benefits to human lives. But these costs and labor are normally invisible and tend to be economically externalized. Such externalized costs and labor should be internalized or compensated by a socio-economic system.

Figure 1.2 shows our starting concept of the landscape management study. In this research we will focus on landscape management labor instead of costs. Treatment of costs is very complex and difficult, because costs include various kinds of contents, such as labor, energy, machinery, transportation and agrichemical materials and so on, and they cannot be easily calculated. On the other hand labor

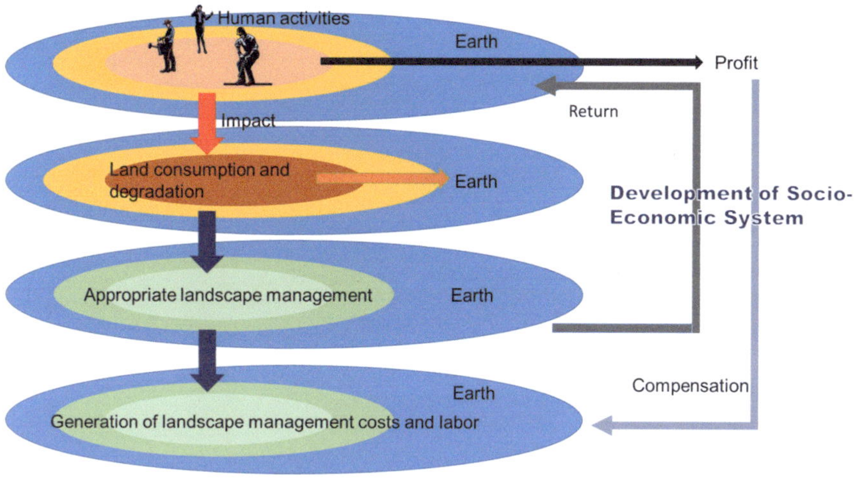

Fig. 1.2 Starting concept of landscape management costs and labor

can be easily grasped or accounted for by observation of landscape management activities or interviews with residents. And in a depopulating society, the landscape management crisis is caused to a great extent by the lack of labor forces. This is a positive reason why we focus on labor instead of costs.

Figure 1.3 shows the more detailed development of the concept of "Landscape Management Labor Accounts". There are many kinds of landscapes from very densely built-up landscapes, less densely built-up landscapes, countryside Satoyama and forest landscapes to natural land landscapes. Economic productivity by property management and human activity is high in the very densely built-up landscapes and becomes lower according to the naturalness of the landscapes. The labor force needed for ecological landscape management might be high in less densely built-up landscapes, countryside Satoyama and forest landscapes, because semi-natural environments in such kinds of landscapes must be maintained ecologically by human labor. But many parts of the labor remain externalized. On the contrary, the economic returns by human activities in the very densely built-up landscapes can be easily visible and internalized. But some parts of these returns or benefits are due to the externalized activities in less densely built-up landscapes or countryside landscapes. Benefits from harvests, tourism, water supply and mitigation of climate can be included. Some of the benefits in very densely built-up landscapes should be redistributed in countryside landscapes. Also the population in very densely built-up landscapes is higher, and there is a possibility for substitution of the shortage of management labor in countryside landscapes on a professional or volunteer basis. Landscape management labor accounts can serve as the method for counting the amount of such kinds of compensation. It is also very important to change the externalized economy to an internalized one by revitalization of the environment for friendly human activities especially by the efforts of residents or residential organizations.

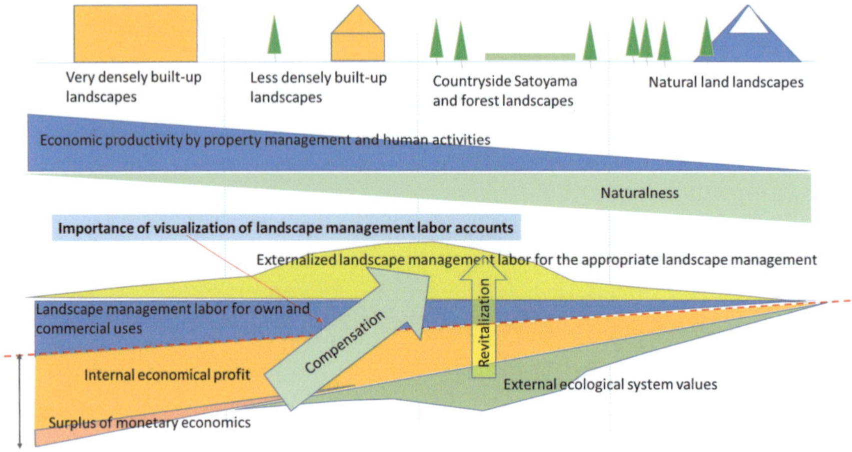

Fig. 1.3 Detailed concept of landscape management labor accounts

Figure 1.4 shows the categories of labor which should be included in landscape management labor accounts. We set three main categories, i.e., labor for creation and maintenance of land forms, labor for creation and maintenance of ecological services and biodiversity and labor for creation and maintenance of ecological cycles of natural and irrigated water. These categories can be divided into two categories respectively, i.e., initial labor and annual labor. Creation of forests or farmland shapes might be included in the first category, maintenance of forests by pruning or of farmland by cultivation might be included in the second category. The creation and management of irrigation channels belongs to the third category.

Table 1.1 is a table of labor categories according to ecological effects of labor. We divided the labor into three main categories, i.e., *ecological landscape management labor*, *non-ecological landscape management labor* and *outside landscape management labor*. The ecological landscape management labor is divided into two

1) Labor for creation and maintenance of ecological land forms

   1-1 Initial labor for a certain ecological land form
   1-2 Annual maintenance labor for a certain ecological land form

2) Labor for creation and maintenance of ecological services and biodiversity

   2-1 Initial labor for creation of ecological services and biodiversity
   2-2 Annual maintenance labor for ecological services and biodiversity

3) Fundamental labor for creation and maintenance of an ecological cycle of natural and irrigated water

   3-1 Initial investigation labor for creation of an ecological cycle of natural and irrigated water
   3-2 Annual maintenance labor for an ecological cycle of natural and irrigated water

Fig. 1.4 Categories of landscape management labor

Table 1.1 Labor categories according to ecological effects and externalization-internalization of labor

Type of labor	Parts of labor	Contents	Examples
Ecological landscape management labor	Fundamental landscape management labor	Fundamental landscape management labor is needed for ecological and sustainable maintenance and conservation of a fundamental land form of a targeted landscape.	Creation and maintenance of paddy field's ribs, stone walls of terraced fields etc.
	General landscape management labor	General landscape management labor is needed generally for ecological and sustainable maintenance and conservation of a targeted landscape.	Mowing, plowing, pruning, seeding, planting, reaping, fertilization, cleaning of irrigation channels etc.
Non-ecological landscape management labor	Non-ecological landscape management labor	Use of non-ecological landscape management labor is inevitable under contemporary social conditions, but has negative ecological impacts.	Disinfection, concrete paving of channels, etc.
Outside landscape management labor	Outside landscape management labor	Outside landscape management labor is done indirectly in other places outside ecological fields.	Maintenance labor of machines, traffic and transportation labor, etc.

sub-categories, i.e., *fundamental landscape management labor* and *general landscape management labor*. Fundamental landscape management labor is labor which is needed for ecological and sustainable maintenance and conservation of a fundamental land form of a targeted landscape, such as creation and maintenance of paddy field's ribs, stone walls of terraced fields and so on. General landscape management labor is labor which is needed generally for ecological and sustainable maintenance and conservation of a targeted landscape, such as mowing, plowing, pruning, seeding, planting, reaping, fertilization, cleaning of irrigation channels and so on. Non-ecological landscape management labor is labor which is inevitable to use to some extent due to contemporary social conditions, but has negative ecological impacts. In these categories, uses of agrichemicals and concrete paving of channels are included. Outside landscape management labor is labor which is done indirectly in other places outside ecological fields. In this book, ecological landscape management labor is mainly handled. Non-ecological landscape management labor can be evaluated as substitute fundamental landscape labor accounts without using non-ecological treatments. We basically do not treat outside landscape management labor because of the difficulties of collecting data, which is a research issue of the future.

Figure 1.5 shows the image of appropriate landscape management and compensation treatments. There is a curve of ecosystem service levels, which is highest

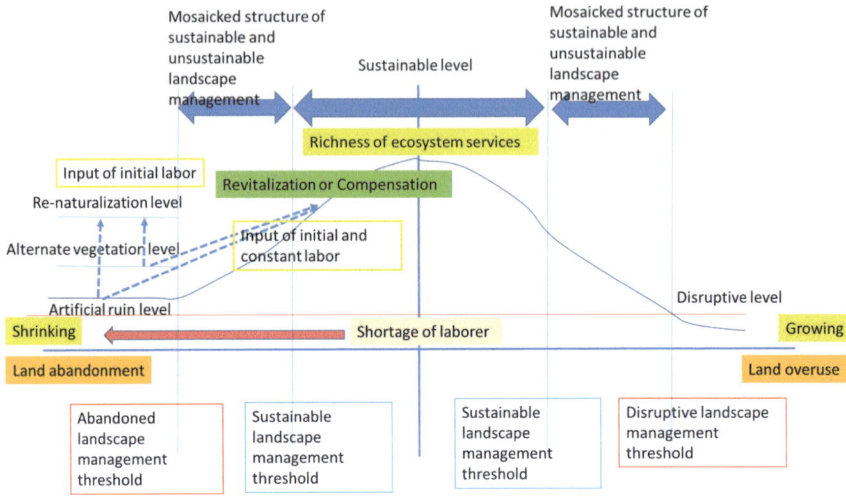

Fig. 1.5 Image of appropriate landscape management and compensation treatments

in the harmonized situation of ecological environments and adequate supporting human labor. These balances decrease according to either the excess or lack of land use pressures in growing or shrinking situations. There might be two kinds of thresholds, namely a sustainable landscape management threshold and an abandoned or disruptive landscape management threshold. All levels of management structures are sustainable between sustainable landscape management thresholds in shrinking and growing sides. Outside zones of the threshold of abandoned or disruptive landscape management, there are mosaics of sustainable and unsustainable situations. If land use levels or pressures reach beyond these second thresholds, then they go to land abandonment or land overuse. In shrinking situations labor is shortened for sustainable landscape management, which creates artificial ruins or undesired alternative vegetation. There are basically two ways of recovery, namely re-naturalization in an abandonment situation or return to a sustainable level. Since re-naturalization forms a ruin or undesired alternate vegetation, appropriate initial labor must be input. By the recovery to a sustainable level through revitalization or compensation treatments, initial labor and constant labor are needed. These might be compensated by some socio-economic systems. This additional labor will be estimated and proposed in the following chapters.

1.2.2 Neighborhood Landscape Complex Unit

Now we will propose the concept of a neighborhood landscape complex unit. Landscape management labor must be considered by combining with community systems, because local landscape management is carried out by public and private organizations, companies or individual residents. Therefore we set up the concept of a neighborhood landscape complex unit. Figure 1.6 shows the typical neighborhood landscape complex units. Figure 1.6a shows a typical Satoyama neighborhood landscape complex unit, which consists of neighborhood landscape elements, such as farmlands, residential sites, surrounding secondary forests and irrigation channels and ponds for supplying water. We must examine how these elements are supporting each other for the sustainable management of lands in a complex, and how much labor is needed for the maintenance of a complex unit as a total community system, because much labor is supplied as *collaborative works of the community*. Besides Satoyama neighborhood landscape complex units, there are other landscape complex units such as a coastal neighborhood landscape complex unit (Fig. 1.6b), a flatland farmland neighborhood landscape complex unit (Fig. 1.6c) and an urban neighborhood landscape complex unit (Fig. 1.6d). According to the difference of community structures, activities and labor supplying systems for the management of landscapes might differ from each other. Thus different neighborhood landscape management complex units should be examined.

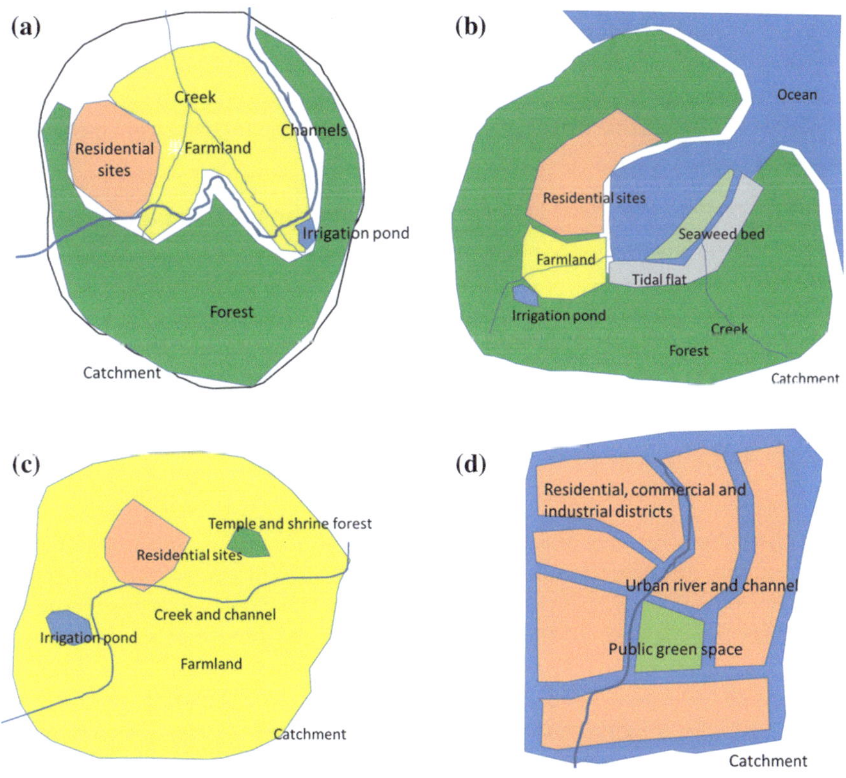

Fig. 1.6 The typical neighborhood landscape complex units. **a** Satoyama neighborhood landscape complex unit. **b** Costal neighborhood landscape complex unit. **c** Flatland farmland neighborhood landscape complex unit. **d** Urban neighborhood landscape complex unit

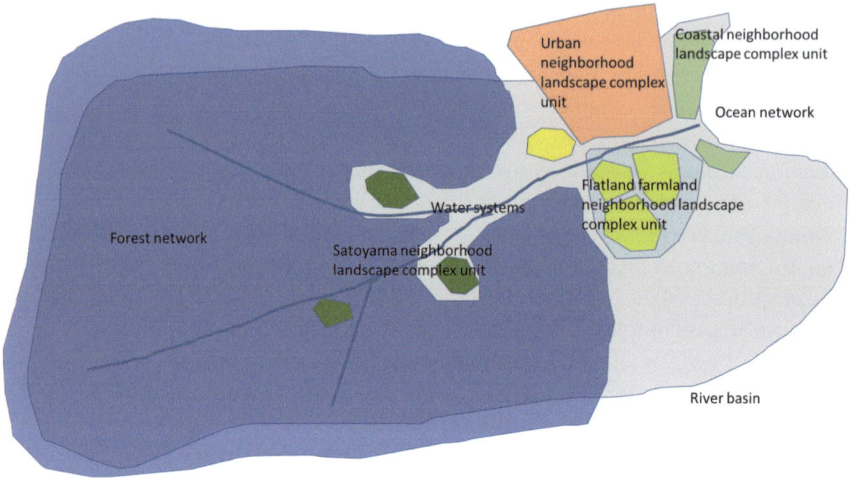

Fig. 1.7 Regional distribution of neighborhood landscape complex units

Figure 1.7 shows the regional distribution of the above mentioned neighborhood landscape complex units in concept. Landscape management labor must also be examined not only at community levels but also at regional levels because the importance of landscape management and planning at the regional level has become higher in recent years.

Figure 1.8 shows a diagram of a socio-economic system of a neighborhood landscape management complex unit. The action of landscape management is carried out by many working bodies, such as land owners, social organizations, community organizations, public bodies, teams of specialists or individuals. The balance of these organizations and individuals is very important to the sound management of complex units and this balance must be examined.

1.2.3 Working Style Types

Regarding the socio-economic system of a neighborhood landscape management complex unit, a concept of *working style types* must be developed. For instance, in order to produce high quality timber safely and efficiently, highly skilled professional laborers are needed for plantation forest management. Also for the management of high quality Japanese gardens, the input of skilled labor forces of professional plantsmen is inevitable. This must be a *professional working style*. There is another working style with professional skills. Many Japanese rice farmers are part-time farmers and work on the weekends, while they work in other places such as factories or shops on the weekdays. This working style can be called a *weekend working style*. On the other hand, homeowners who are amateur gardeners might work on holidays to manage their house gardens. This can be called an

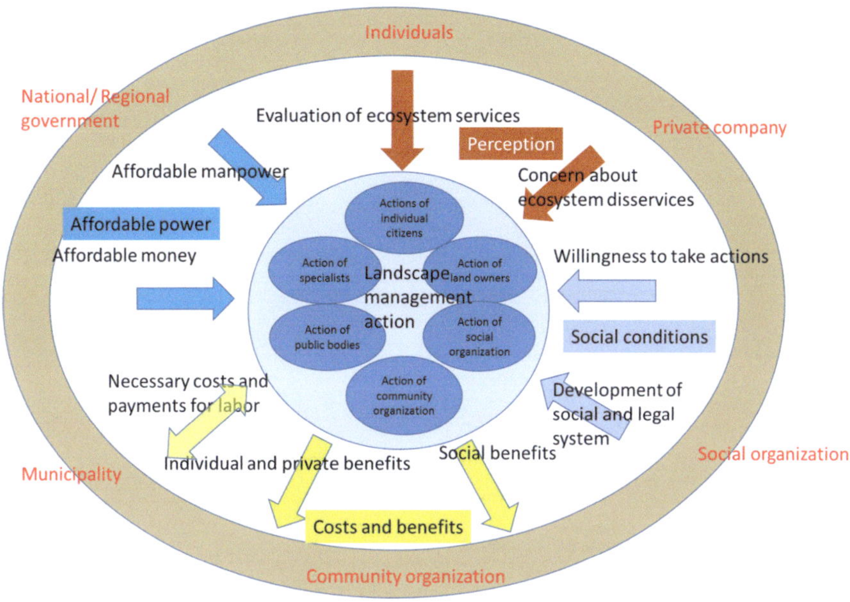

Fig. 1.8 Socio-economic system of a neighborhood landscape complex unit

amateur working style. Otherwise, there are some people who will work in the fields for their health management or self-sufficiency. They don't work so hard but stay in the fields for a relatively long time. This can be called a *self-realization working style*. In this book, it will not be evaluated as to which working style is superior. Rather, it is important to think about an appropriate mixture of different working styles according to the characteristics of a targeted landscape complex unit. The working styles must fit the community structure of a landscape complex unit.

1.2.4 Importance of Dealing with All Kinds of Landscape Elements on the Same Level

Figure 1.9 is a photo montage which shows different kinds of landscape elements existing in contemporary Japan. In an urban area, there are parks with lawns and trees, roadside trees, greens on canal banks, hedges and green mosaics in residential districts. In peri-urban areas, there are many allotment gardens. Besides, some houses continue onto green foothills. In countrysides there are many farmland landscapes with paddy fields, orchards and secondary forests. In hilly districts' paddy fields, forests and residential sites there exist Satoyama landscapes. Forests also have many different features, such as matured plantations, abandoned plantations, bamboo forests and so on. In coastal areas there are many wetlands, which provide rich biodiversity.

Fig. 1.9 Different kinds of landscape elements in Japan

These are all important landscape elements. In order for these landscapes to remain as they are, it is necessary to manage them appropriately, for which an appropriate mixture of working styles is needed. Also they must be connected to each other in an integrated network as hubs and links. But nowadays these elements are managed separately in a different manner and by different management bodies under different policies.

The concept of landscape management labor accounts aims also to provide an identical horizon for creating integrated academic and political perspectives on sustainable and efficient management of all kinds of landscapes.

1.2.5 Balance Between Biodiversity, Productivity and Labor Efficiency

There is another important viewpoint regarding landscape labor accounts. It is the balance between biodiversity, productivity and labor intensity. It is usual to think that human intervention causes biodiversity degradation. This may not necessarily be the case, for instance, in traditional paddy field farming. Natuhara (2013) pointed out the richness of biodiversity of traditional paddy fields as substitutes for natural

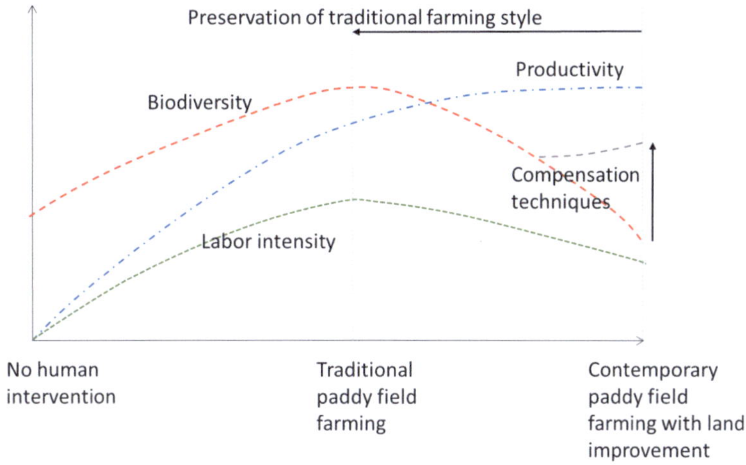

Fig. 1.10 Balance between biodiversity, productivity and labor intensity

wetlands and the loss of biodiversity by land consolidation and modernization of irrigation systems.

Figure 1.10 shows the balance between biodiversity, productivity and labor intensity of this case. In order to get more productive efficiency, land improvement, or readjustment projects were carried out in the latter half of the 20th century in Japanese paddy fields. Semi-natural open channels with lush grasses changed to feed water valves because of the rationalization of the water supply, and the level between paddy fields and drainages became deeper due to the rapid drying of paddy fields. Such modern improvements promise high productivity with less labor intensity, but bring less biodiversity. For the compensation of the loss of biodiversity, some kinds of techniques, such as fish ladders and catchment ditches, have been employed in recent years. This is one method to recover biodiversity, but a more positive improving method might be proposed. That is the preservation of traditional paddy field farming styles. But the preservation of old farming styles forces farmers to use more labor. It is very important that local farmers understand fully about the importance of preservation of biodiversity through traditional farming and accept the extra amount of labor. For such understanding the concept of landscape management labor accounts might play a large role.

1.2.6 Landscape Labor Accounts and Ecosystem Services

The concept of landscape management labor accounts does not directly link to the concept of ecosystem services, but both have a strong relation. The concept of ecosystem services comes from the supply side perspective, which is based on the idea that nature can give various kinds of benefits to human society. On the contrary, the concept of landscape management labor accounts requires humans to maintain landscapes in a good condition. Well maintained landscapes might provide rich ecosystem services unless human intervention exceeds the capacity of nature. In countries like Japan, those lands are historically cultivated and densely populated, and rich ecosystem services have been created by the interaction between nature and human activities. The Satoyama landscape mentioned in this book is one of such cases. Well maintained agricultural landscapes can also provide rich harvests for human society and beautiful countrysides for travelers. On the contrary, unmanaged or abandoned landscapes cause various kinds of ecosystem disservices, discomfort caused by fallen leaves on streets, generation of mosquitos, agricultural damage caused by animals and so on.

This book starts from a pessimistic view, that the human power needed to manage such landscapes might not be supplied by an aging and depopulated society in the future. In such a shrinking society, the styles of managing landscapes might be changed from those of a growing society. How we can find the appropriate solution has become an urgent issue. The concept of landscape management labor accounts can act as a basic indicator for evaluating the management situation of all kinds of landscapes. In shrinking conditions it will be needed to develop methods to manage landscapes with rich ecosystem services while using less management power. However, some landscapes might have to be abandoned, possibly leaving high ecosystem services. For such an evaluation, the concept of landscape management labor accounts can play an important role.

But finally, it must be added, that there are some landscapes which can provide the richest ecosystem services without needing human intervention. In this sense, landscape management labor accounts and ecosystem services do not connect directly. It is necessary to develop research on landscape management labor accounts, giving attention to this aspect.

1.2.7 Integration from Neighborhood Landscape Complex Unit to a National Scale

Finally the method of statistically integrating landscape management labor accounts from neighborhood landscape complex units to a national scale is proposed. There are two viewpoints; the first is the calculation of labor accounts based on the area of each landscape element such as a paddy field, secondary forest, private garden, and so on. The other is the calculation of labor accounts based on the population of each working style type such as a professional farmer, volunteer, and so on Fig. 1.11.

In the scale of neighborhood landscape complex units, which is the basic scale of the landscape management, the total labor accounts in each complex unit according to the above two factors are calculated, and they are compared to each other. According to this, we can estimate how many landscape elements can be managed

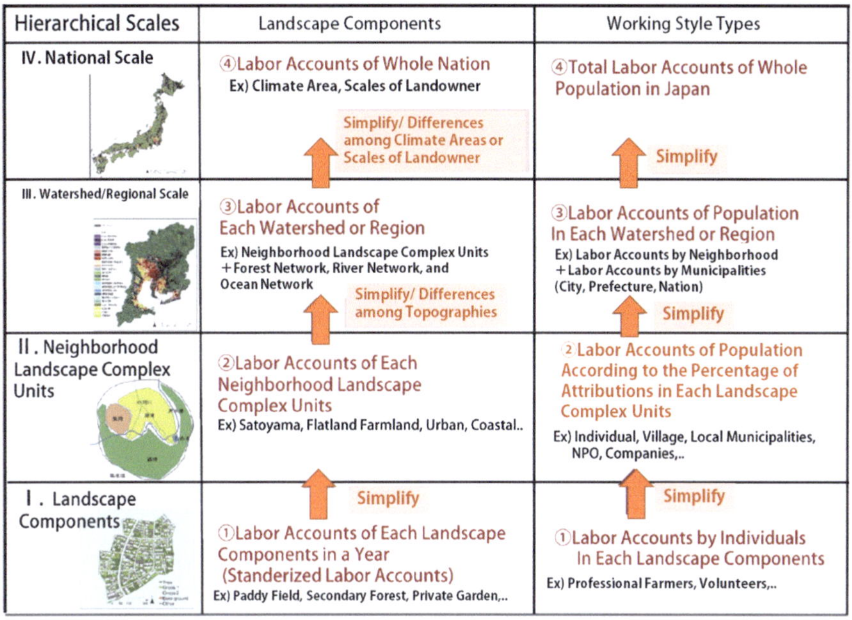

Fig. 1.11 Integration of labor accounts in hierarchical scales

or abandoned by the future population and the ratio of working style types and whether they are appropriate for ecologically sustainable landscapes.

In the scale of a watershed or region, the total labor accounts of the above two factors are also totaled. In this scale, the neighborhood landscape complex units in each watershed are connected by the ocean, river, and forest network which are managed mainly by the basic municipalities such as cities, prefectures, and nation, or by large company, corporation, association, NPO and so on. In comparison between landscape management labor accounts in the neighborhood landscape complex units in the upstream side and downstream side in a watershed, we can evaluate and estimate how many people potentially will move to manage the landscape in the neighborhood landscape complex units which need more management for the appropriate landscape.

On the national scale, the total labor accounts are totaled by considering the local difference among the regions or watersheds, which are natural factors such as climate and soil, and human factors such as the landholding scale of farmers. And the estimations of present and future population are compared to the appropriate landscapes in each region.

Above we have shown the important concepts for the consideration of landscape management labor accounts. In the following section we will show the detailed individual issues on land uses and landscape management in typical landscape types. They have a lifestyle to go to their fields as a daily routine.

References

Hindmarch C, Pienkowski M (2000) Land management: the hidden costs, the British ecological society. Ecological issues series, Blackwell Science Ltd., UK

National Institute of Population and Social Security Research (2012) Population projections for Japan. http://www.ipss.go.jp/site-ad/index_english/esuikei/ppfj2012.pdf. Accessed 2015

Natuhara N (2013) Ecosystem services by paddy fields as substitutes of natural wetlands in Japan. Ecol Eng 56:97–106

United Nations Environment Programme (2012) Fifth global environment outlook-5 (Geo-5). Environment for the future we want, Progress Press LYD., Malta

Chapter 2
Japanese Basic Landscape Types, and Change in Population and Urban Land Use

Hiroyuki Shimizu

Abstract This chapter illustrates typical landscape types extracted by principal component analysis and cluster analysis. It also discusses the distribution patterns and characteristics of these landscape types and examines the relationship between land use and population changes. The following seven landscape types were extracted: urban landscape, urban paddy field mixed landscape, paddy field landscape, other field landscape, paddy field satoyama landscape, other field satoyama landscape, and nature landscape. Then, through the combination analysis of land use and population changes in recent years, shrinking, compacting, stability, scattering, and expanding tendencies are observed. Not only in Nagoya but also in the periphery of the Tokyo Metropolitan Area, a mosaicked structure of the above tendencies was clearly observed, with satoyama landscape types most dramatically confronting the crisis of disappearance.

Keywords Population · Land use · Change · Landscapes type · Population estimation

2.1 Overview of Japanese Landscapes and Land Use

Japan is an elongated island nation, where a steep sloped mountainous area is spreading as a backbone in the center. Figure 2.1 shows the land use of the national land in 2009. These mountainous areas are covered mostly by forest and created a forest landscape (Fig. 2.2). Even though steep sloped mountainous areas cover the main parts of Japanese lands, there are some flat plains which were made by the

The original version of this chapter was revised: In Fig. 2.49 caption: 1975 was changed to 2005. The erratum to this chapter is available at 10.1007/978-981-10-2278-4_26

H. Shimizu (✉)
Graduate School of Environmental Studies, Nagoya University,
Furo-cho, Chikusa-ku, Nagoya, Aichi 464-8601, Japan
e-mail: shimkankyou@msn.com

H. Shimizu et al. (eds.), *Labor Forces and Landscape Management*,
DOI 10.1007/978-981-10-2278-4_2

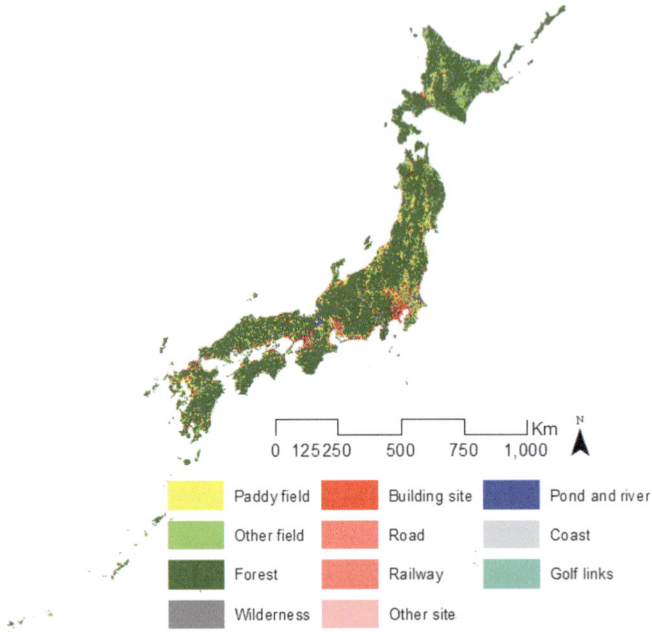

Fig. 2.1 Land use of the national land in 2009

Fig. 2.2 Forest landscape in a mountainous area

Fig. 2.3 Northern edge of Nobi Plain

Fig. 2.4 Small plain in Chita peninsula

Fig. 2.5 Farmland landscape in Toyama Prefecture

Fig. 2.6 Urban-farmland mixed landscape

Fig. 2.7 Satoyama landscape

marine transgression in the ancient age. These plains are highly cultivated and most of Japanese population is living there. Figure 2.3 shows an edge of urban area to a forest area. High densely populated urbanized areas, i.e. urban landscapes, are spreading up to mountainous areas. Even in a small plain among hills, high densely cultivated countrysides are spreading everywhere (Fig. 2.4). Farmland landscapes are also spread in flatlands (Fig. 2.5). In big plains with big cities like Tokyo, Osaka or Nagoya, endless urban sprawls with fields, houses and factories, are remarkable. This can be termed an urban-farmland mixed landscape (Fig. 2.6). This mixed landscape is one of the typical landscapes of Japan. In relatively gentle sloped

Fig. 2.8 Land use change between 1976 and 2009

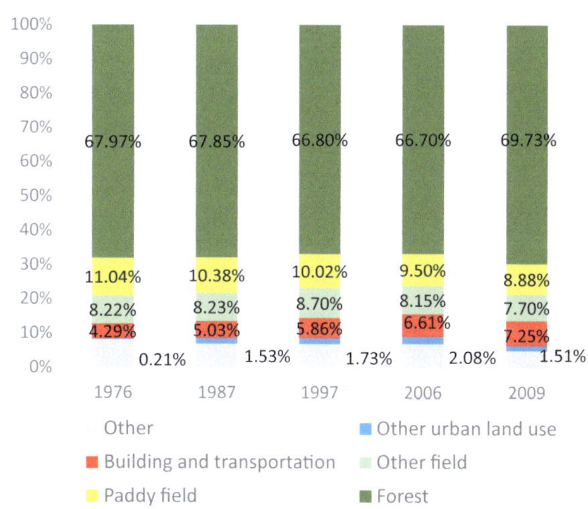

mountainous areas there are many small villages with fields and secondary forests, which consist of Japanese typical landscapes, Satoyama landscapes (Fig. 2.7).

Figure 2.8 shows land use change between 1976 and 2009. Forests cover constantly over 66 % of the total national land. The significant land use change occurs between farmlands and urban land uses. Farmlands decrease constantly and changed to urban land uses. These changes occur mainly in flatlands. As a result, urban-farmland mixed landscapes are spreading.

2.2 Extract of Japanese Basic Landscape Types and Characteristics

This section shows typical landscape types extracted by principal component analysis and cluster analysis, and distribution patterns and characteristics of them. The land use data in 1 km meshes in 2009 (*National Land Numerical Information, Land utilization tertiary mesh Data*) provided by the National Land Numerical Information Download Service (National Land Information Division, National Spatial Planning and Regional Policy Bureau, MILT of Japan 2009) is used for the analysis.

As a preparation for the analysis, Normalized Difference Index (NDI) between different land uses are prepared. NDI is defined as follows:

A as an area ratio of a certain land use in a certain mesh

B as an area ratio of another land use in a same mesh

$$\mathrm{NDI} = (\mathrm{B} - \mathrm{A})/(\mathrm{A} + \mathrm{B})$$

Forest and urban land use NDI, forest and total field NDI, total field and urban land use NDI and paddy field and other field NDI are prepared. A principle component analysis is carried out by using variables of forest ratio, total field ratio, urban land use ratio, forest and urban land use NDI, forest and total field NDI, total field and urban land use NDI and paddy field and other field NDI. Table 2.1 shows correlations and significant probabilities of variables. Significant probabilities of all the variables are significant at the 1 % level, and three factors shown in Table 2.2 are extracted.

According to the results of the principle component analysis, a cluster analysis is carried on, and the seven basic landscape types are extracted, namely paddy field type, urban paddy field mixed type, paddy field Satoyama type, other field type, urban type, other field Satoyama type and nature type. Table 2.3 shows seven basic landscape types extracted by the cluster analysis. Table 2.4 shows the percentages of seven basic landscape types throughout Japan. Figure 2.9 shows the spatial distribution of the seven basic landscape types.

Areas of the paddy field Satoyama and other field Satoyama types, which correspond to Satoyama landscapes, account for 28.4 %, but the population in them is only 5.9 %. On the contrary, areas of the urban type account for only 5.7 %, but the

Table 2.1 Correlations and significant probabilities of variables

		Foratio09	Ubratio09	Firatio09	FoFiNDI09	FoUbNDI09	FiUbNDI09	PaOfNDI09
Correlation	Foratio09	1.00	−0.63	−0.73	0.90	0.85	−0.03	−0.04
	Ubratio09	−0.63	1.00	0.10	−0.42	−0.80	−0.50	0.07
	Firatio09	−0.73	0.10	1.00	−0.88	−0.48	0.48	0.01
	FoFiNDI09	0.90	−0.42	−0.88	1.00	0.76	−0.25	−0.06
	FoUbNDI09	0.85	−0.80	−0.48	0.76	1.00	0.25	−0.15
	FiUbNDI09	−0.03	−0.50	0.48	−0.25	0.25	1.00	−0.04
	PaOfNDI09	−0.04	0.07	0.01	−0.06	−0.15	−0.04	1.00
Significance probability	Foratio09		0.000	0.000	0.000	0.000	0.000	0.000
	Ubratio09	0.000		0.000	0.000	0.000	0.000	0.000
	Firatio09	0.000	0.000		0.000	0.000	0.000	0.000
	FoFiNDI09	0.000	0.000	0.000		0.000	0.000	0.000
	FoUbNDI09	0.000	0.000	0.000	0.000		0.000	0.000
	FiUbNDI09	0.000	0.000	0.000	0.000	0.000		0.000
	PaOfNDI09	0.000	0.000	0.000	0.000	0.000	0.000	

Table 2.2 Three factors extracted by principle component analysis

Component matrix after rotation	Component		
	Forest-non forest	Non urban-urban	Paddy field-other field
Foratio09	0.942	0.237	−0.000
Ubratio09	−0.478	−0.812	0.029
Firatio09	−0.888	0.368	−0.004
FoFiNDI09	0.979	−0.027	−0.029
FoUbNDI09	0.792	0.537	−0.120
FiUbNDI09	−0.259	0.894	−0.016
PaOfNDI09	−0.030	−0.032	0.999

Table 2.3 Seven basic landscape types extracted by cluster analysis

Cluster centers	Clusters						
	Paddy field satoyama type	Paddy field type	Other field satoyama type	Urban type	Nature type	Urban paddy field mixed type	Other field type
Forest-non forest	0.168	−1.936	0.084	−0.787	0.727	−1.222	−1.574
Non urban-urban	0.667	0.519	0.587	−2.742	−0.135	−1.721	0.895
Paddy field-other field	1.322	1.120	−1.417	−0.741	−0.026	1.131	−1.411

population of them reaches to 54.8 %. Areas of the urban paddy field mixed type, where urban land uses and paddy field land uses are mixed, are only 4.2 %, but the population accounts for 24.0 %. Therefore, 79 % of the Japanese population lives in the urban paddy field mixed and urban types, with an area size of less than 9.9 % of all of Japan.

Table 2.4 Percentages of seven basic landscape types in all of Japan

Landscape types	Mesh number	Mesh number ratio (%)	Population	Population ratio (%)
Paddy field satoyama type	59,411	15.7	46,312,131	3.6
Paddy field type	30,039	7.9	11,332,782	8.9
Other field satoyama type	47,903	12.7	2,884,748	2.3
Urban type	21,462	5.7	70,094,360	54.8
Nature type	174,449	46.1	3,231,294	2.5
Urban paddy field mixed type	15,760	4.2	30,658,752	24.0
Other field type	29,522	7.8	5,139,252	4.0
Sum	378,546	100.0	12,792,419	100.0

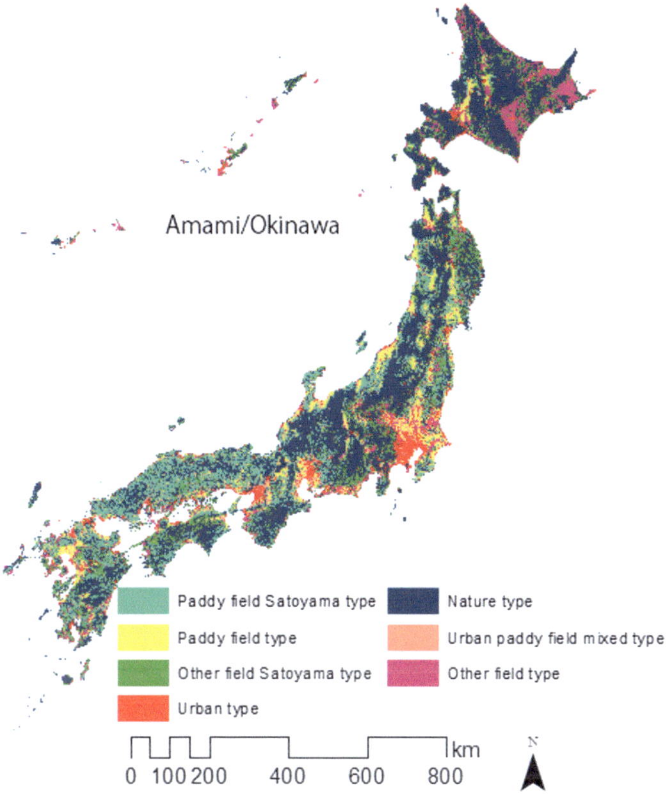

Fig. 2.9 Spatial distribution of seven basic landscape types

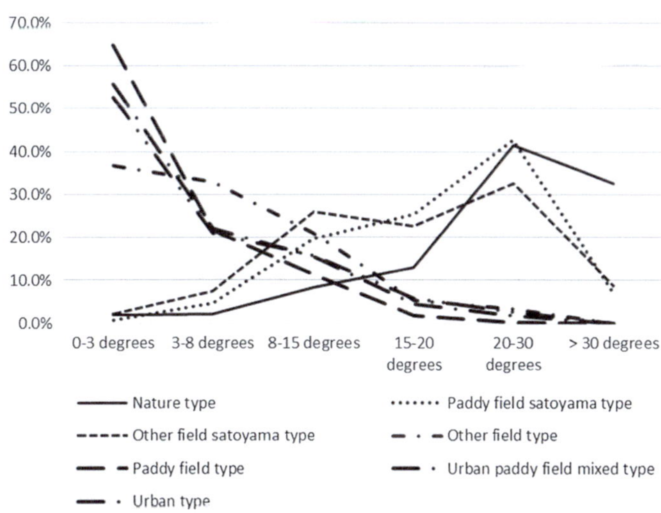

Fig. 2.10 Seven basic landscape types and land gradients

Figure 2.10 shows the relation of these seven basic landscape types and average land gradients of meshes. Specialization of the paddy field, urban paddy field mixed and urban types to the areas of the flat or gentle sloped areas of average land gradients between 0° and 15° is remarkable. Other field type shows the same tendency, but it spreads much more in sloped areas.

On the contrary, the distribution of the paddy field Satoyama and other field Satoyama types shifts to the area of average land gradients between 8° and 30°. Also, specialization of the nature type to the areas of average land gradients over 20° is remarkable.

2.3 Distribution of Japanese Basic Landscape Types in Wide Regions

Figure 2.11 shows the distribution of landscape types in Hokkaido in a subarctic zone. Landscapes, especially in the eastern part, are very similar to Central Europe with hilly farmlands, where large-scale crops cultivations have been developed.

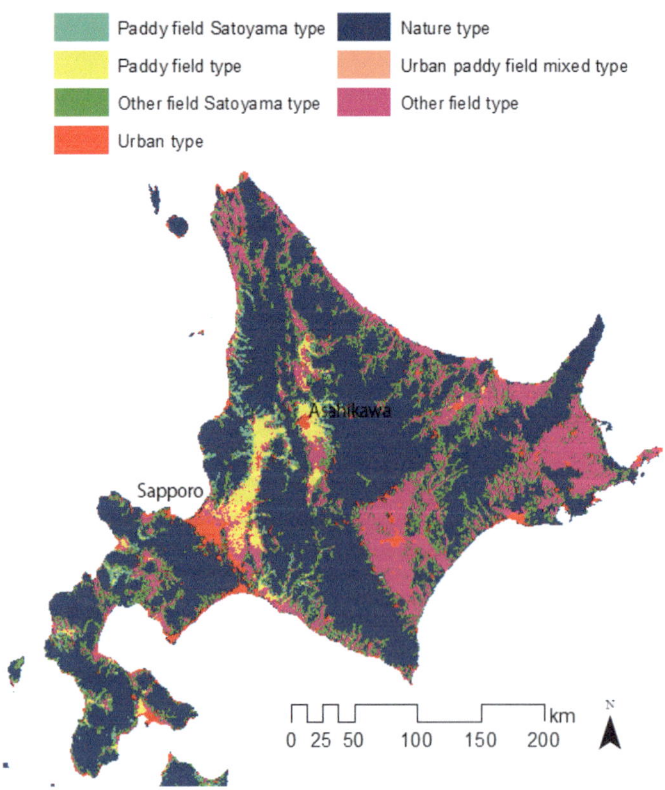

Fig. 2.11 Basic landscape types in Hokkaido

Most population is agglomerated in Sapporo, a prefectural capital. In the northeast part of Sapporo, the paddy field type is spreading widely, where good quality rice is cultivated.

Figure 2.12 shows the landscape types in the Tohoku, Kanto, Hokuriku and Tokai wide regions. The Tohoku and Hokuriku wide regions are the most important producing areas of rice, especially in the Shinano River basin in the southern part of Niigata and in the Kitakami River basin in the northern part of Sendai, where the

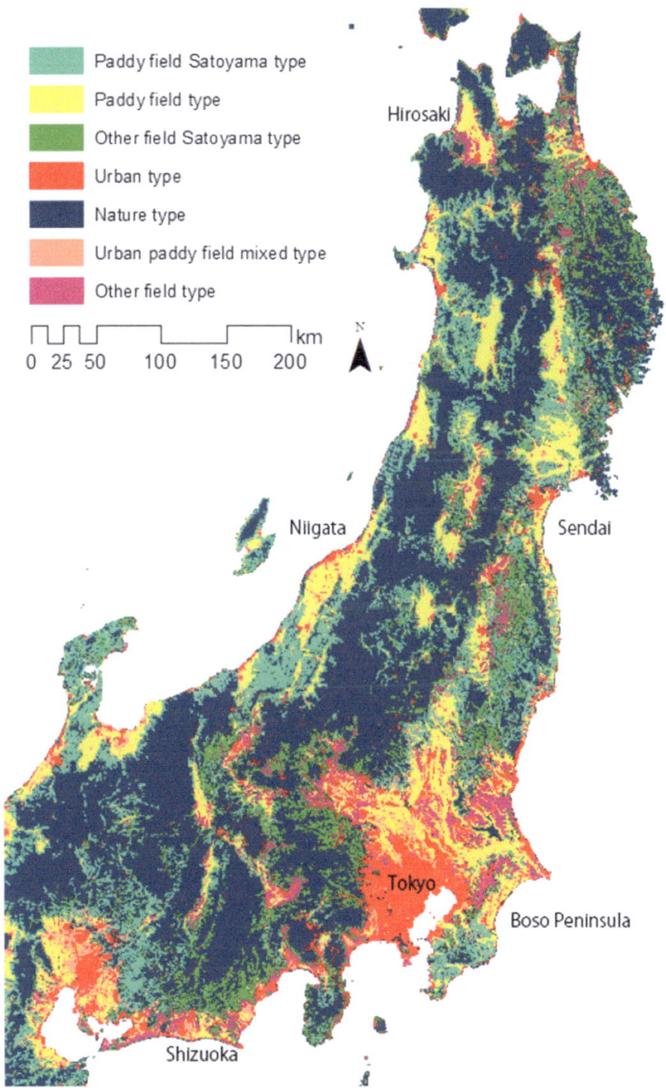

Fig. 2.12 Basic landscape types in Tohoku, Kanto, Hokuriku and Tokai wide regions

paddy field and paddy field Satoyama types are spreading. The urban type is spreading in and around Tokyo widely. In the northern part of Tokyo, the Tone River flows from west to east, along which the paddy field type is spreading in low lands. In plateaus of the northern part of the Boso Peninsula, the other field type is spreading. In the north western part of Tokyo, between Tokyo and Nagano, which is a relatively cool highland, the typical other field type is widely spreading. These are very important districts of vegetable and fruit production. In the plateaus of Shizuoka mandarin orange tree and tea plantations are spreading widely, which create a special other field type.

Figure 2.13 shows the landscape types in Kansai, Shikoku, Chugoku and Kyushu wide regions. A very big Median Tectonic Line is crossing these wide regions from Kumamoto, Matsuyama to Wakayama. It is very interesting to know that there is a big difference in landscape types between the northern and southern parts of the line. In the southern part of the line, the nature and other field Satoyama types are spreading widely; on the contrary, in the northern part the paddy field Satoyama and paddy field types are spreading dominantly. The southern part of Wakayama and the western end of Shikoku are very famous mandarin orange

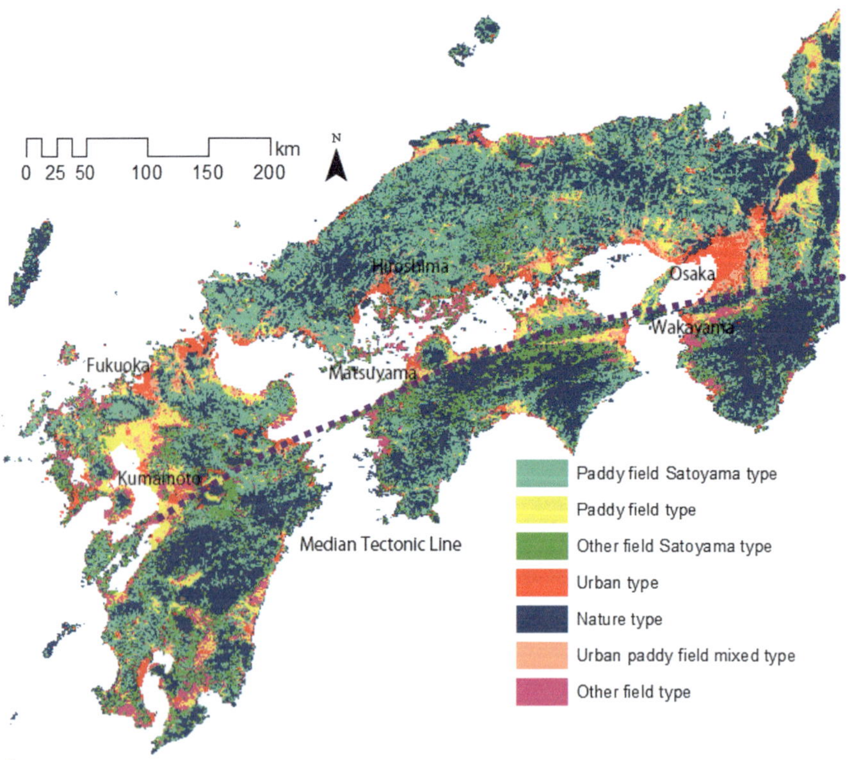

Fig. 2.13 Basic landscape types in Kansai, Shikoku, Chugoku and Kyushu wide regions

plantation districts and create the typical other field type. It is also characteristic that the paddy field Satoyama type is spreading widely in the mountainous area in the Chugoku wide region.

Figure 2.14 shows the detailed distribution pattern of 7 landscape types around the area of Ise Bay, where Nagoya is located in the center. In the southeastern part and in the western part of Nagoya, the urban paddy field mixed and urban types are widely spreading and are mixed with the paddy field type. Nagoya and the surrounding districts are one of the most developed industry areas and a strong agglomeration of population and urban expansions are observed. Matsusaka, located in the western coast of Ise Bay, is a relative small regional city but historically very famous and has a wide area which consists of all types of landscapes.

As described above, Japan is not homogeneous country in its landscapes. There are characteristic regional differences. Especially Satoyama landscapes occupy 28 % of Japanese land and consists an important basic landscape, but the population is only less than 6 %. On the contrary in the urban and urban paddy field

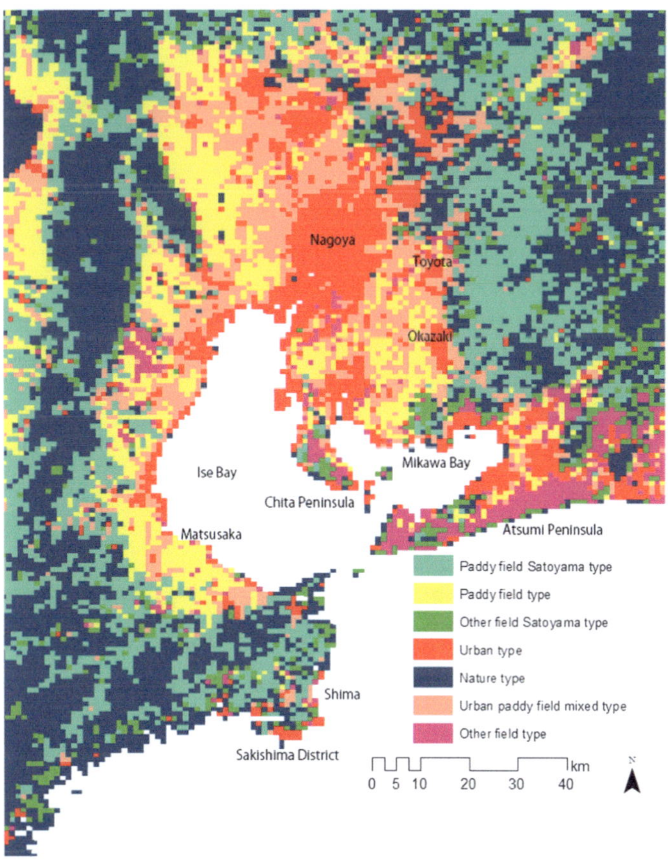

Fig. 2.14 Basic landscape around Ise Bay area

mixed types, which occupy only 10 % of the national land, lives 79 % of the population. This might be a reason that there is less public attention to the situation of Satoyama landscapes.

2.4 Overview of Changes in Population, Urban Land Use

Expansion of urban land use has affected the wide range of ecological systems. Urban landscape changes might be related to population changes. This section shows the characteristics of spatial distribution patterns and their change of urban land use and population throughout the national land of Japan by using 1 km mesh analysis on GIS. The data employed are from the Statistics Bureau of the Ministry of Internal Affairs and Communication (E-Stat 2015), and the University of Tokyo (CSIS 2015).

Japan consists mostly of steeply sloped mountainous areas in the center (Fig. 2.15). There are, however, some flat diluvial plains where most of the population is agglomerated. Big cities such as Tokyo/Yokohama, Osaka, Nagoya and so on are located in such plains (Fig. 2.16).

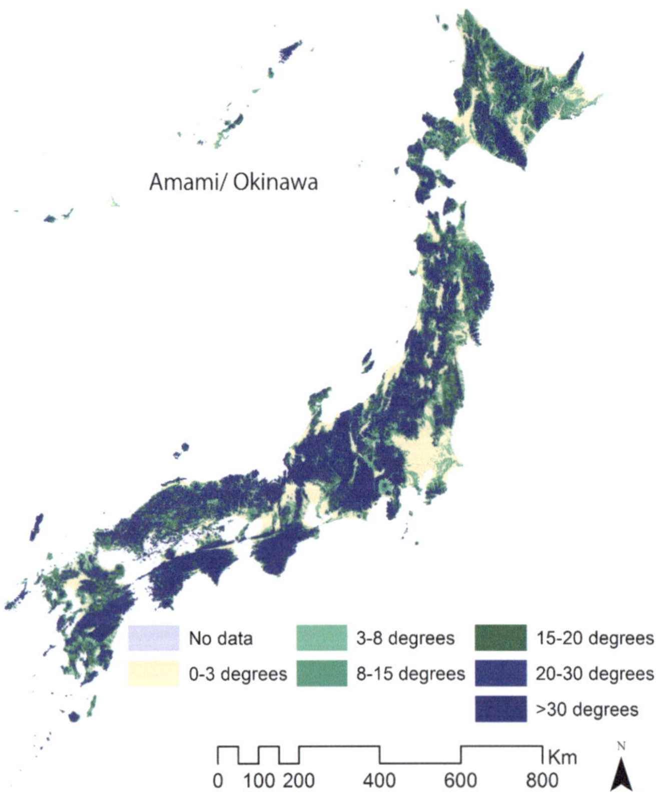

Fig. 2.15 Average land gradients in 1 km meshes

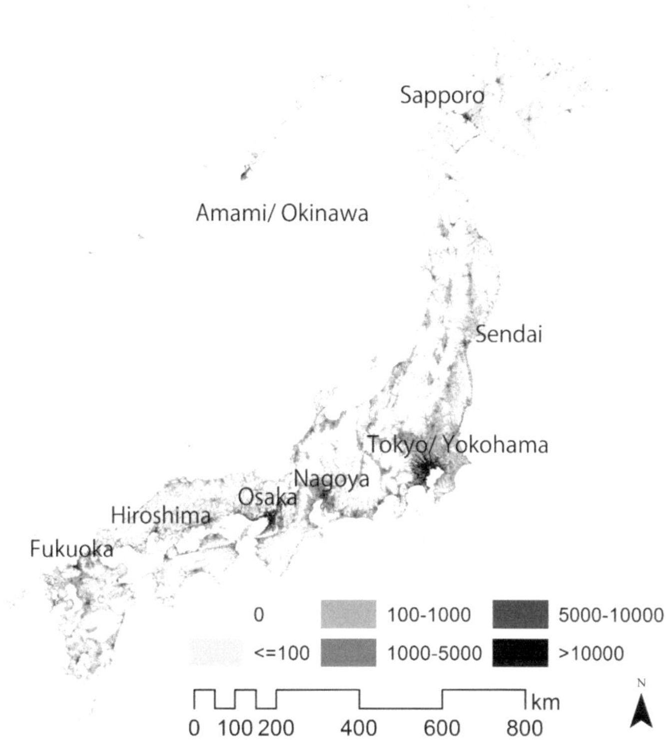

Fig. 2.16 Population distribution in 2010

Japan is now in the phase of population shrinking which began in 2008. Figure 2.17 shows the population changes between 1975 and 2010. The threshold between decrease or increase and being stable is ±5 % change. Although the distribution pattern shows an increasing trend as a whole, there are many areas with a shrinking population in the countryside. Figure 2.18 shows the recent population changes between 2005 and 2010. The threshold is set to 1/7 of the threshold of the change between 1975 and 2010 in correspondence with the period length. It is obvious that the areas of population decrease expand widely in the countryside, although there is a tendency toward increasing population in densely populated urban areas of big cities, such as Tokyo, Osaka and Nagoya and so on.

Figure 2.19 shows the relation of the population changes between 1975 and 2010, and the land gradients. It is obvious that the population increase occurs in the relative flat areas under 8°. But it can't be ignored that the population decrease already occurs in all categories of land gradient ranks. Figure 2.20 shows the relation of the recent population changes between 2005 and 2010, and land gradients. It is remarkable that decreasing and increasing trends are reversed from the change between 1975 and 2010 in their ratios and the decreasing trend becomes dominant.

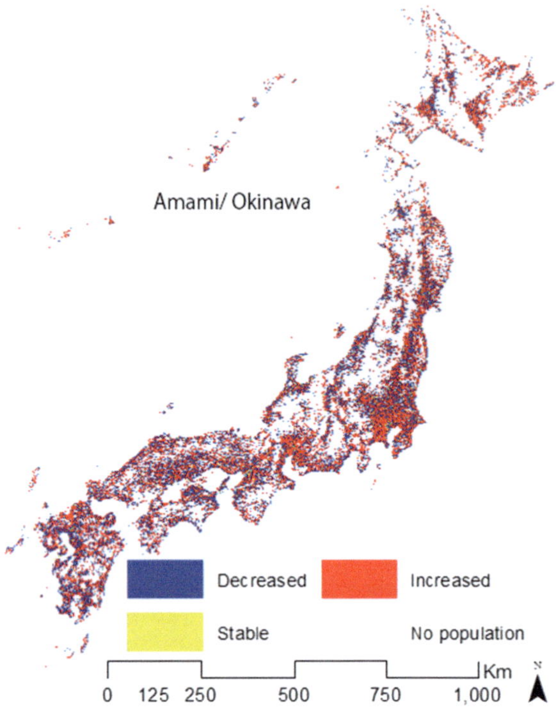

Fig. 2.17 Population changes between 1975 and 2010 (Threshold 5 %)

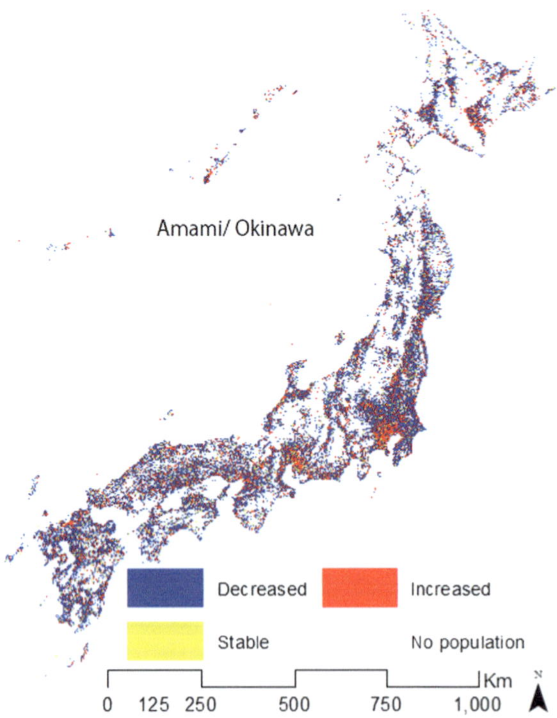

Fig. 2.18 Population changes between 2005 and 2010 (Threshold 0.7 %)

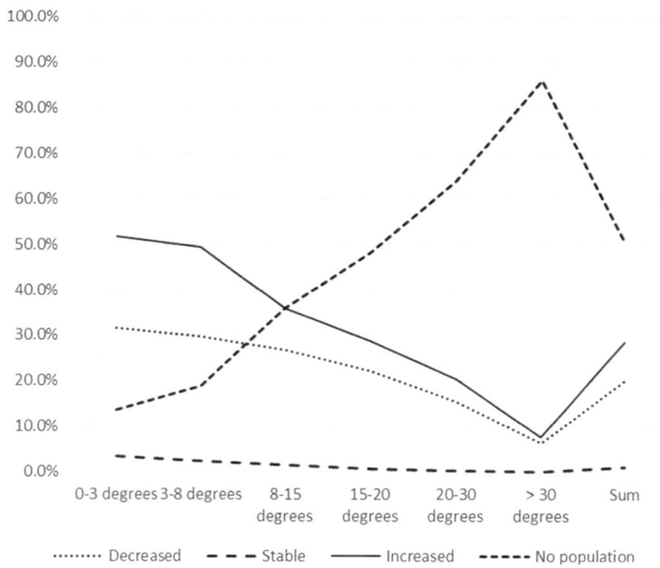

Fig. 2.19 Population changes between 1975 and 2010 and land gradients (Threshold 5 %)

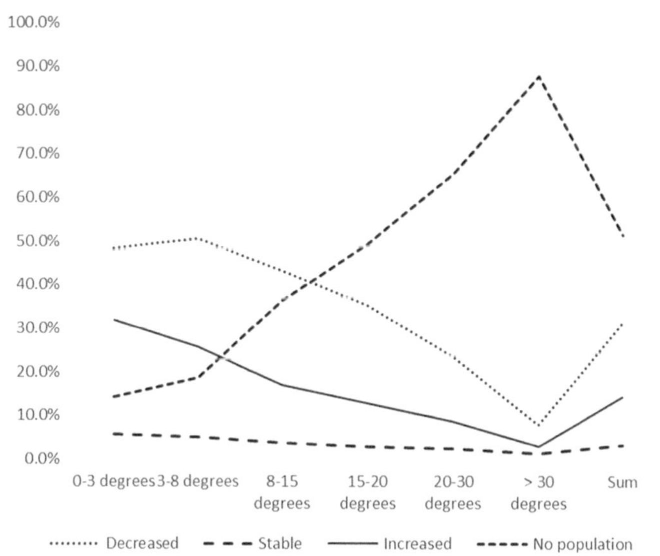

Fig. 2.20 Population between 2005 and 2010 changes and land gradients (Threshold 0.7 %)

Figure 2.21 shows the urban land use changes between 1976 and 2009. The threshold between decrease or increase and being stable is ±5 % change. High urban land use increases can be observed in the densely populated areas, such as Tokyo, Nagoya, Osaka and so on. But the decrease in urban land use is already

Fig. 2.21 Urban land use changes between 1976 and 2009 (Threshold 5 %)

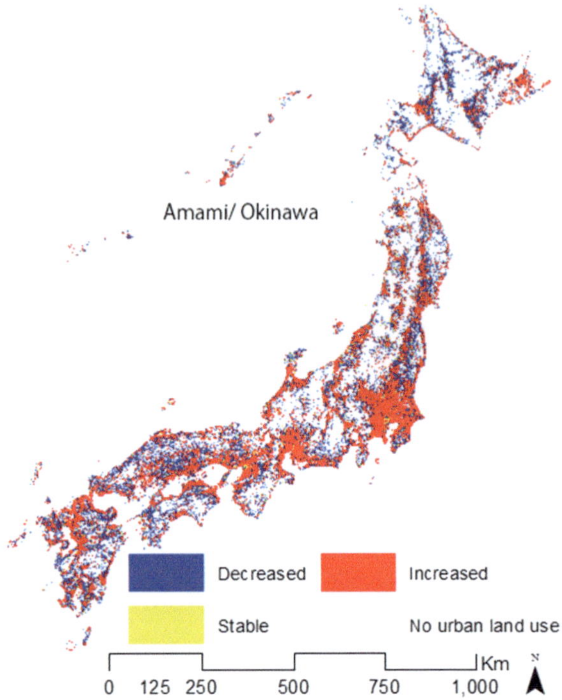

Fig. 2.22 Urban land use changes between 2006 and 2009 (Threshold 0.5 %)

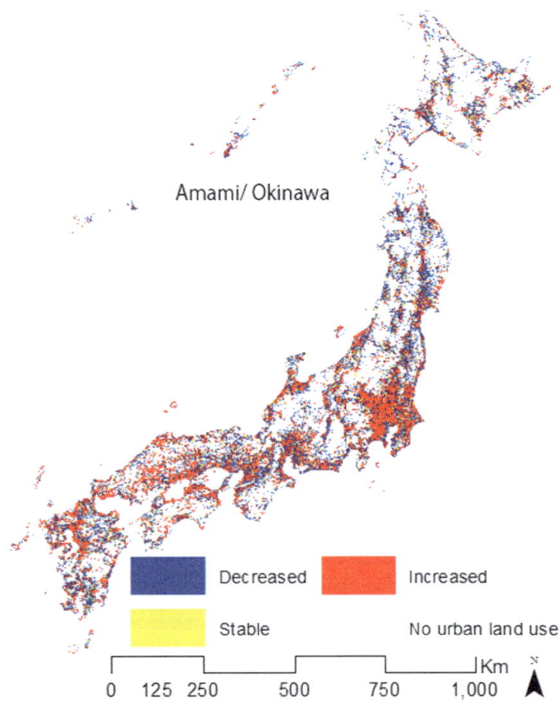

shown, mostly scattered in the countryside and mainly in the mountainous areas. Figure 2.22 is the recent urban land use changes between 2006 and 2009. The threshold of the changes is set to 1/11 of the threshold of the change between 1976 and 2009 in correspondence with the period length. Compared to the recent population decrease, the recent urban land use decrease is not so strong. That is, the urban land use increase might continue in some areas, where the population decrease has already begun.

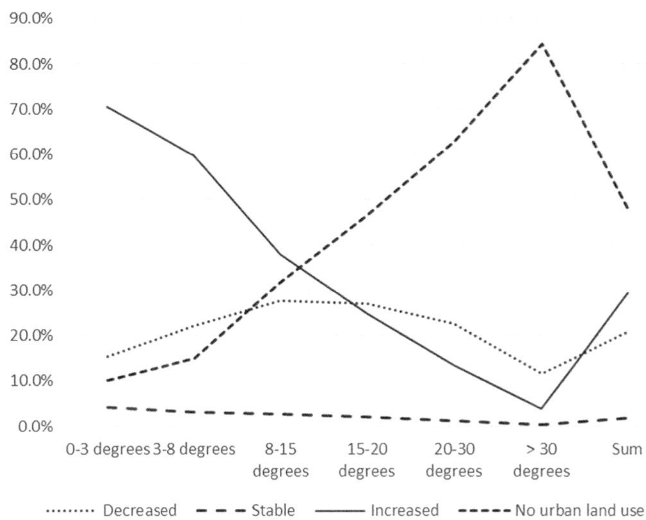

Fig. 2.23 Urban land use changes between 1976 and 2009 and land gradients (Threshold 5 %)

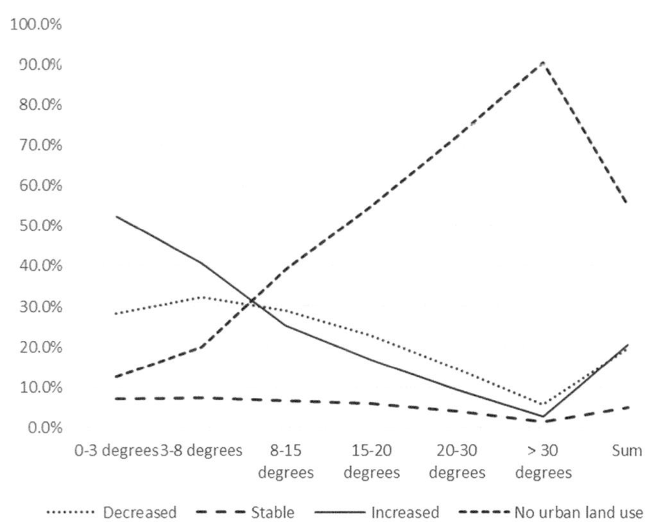

Fig. 2.24 Urban land use changes between 2006 and 2009 and land gradients (Threshold 0.5 %)

Figure 2.23 shows the relation of the urban land use changes between 1976 and 2009, and the land gradients. Compared to the population change, the increase of urban land use is higher than the decrease of urban land use in the areas of land gradients under 8°. Figure 2.24 shows the relation of the urban land use changes between 2006 and 2009, and land gradients. In the change of population, the decrease exceeds the increase between 2005 and 2006, and the increase of the urban land use remains higher than the decrease in the areas of land gradients under 8°. It might be said that the tendency of the decrease and increase of population and land uses differs according to the land gradients, and urban land use increases cannot stop easily, even when the decrease of population begins, especially in the flatlands.

2.5 Estimated Population Changes Between 2010 and 2050, and Basic Landscape Types

The Japanese Government has provided an estimate of population distribution in 1 km meshes. According to the data, estimated population changes in basic landscape types are examined in this section.

Figure 2.25 is the special distribution of estimated population changes between 2010 and 2050. The threshold between decrease or increase and being stable is ±5 % change. Areas of increasing population are very limited in some parts of

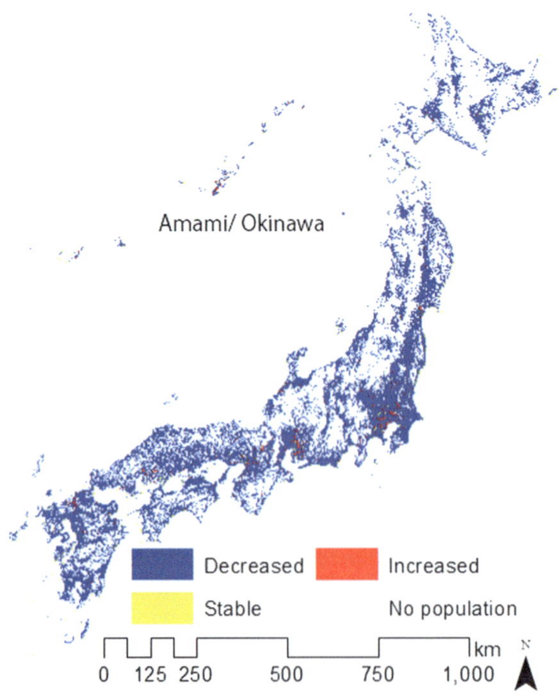

Fig. 2.25 Estimated population changes between 2010 and 2050 (Threshold 5 %)

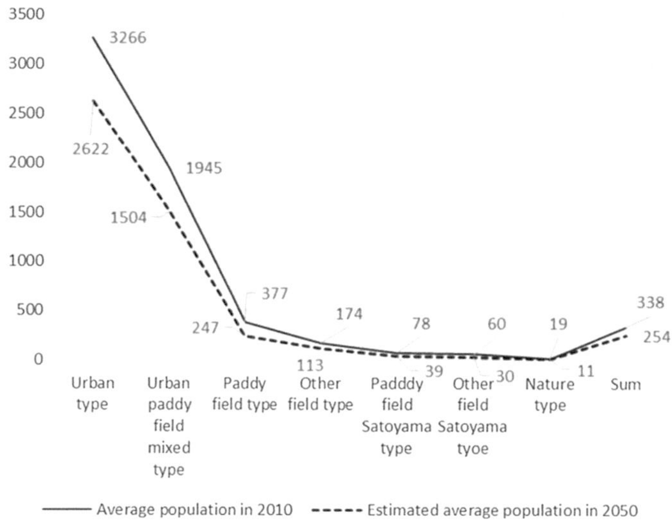

Fig. 2.26 Average mesh population in 2010, estimated average mesh population in 2050 in each basic landscape type

highly agglomerated urban areas around Tokyo, Nagoya Osaka, Fukuoka and so on. Almost all areas show a reducing trend. The extreme lack of labor for the appropriate landscape management in the future will be of strong concern. Figure 2.26 shows the average mesh population in 2010 and estimated average mesh population in 2050 in each basic landscape type.

In the highly populated landscape types such as urban type and urban paddy field mixed type, the population decrease is around 20 %. But in low populated landscape types such as paddy field Satoyama type and other field Satoyama type, the decrease in population could reach to 50 %. It is an enormous decrease and these landscape types will face a crisis of a vanishing population.

Figure 2.27 shows population changes between 1975 and 2010 in each landscape type. On the whole, the tendency of increase is stronger than the tendency of decrease in all landscape types. But in the population changes between 2005 and 2010 shown in Fig. 2.28, the decreasing tendency is greater than the increasing tendency in all landscape types. Nowadays the already shrinking phenomenon has become dominant. Figure 2.29 shows estimated population changes between 2010 and 2050 in each landscape type. In most areas of all landscape types, the decreasing tendency greatly exceeds the increasing tendency. It is easily recognized that the Satoyama landscapes in the countryside suffer from strong population decrease, but indeed in urban periphery landscapes such as urban paddy field mixed type and paddy field type, population decreases will also progress. Until now, Japan was not hit by such strong population decrease. Naturally, it is necessary as ever to consider the different measures that occur under such a strong decrease.

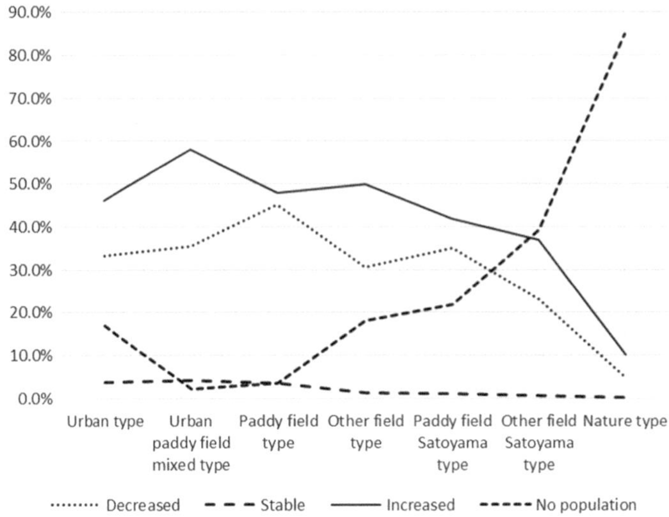

Fig. 2.27 Population changes between 1975 and 2010 in each landscape type (Threshold 5 %)

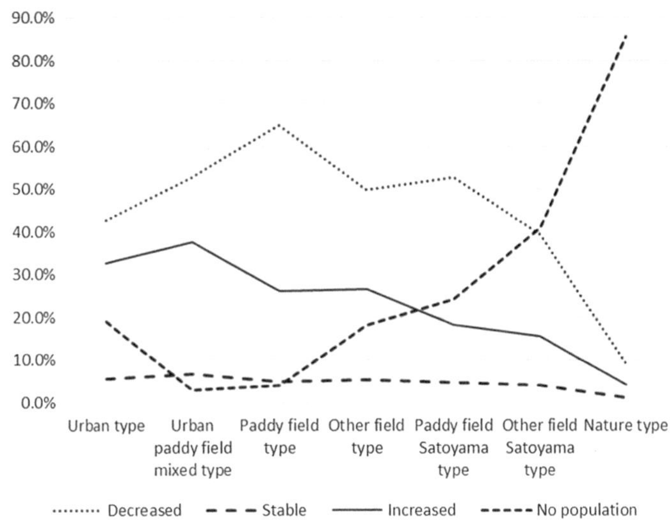

Fig. 2.28 Population changes between 2005 and 2010 in each landscape type (Threshold 0.7 %)

Figure 2.30 shows the urban land use changes between 1976 and 2009 in each landscape type. Urban land use increase exceeds urban land use decrease in the landscape types of urban, urban paddy field mixed, paddy field and other field types. Especially in the urban and urban paddy field mixed types, the decrease in urban land use is at a very low level. On the contrary, the decrease in urban land use is greatest in the paddy field Satoyama and other field Satoyama types. This means

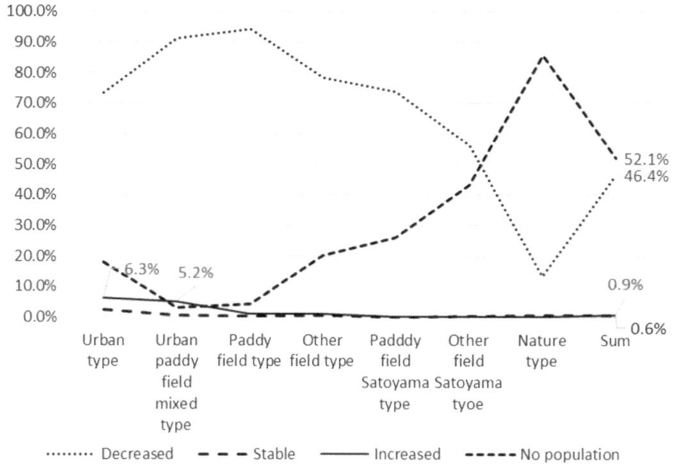

Fig. 2.29 Estimated population changes between 2010 and 2050 in each landscape type (Threshold 5 %)

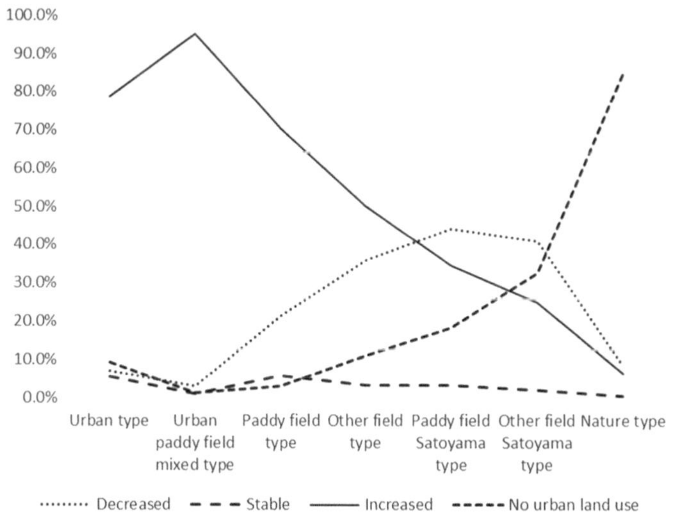

Fig. 2.30 Urban land use changes between 1976 and 2009 in each landscape type (Threshold 5 %)

that, in Satoyama areas, the decrease in urban land use has already begun in population growth periods. Figure 2.31 shows the urban land use changes between 2006 and 2009 in each landscape type. In the areas of urban, urban paddy field mixed and paddy field type, urban land use increase continues still, but remains at lower levels. In the area of other field types, the tendency to decrease changed to exceed increase.

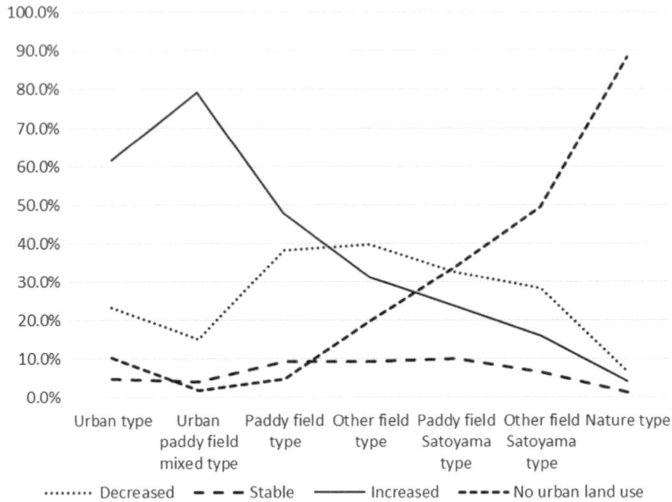

Fig. 2.31 Urban land use changes between 2006 and 2009 in each landscape type (Threshold 0.5 %)

There is no estimation of urban land use accounts in 2050. The relation between urban land use areas and population in 2050 can be estimated. Figure 2.32 shows the urban land use area per population in 2010 in each landscape type. It is interesting to see that in the other field type, the other field area per population is extremely high. It might be said that fields without paddy fields could be managed in a wider area than paddy fields, which need more labor-intensive methods.

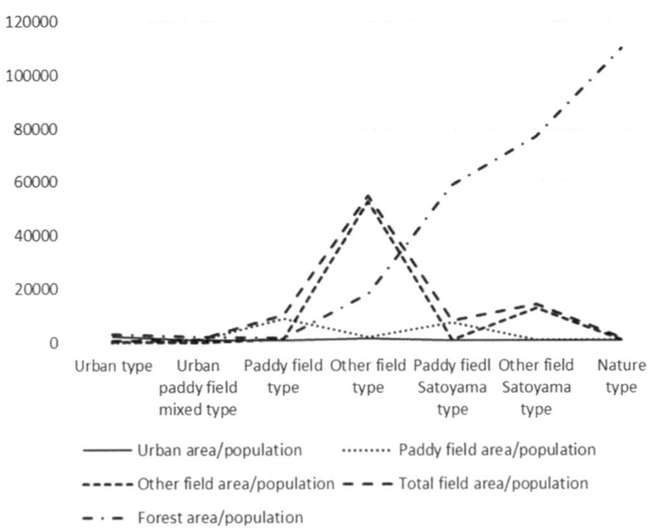

Fig. 2.32 Land use areas per population in 2010 in each landscape type

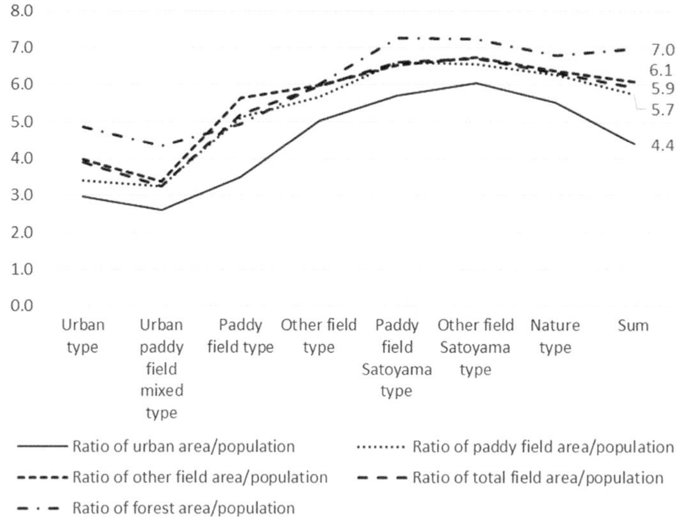

Fig. 2.33 Ratios between urban land use areas per population in 2010 and 2050 in each landscape type

Figure 2.33 shows ratios between urban land use area per population in 2010 and 2050 in each landscape type. Because of strong population decrease, managing area sizes per population of paddy field, other fields and forests will increase more than 6 times in 2050 than that of today, in the assumption that the managing area in 2050 is the same as in 2010.

As seen above, the decrease in Japanese population will hit all categories of landscape types, and land use changes will go into a very different situation than ever before. Especially rich green environments or rich ecological landscapes, which are created and maintained with collaboration between nature and human beings, will confront the lack of management labor. Each landscape type has its own management style, and impact on the environment will differ in each landscape type. It is very important to adopt appropriate ways of landscape management, thinking about the difference of working styles in each landscape type. In part III landscape management in different landscape types will be observed concretely in case studies.

2.6 Combined Analysis Between Integrated Categories of Population and Urban Land Use Changes, and Landscape Types

Usually shrinking or growing phenomena are observed as a matter of population. But the increase or decrease of urban land uses is also an important factor of the phenomena. The decrease of population is not necessarily linked to a decrease in urban land use. And urban sprawl seems to be progressing, even in areas of population decrease.

In this report, the combined changes of population and urban land uses are calculated and examined. As available population and urban land use data are from 2005 and 2010, and from 2006 and 2009 respectively, it is necessary for the land use data to be corrected in their time periods. For the one year correction towards the previous year, the addition of 1/9 of the land use change between 1997 and 2006 is used, and the correction towards the later year, a one year extrapolation of the change between 2006 and 2009 is used. The threshold between decrease or increase and being stable is ±5 % change.

As the integrated categories of urban land use and population changes, five types are introduced (Fig. 2.34). That is, "shrinking", as the population stabilizes or decreases and urban land use decreases, "scattering" as the population decreases and urban land use stabilizes or increases, "remaining stable" as the population and urban land use stabilize, "expanding" as the population stabilizes or increases and urban land use increases, or "compacting" as the population decreases and urban land use decreases or stabilizes.

Figure 2.35 shows the spatial distribution of the combination changes of population and urban land use between 1975 and 2010 with the change threshold 5 %. Figure 2.36 shows the spatial distribution of combination changes of the population and urban land use between 2005 and 2010 with the change threshold 0.7 %. When comparing the two figures, the expanding tendency weakened and the scattering tendency between 2005 and 2010 became stronger than between 1975 and 2010, even though the expanding phenomena is seen in urbanizing areas.

Figure 2.37 shows the combination changes of population and urban land use between 1975 and 2010, and 2005 and 2010. Compared to the change between 1975 and 2010, scattering and shrinking occur relatively high in every category between 2005 and 2010, and the spatial character observed in Figs. 2.35 and 2.36 is confirmed.

Figure 2.38 shows the combination changes of population and urban land use between 1975 and 2010 in each landscape type. In urban, urban paddy field mixed, paddy field and other field types, the rate of expanding exceeds the rate of scattering. But in the combination changes between 2005 and 2010 (Fig. 2.39), scattering exceeds expanding in all types and shrinking exceeds expanding in the paddy

Types	Population decreased	Population remaining stable	Population increased
Urban land use decreased	Shrinking	Shrinking	Compacting
Urban land use remaining stable	Scattering	Remaining Stable	Compacting
Urban land use increased	Scattering	Expanding	Expanding

Fig. 2.34 Integrated categories of urban land use and population changes

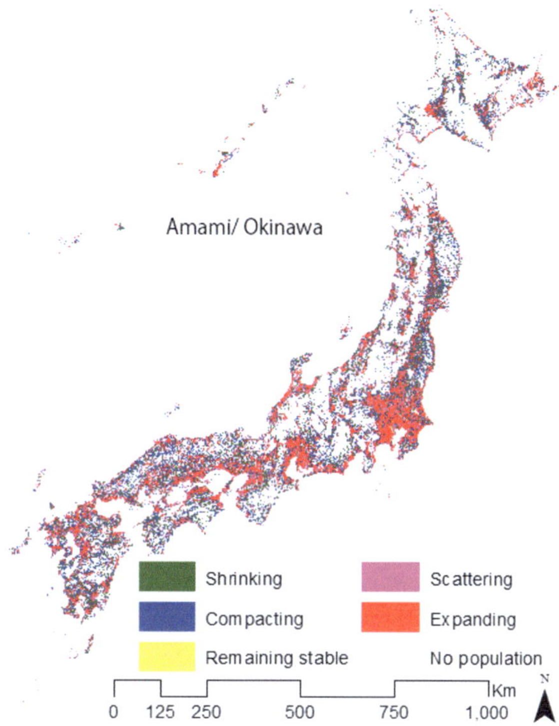

Fig. 2.35 Spatial distribution of combination changes of population and urban land use between 1975 and 2010 (Threshold 5 %)

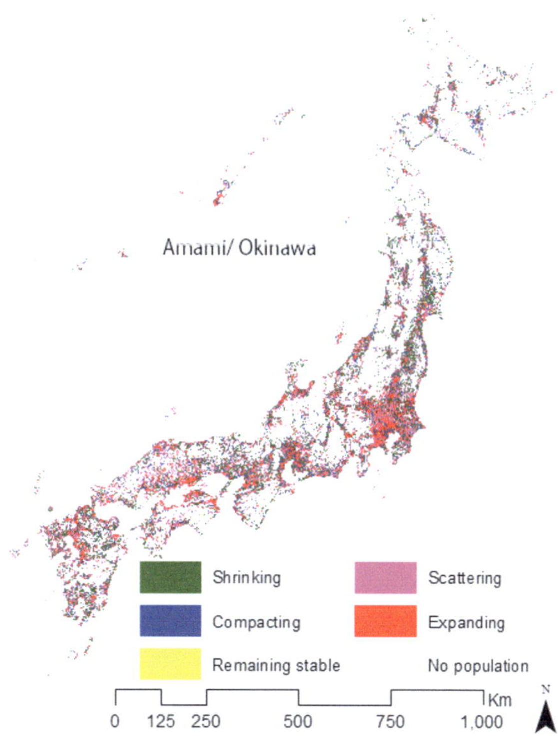

Fig. 2.36 Spatial distribution of combination changes of population and urban land use between 2005 and 2010 (Threshold 0.7 %)

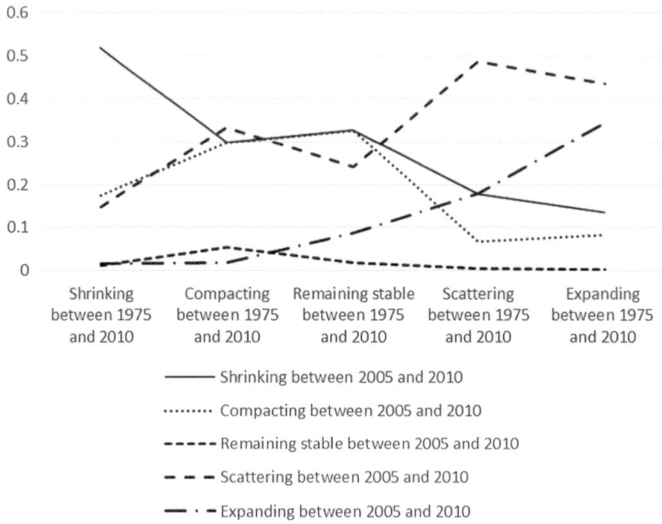

Fig. 2.37 Combination changes of population and urban land use between 1975 and 2010, and between 2005 and 2010

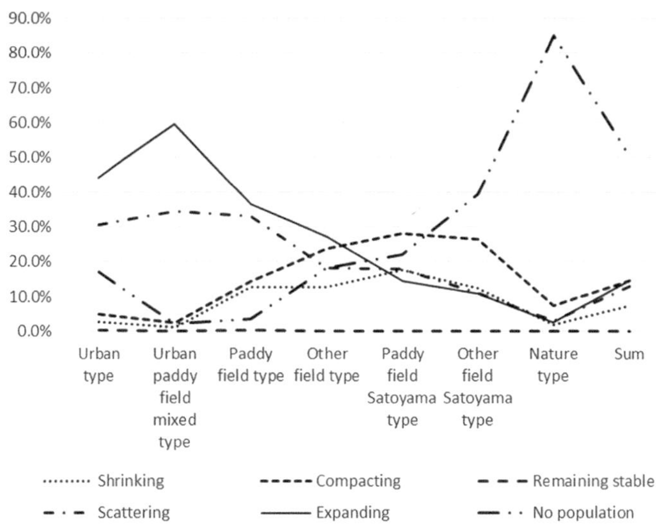

Fig. 2.38 Combination changes of population and urban land use between 1975 and 2010 in each landscape type (Threshold 5 %)

field, other field, paddy field Satoyama, other field Satoyama and nature types. This shows the dilution of urban land use is progressing in all types of landscape.

Figure 2.40 shows the population ratio of 2050 against 2010 in each landscape type. In landscape types of the countryside such as other field, paddy field

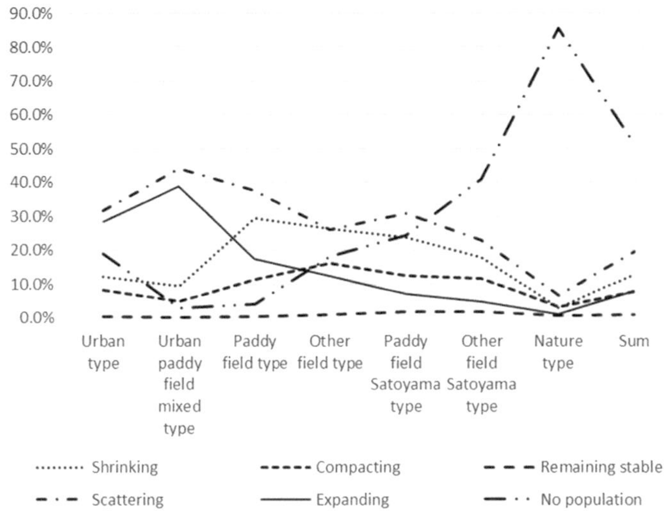

Fig. 2.39 Combination changes of population and urban land use between 2005 and 2010 in each landscape type (Threshold 0.7 %)

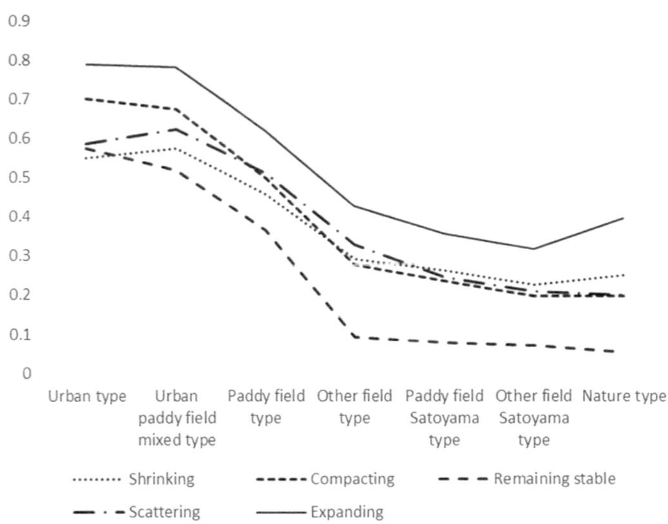

Fig. 2.40 Population ratio of 2050 against 2010 in each landscape type

Satoyama, other field Satoyama and nature types, population decline will progress at an enormous speed, and even in the expanding types, the population in 2050 will decrease to almost under 40 % of that in 2010. The other type population will decrease to almost under 30 %. Even in the urban related landscape types, such as urban, urban paddy field mixed and paddy field types, a relatively strong decrease in population is observed. In such a strong population decrease, cultural landscapes of these landscape types, which are created by the collaboration of human beings and nature, might not be maintained in the same manner and size of today.

Now more detailed special distribution of combination changes of population and urban land use will be examined. Figures 2.41 and 2.42 show the spatial distribution of combination changes of population and urban land use around Nagoya city with a population of 2,276,000 between 1975 and 2010, and between 2005 and 2010 respectively. Comparing these two figures, the center of Nagoya shows scattering changes to compacting, but the expanding phenomenon is weakened and turned to scattering and shrinking.

Figure 2.43 shows the distribution of combination changes of population and urban land use between 2005 and 2010 according to the distance from the center of Nagoya. Expanding occurs mainly within the area of less than 40 km from the center, and scattering occurs mostly in the area between 20 and 40 km from the center. Figure 2.44 shows the combination changes of population and urban land use between 2005 and 2010 around Nagoya in each landscape type. Expanding can be seen dominantly in urban and urban paddy field mixed landscape types, but in paddy field, other field, paddy field Satoyama, and other field Satoyama types, the shrinking phenomenon becomes dominant.

Figure 2.45 shows the distribution of landscape types around Nagoya according to the distance from the center of Nagoya. Between 20 and 40 km urban related landscape types such as urban and urban paddy field mixed landscapes and

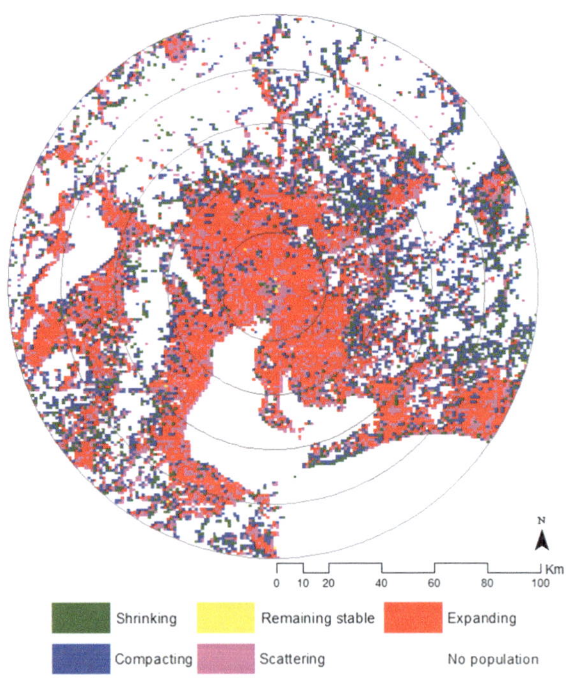

Fig. 2.41 Spatial distribution of combination changes of population and urban land use between 1975 and 2010 around Nagoya

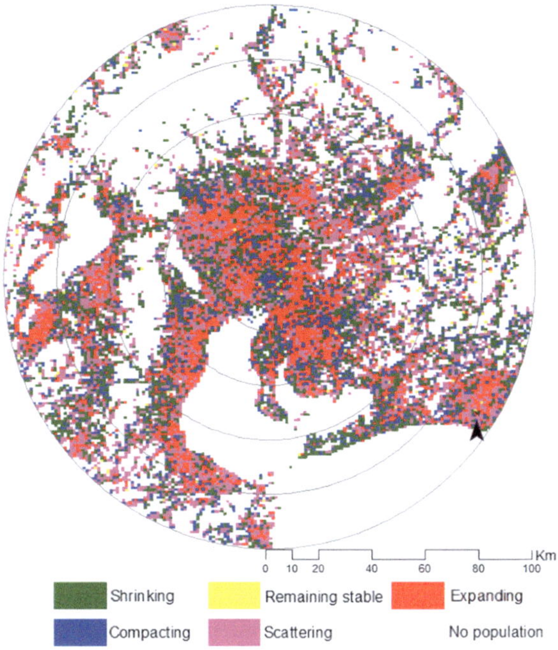

Fig. 2.42 Spatial distribution of combination changes of population and urban land use between 2005 and 2010 around Nagoya

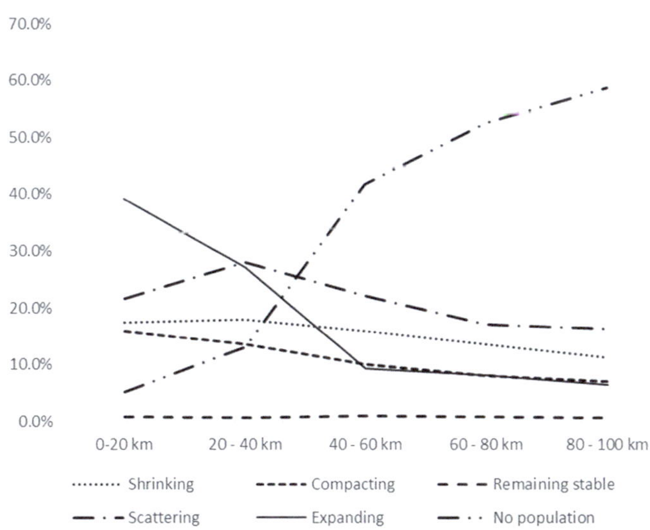

Fig. 2.43 Distribution of combination changes of population and urban land use between 2005 and 2010 around Nagoya

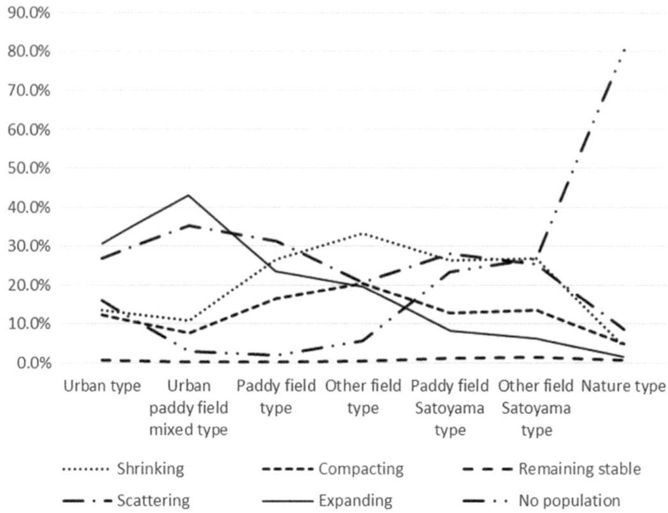

Fig. 2.44 Combination changes of population and urban land use between 2005 and 2010 around Nagoya in each landscape type

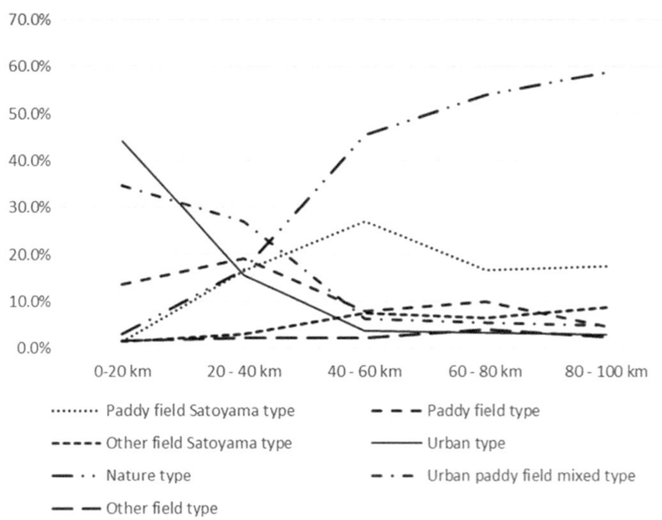

Fig. 2.45 Distribution of landscape types around Nagoya

countryside landscape types such as paddy field and paddy field Satoyama landscape types are mixed strongly and show sprawled urban peripheries. In the area more than 40 km from the center, paddy field and the Satoyama landscape become the dominant landscape types.

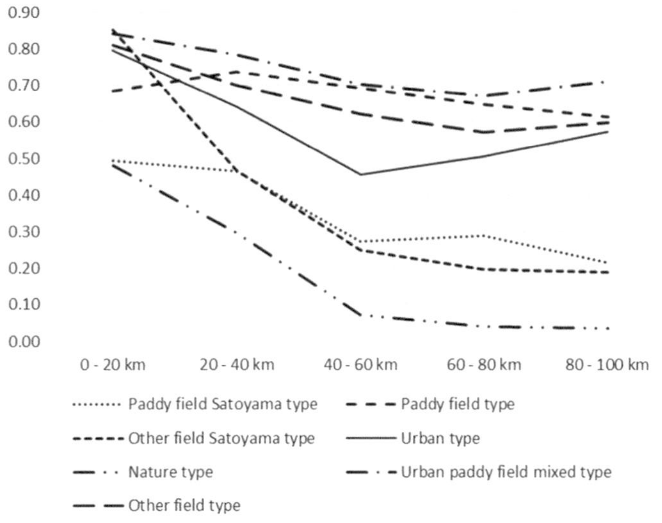

Fig. 2.46 Estimated population change ratio of each landscape type around Nagoya between 2010 and 2050

Figure 2.46 shows the estimated population change ratio of each landscape type around Nagoya between 2010 and 2050. The population decrease of paddy field Satoyama, other field Satoyama and nature types are extremely stronger than other landscape types, especially in the areas more than 40 km from the center. Figure 2.47 shows the population ratio of 2050 against 2010 in each landscape type

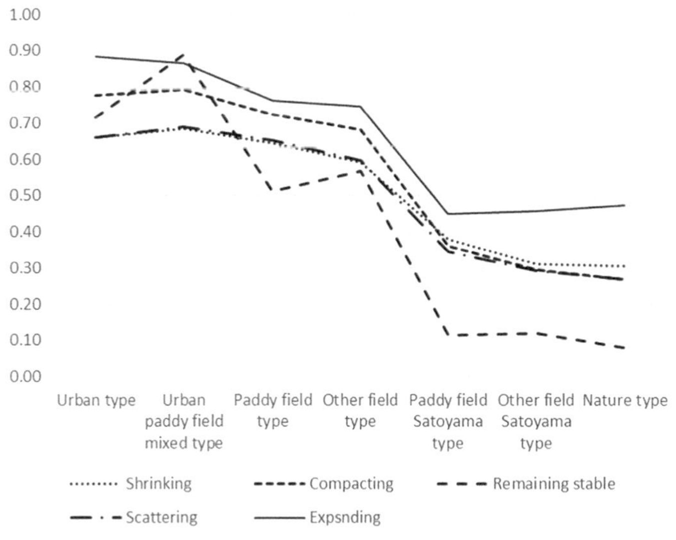

Fig. 2.47 Population ratio of 2050 against 2010 in each landscape type around Nagoya

around Nagoya. Strong population decline can be observed here also in paddy field Satoyama and other field Satoyama landscape types. Nagoya is one of the most active industrial districts, but shrinking and scattering phenomena around Nagoya are progressing and mosaicked structure of shrinking, scattering, compacting and expanding can be observed clearly. And in the periphery of such active areas, Satoyama landscape types are confronting the crisis of disappearance through population decline.

Figures 2.48 and 2.49 show the spatial distribution of combination changes of population and urban land use around Tokyo city between 1975 and 2010, and between 2005 and 2010 respectively. As seen in the comparison of Nagoya, the city center changes from scattering to compacting, and in the peripheries expanding changes to scattering or shrinking. Especially in the eastern and northern part of the periphery shrinking phenomenon is strongly observed.

Figure 2.50 shows the distribution of combination changes of population and urban land use between 2005 and 2010 according to the distance from the center of Tokyo. Expanding occurs mainly within the area of less than 60 km from the center, and scattering occurs mostly in the area between 40 and 80 km from the center. Figure 2.51 shows the combination changes of population and urban land use between 2005 and 2010 around Tokyo in each landscape type. Expanding can be seen dominantly in urban and urban paddy field mixed landscape types, but in paddy field and other field types shrinking phenomenon becomes higher, and in paddy field Satoyama, and other field Satoyama types, shrinking becomes dominant.

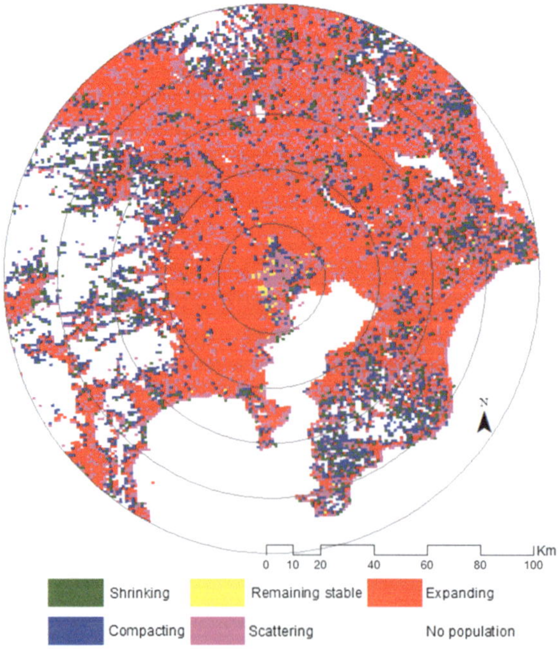

Fig. 2.48 Spatial distribution of combination changes of population and urban land use between 1975 and 2010 around Tokyo

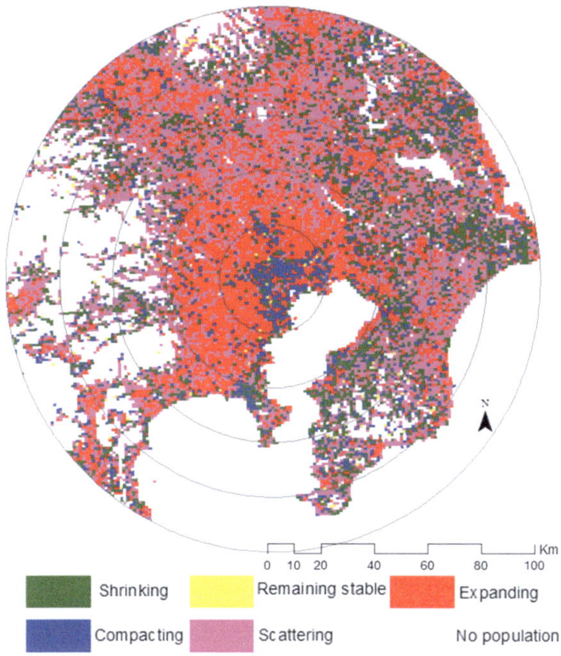

Fig. 2.49 Spatial distribution of combination changes of population and urban land use between 2005 and 2010 around Tokyo

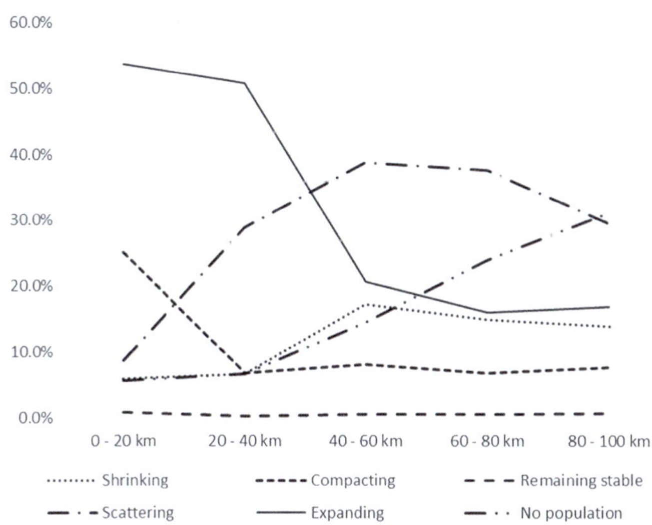

Fig. 2.50 Distribution of combination changes of population and urban land use between 2005 and 2010 around Tokyo

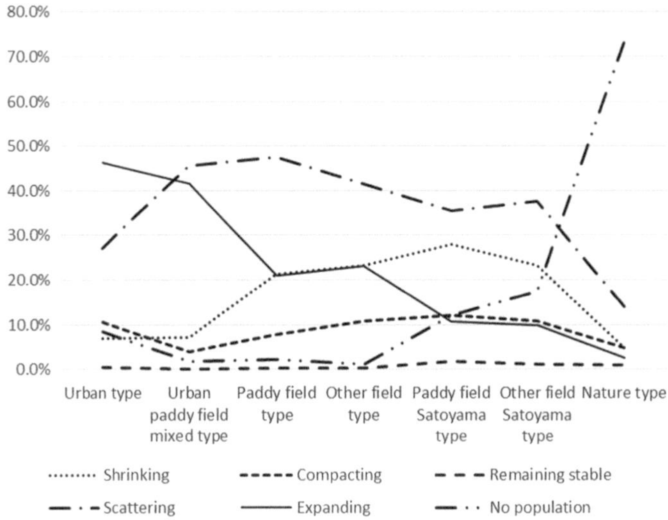

Fig. 2.51 Combination changes of population and urban land use between 2005 and 2010 around Tokyo in each landscape type

Figure 2.52 shows the distribution of landscape types around Tokyo according to the distance from the center of Tokyo. Between 40 and 60 km urban related landscape types such as urban and urban paddy field mixed landscapes and countryside landscape types such as paddy field, other field Satoyama and paddy field Satoyama landscape types are mixed strongly and show sprawled urban

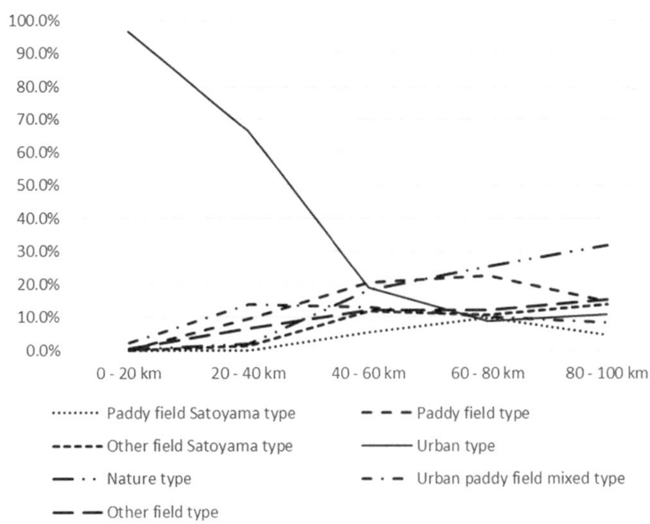

Fig. 2.52 Distribution of landscape types around Tokyo

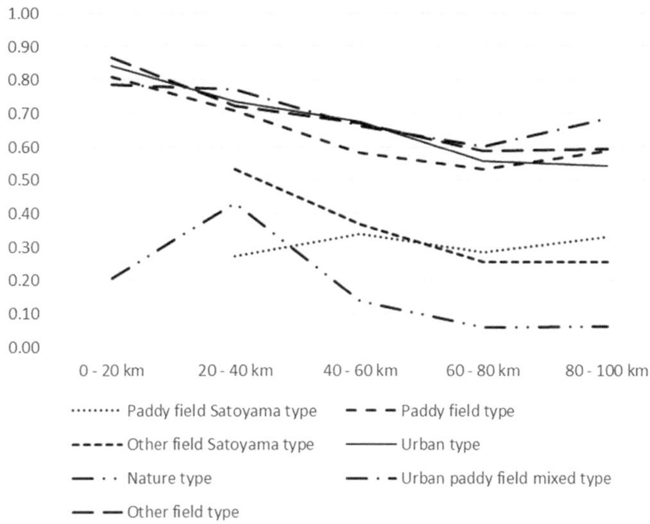

Fig. 2.53 Estimated population change ratio of each landscape type around Tokyo between 2010 and 2050

peripheries. On the contrary, in the area more than 60 km from the center, urban types decrease to nearly 10 %.

Figure 2.53 shows the estimated population change ratio of each landscape type around Tokyo between 2010 and 2050. The population decrease of paddy field Satoyama, other field Satoyama and nature types is much stronger than other landscape types in the areas more than 20 km from the center. Figure 2.54 shows

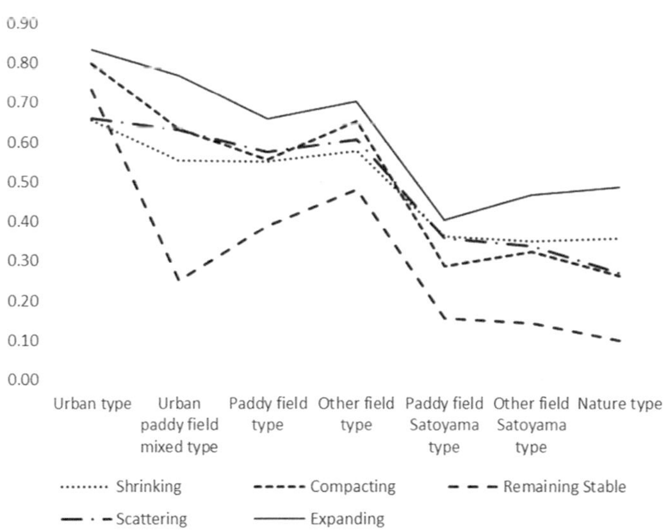

Fig. 2.54 Population ratio of 2050 against 2010 in each landscape type around Tokyo

the population ratio of 2050 against 2010 in each landscape type around Tokyo. A strong population decrease can also be observed here in paddy field Satoyama and other field Satoyama landscape types. Not only Nagoya but also in the periphery of the metropolitan area of Tokyo, mosaicked structure of shrinking, scattering, compacting and expanding are observed more clearly and Satoyama landscape types are confronting the crisis of disappearance.

2.7 Conclusion

The following facts are clarified. There are many scattering areas around big cities where urban land use is increasing and the population is decreasing. And a mosaic structure of shrinking, expanding, compacting and scattering in the peripheral areas away from the city center becomes clear.

It is also clarified that the land gradient regulates the development of urban land use, and land use of paddy and crop fields can easily be changed to urban land use. Even when there is a tendency for the population to shrink, land use "scattering" and partially "expanding" might continue in flatlands.

In the typical country landscape types such as paddy field Satoyama and other field Satoyama types, population decrease will progress in rapid speed in the future, and in the long term continued traditional cultural landscapes, such as Satoyama, might continuously decline.

But not only in Satoyama, but also in urban type landscapes, population decrease will progress in a certain amount, and ecological and cultural management of urban green spaces will be in crisis because of the lack of maintenance budget and labor forces.

These phenomena will lead to the unsustainability of landscape management all over the land. For protecting from such crisis, we need to know the exact amount of the management ability of all kinds of landscapes and landscape elements. The concept of "Landscape management labor accounts", proposed in this book, can serve for this purpose. This research was carried out in 2015 by the help of the Center for Spatial Information Science, the University of Tokyo.

References

Center for Spatial Information Science (CSIS) (2015) Grid square statistics of 1975 population census. http://www.csis.u-tokyo.ac.jp/english/index.html. Accessed 2015

E-Stat (2015) Grid square statistics of 2005 and 2010 population census. Ministry of Internal Affairs and Communication. http://e-stat.go.jp/SG2/eStatGIS/page/download.html. Accessed in 2015

National Land Information Division, National Spatial Planning and Regional Policy Bureau, Ministry of Land, Infrastructure, Transportation and Tourism (MILT) of Japan (2009) National land numerical information, land utilization tertiary mesh data. http://nlftp.mlit.go.jp/ksj-e/gml/datalist/KsjTmplt-L03-a.html. Accessed 2015

Part II
What is Happening in Typical Landscapes in Japan?

Chapter 3
Urban Landscape: Urban Planning Policies and Institutional Framework

Akito Murayama

Abstract This chapter reviews the urban landscape policies and institutional frameworks related to the urban landscape in Japan, introduces an example of a "green" urban planning scheme for Nagoya City and presents the issues of urban landscape in other Japanese cities. The need for a regional and spatial scheme is emphasized, where the calculation of conservation cost plays an important role, as well as the need for green infrastructure development in urban areas, including urban centers, inner-city areas, and the suburbs through interventions of private lands due to scarce public financial resources.

Keywords Planning system · Green planning · Green infrastructure · Nagoya city

3.1 Urban Landscape Policies and Institutional Framework

3.1.1 National Policies and Institutional Framework

The "Grand Design of National Spatial Development towards 2050, Japan" was developed by the Ministry of Land, Infrastructure, Transport and Tourism (2014, July) to define a spatial development vision that would respond to the challenges Japan faces, including population decline; a low fertility rate; a hyper-aging society; competition both among cities and globally; mega disasters combined with aging infrastructure; threats to food, water, and energy supplies; global environmental problems; and technological innovation. "Compact" and "Network" are basic concepts applied to reshaping spatial structures in a manner that improves the

---

A. Murayama (✉)
Urban Land Use Planning Unit, Department of Urban Engineering,
School of Engineering, The University of Tokyo, 7-3-1 Hongo,
Bunkyo-Ku, Tokyo 113-8656, Japan
e-mail: murayama@up.t.u-tokyo.ac.jp

H. Shimizu et al. (eds.), *Labor Forces and Landscape Management*,
DOI 10.1007/978-981-10-2278-4_3

productivity of the entire country. Under the key concepts of "diversity," "connectivity," and "resilience," twelve basic strategies were laid out.

After the development of the "Grand Design," the process of revising the National Spatial Strategy, including national and regional plans, began. In August 2015, the National Spatial Strategy Plan was revised. Its basic vision was as a "National Spatial Structure to Promote Convection," and directions were presented to achieve this vision, some of which are deeply related to the urban landscape of Japan. First, in order to respond to population decline, reshaping and networking "compact cities" became a desirable direction. This idea focuses on reshaping smaller urban areas with sufficient population density to maintain a certain level of public/private services and to improve productivity. As a result, urban areas without services will experience further population decline and vacant properties, such as buildings and parcels. Second, appropriate management of land to shape a safe, secure, and sustainable nation is called for. Conservation and utilization of agricultural land and forests, conservation and regeneration of the natural environment, change from ownership to utilization of underutilized parcels and vacant buildings, and selective use of national land are listed as steps toward achieving this goal. Third, collaboration is called for between this National Spatial Strategy and National Land Use Plans.

National Land Use Plans are developed at the three levels of national, prefectural, and municipal under the National Land Use Planning Act, which aims to ensure the comprehensive and systematic use of national land (Ministry of Land, Infrastructure, Transport and Tourism 2006). The national plan is drafted by the Ministry of Land, Infrastructure, Transport and Tourism, with input from prefectural governors, and decided by the cabinet. The prefectural plan is drafted by a governor, with the input of mayors and the council, and is adopted by the relevant assembly. Municipal plans are drafted based on the prefectural plan by a mayor, with input from local citizens, and adopted by the relevant assembly. The contents of the plans at the three levels are coordinated. Each plan consists of the following sections: (1) the basic concept concerning land use, (2) the size targets and outlines for five categories of land corresponding to each objective, and (3) an outline of the measures needed to achieve these.

Based on the National Land Use Plans, the Land Use Master Plan is developed at the prefectural level to adjust land use at the same and to establish the basic land-use direction. The Land Use Master Plan is composed of a written proposal as well as a planning map on a scale of 1:50,000 that shows the boundaries of the Urban, Agricultural, Forestry, Natural Park, and Natural Reserve Areas. To promote appropriate and reasonable land use based on the Land Use Master Plan, measures are implemented based on individual laws and regulations for each of the five areas. Urban Areas are regulated by the City Planning Law, Agricultural Areas by the Law Concerning the Improvement of Agricultural Promotion Areas, Forestry Areas by the Forest Law, Natural Park Areas by the Natural Park Law, and Natural Reserve Areas by the Natural Conservation Law.

3.1.2 Local Urban Planning System

The City Planning Law regulates land use in Urban Areas (Ministry of Land, Infrastructure, Transport and Tourism 2003). The city planning system under the law consists of the following parts:

- Designation of City Planning Area and Quasi-City Planning Area.
- Development of Master Plan for City Planning Area (by the prefectural government) and Municipal Master Plan (by the municipal government).
- Land use regulations (implemented through Land Development Permission, Building Confirmation, and Notification [District Plan, etc.]).
  - Division of Urbanization Promotion Area and Urbanization Control Area.
  - Zones and Districts (Land Use Zone, Special Land Use District, Special Land Use Restriction Zone, Height Control District, Specified Block, Special District for Urban Renaissance, Efficient Land Utilization District, etc.).
  - District Plan.
- Urban facilities (implemented through public projects).
  - Transport facilities (roads, etc.).
  - Public spaces (parks and green spaces, etc.).
  - Supply and treatment facilities (sewage, etc.).
  - Others.
- Urban development projects (implemented through public/private projects).
  - Urban development projects (Land readjustment project, urban redevelopment project, and others).
  - Urban Development Expediting District.
  - Urban Development Promotion District.
  - Promotion District for Reconstruction of a Disaster Struck Urban Area.

Land Use Zones under the City Planning Law control building coverage, volume, and height of buildings as well as their use under the provisions of the Building Standard Law. These regulations are designed to prevent a mixture of buildings used for different purposes in one area and to ensure a suitable environment for the specified type of land use. The volume of buildings is regulated using Floor Area Ratio (total floor area/site area) and Building Coverage Ratio (building area/site area). The lower the Building Coverage Ratio, the higher the ratio of open space. In addition, to prevent overcrowding and encourage efficient land use, a special bonus is awarded to the total floor area ratio of "excellent projects," which include the development of public infrastructures, open spaces, and housing in their projects. The District Plan (Redevelopment Promotion District), Specified Block, Efficient Land Utilization District, Special District for Urban Renaissance, and Permission System for Comprehensive Building Design are examples of such incentive systems.

In addition to the City Planning Law, there are several other laws that are related to open/green spaces. Open/green spaces are divided into Facility Open/Green Space and Zoning Open/Green Space. As shown in Table 3.1, the former is provided through the development of urban facilities (both public and private) and the latter is provided through the zoning systems of several laws.

Recently, the Siting Optimization Planning System under the Act on Special Measures Concerning Urban Reconstruction was introduced to promote "Networked-Compact Cities" by attracting urban functions and residents to areas that are well served by public transit. Local governments can designate areas to attract urban functions and residents within the Urbanization Promotion Area. The idea of this planning system is to maintain urban functions, such as welfare, medical, commercial, etc., as well as population density in a city center and areas around transit stations while controlling developments and managing de-intensified areas or those outside the designated areas. The greatest challenge lies in drawing the lines of the Community Amenity Advancement District and Residential Advancement District within the Urbanization Promotion Area that result in defining the area outside the dwelling attraction area where people are likely to experience a decline in population and an increase in vacant properties, i.e., potential for more green/open space if planned, designed, and managed properly (Murayama 2016).

Table 3.1 Categorization of green/open space in urban areas

Open/green space		
Facility open/green space		
Urban parks		
Non-urban parks	Public facility open/green space	Public space, pedestrian way, public-operated community garden, public university facility, river bank green space, harbor green space, road environmental facility, etc.
	Private facility open/green space	Civic green space, community garden, private university facility, grounds of private business, temple, shrine, green roof of private building, private zoo, private botanical garden
Zoning open/green space		
Area by law	Green conservation area, special green conservation area, scenic zone, productive green space, landscape district, natural park, river bank, protection forest, agricultural promotion area, etc.	
Agreement	Green space agreement, landscape agreement	
Ordinance	Green area and greening by ordinance, guideline, and contract	

3.2 An Example of a "Green" Urban Planning Scheme: The Case of Nagoya City

3.2.1 Brief History of Urban Green Infrastructure Development in Nagoya City

Modern urban planning in Japan started in 1919 with the establishment of the Former City Planning Law. Nagoya City is recognized as one of the cities successful in practicing modern urban planning with the intensive implementation of the Land Readjustment Project, a basic type of project in Japanese urban planning to develop modern infrastructure, including streets, parks, water works, and sewage systems, by having private land owners provide parts of their parcels for public use and reorganizing the shape of private parcels for efficient land use. Nagoya City, in response to population growth, has successfully developed planned urban areas through this Land Readjustment Project. As much as 70 % of the city's land was developed by Land Readjustment Projects. In addition, urban facilities, including transport facilities, parks, and supply and treatment facilities, were developed through public projects. The Greater Nagoya Future Concept Map in 1965 included the ambitious proposal to develop a wide greenbelt around the rivers and the periphery of Nagoya City. However, the greenbelt was never developed because many private landowners were opposed to the strict regulation of their lands. As a result, only a limited number of urban parks were developed as the hubs of green infrastructure. Several large rivers function as links or corridors of green infrastructure.

In 2010, 23.3 % of Nagoya City was covered by "green," including forest, grass, and meadow, farmland, and water surface. Through urban development, green coverage has been replaced by grey coverage such as asphalt and concrete. In fact, green coverage ratio of the city was 29.8 % in 1990, and 2,136 ha of green coverage was lost in the following 20 years. It should be noted that there are green spaces not only in public lands, such as parks and streets, but also in private parcels, including single-family home parcels, multi-family building parcels, privately-owned urban forests, temples, and shrines. These are the mosaics of green infrastructure.

Nagoya City might experience a population decline in the future. Even if the population stabilizes or increases, there are various transformations of urban areas due to private development and demolishment activities. In the urban center, there are opportunities to develop green area through redevelopment or major infrastructure projects. Inner-city areas might experience de-intensification, and there might be opportunities to insert urban green infrastructures into vacant parcels. Suburban areas without convenient public transit systems might also experience de-intensification, leaving more land for green spaces.

3.2.2 Nagoya City's "Green" 2050 Urban Strategies

The "Water Circulation Restoration 2050 Nagoya Strategy" (Nagoya City 2009b, March), "Low Carbon City 2050 Nagoya Strategy" (Nagoya City 2009a, November), and "Biodiversity 2050 Nagoya Strategy" (Nagoya City 2010, March) are the three environmental, or "green," strategies developed by the Environmental Affairs Bureau of Nagoya City. These strategies present the most long-range vision and strategies regarding the city's physical environment.

3.2.2.1 Water Circulation Restoration 2050 Nagoya Strategy

The following problems occur because of the decrease of green areas and water surfaces resulting from excessive urban development.

- As species decreases, opportunities for people to get in touch with them decrease. It becomes difficult for people to enjoy seasons and for children to be educated about the environment.
- As it becomes difficult for rainwater to permeate the ground, there will be an extreme difference in the river flow and, in the case of isolated rains, an increase in the chance of water disaster. In addition, ground water will not be fostered well, and the amount of ground water will decrease. The level of biodiversity falls as swamps dry.
- The amount of water evaporation from the city decreases, and the heat island phenomena will worsen.
- The development and maintenance of more levees, rainwater pipes, and pumps are essential to the prevention of water disasters. By worsening the heat island phenomena, dependency on air-conditioning systems will rise, thus increasing the city's energy consumption.

The restoration of water circulation in the city and region is needed to mitigate these problems.

Nagoya City developed the "Water Circulation Restoration 2050 Nagoya Strategy," presenting vision, an action plan, issues, and indicators to evaluate the actions necessary for the restoration of water circulation. The representative indicator to evaluate the actions is water balance (when rainwater goes out of the city, how does it divide into "evapotranspiration," "permeation and retention," and "direct runoff." Based on the calculation of water balance in 1965 before urbanization and in 2001 after urbanization, the ideal approach is to increase "evapotranspiration" and "permeation and retention." Considering the reality of the actions, the goals were set to increase "evapotranspiration" from 24 % in 2001 to 31 % in 2050 and "permeation and retention" from 14 to 33 % in the same timeframe. As a result, "direct run off" will decrease from 62 to 36 %. In order to realize these goals, the city must increase its green coverage ratio by around 10 % through the implementation of a green zoning system, conservation of forests, transformation of some urban land uses to natural land uses by urban area compaction, change of paved and roof surfaces of residential/commercial land to better materials with higher water-retention abilities, increase of the ratio of permeable sidewalks to 50 %, and the transformation of 70 % of surface parking spaces to permeable pavement. These are sizeable challenges.

3.2.2.2 Low Carbon City 2050 Nagoya Strategy and Action Plan

The "Low Carbon City 2050 Nagoya Strategy" adopted in 2009 sets the goals of reducing carbon dioxide emission to 25 % below the 1990 level by 2020 and 80 % below the 1990 level by 2050. The city is trying to achieve this "–80 %" through *Machizukuri* (passive design, shift to pedestrians and bikes, and area energy use) (−17 %), product innovation (eco-electronics, facilities and buildings, eco-automobiles) (−31 %), energy supply (de-fossilization of energy suppliers, introduction of alternative energy) (−22 %), and social system (eco-lifestyle) (−11 %). This strategy shows an ambitious image of future urban structure. Urban structures proposed in the strategy include citywide Green/Water Corridors along major rivers and the Higashiyama Hills, Wind Corridors along rivers and canals, and the idea of a compact urban forms and intensified low-carbon station areas with various urban and residential functions surrounded by low-density residential areas and assembled open spaces. Assembled open spaces are well connected with rivers and existing parks and forests, thus creating a green infrastructure.

The action plan for the strategy, which was adopted in 2011, sets 23 sustainability targets for 2020 in the following fields: station area living, automobile, home, green and water, energy, and work. The plan also sets key implementation measures: low-carbon model district, top-runner business support, CO_2 reduction at home, and environmental education/human development. In the designated low-carbon model district, various environmental measures will be implemented in an integrated way, including the introduction of green infrastructure, district energy management, urban heat waste, unused energy, a car-sharing system using low-carbon cars and green buildings, facilities, and equipment.

3.2.2.3 Biodiversity 2050 Nagoya Strategy

The "Biodiversity 2050 Nagoya Strategy" is a policy to realize a sustainable Nagoya in harmony with various species. There are four strategies. The first is to create a healthy city supported by nature through the increase of the habitat of species to 40 % (i.e., increase the green coverage ratio from 25 to 40 %) and increase opportunities for people to touch species in familiar places and creation of corridors of soil, water, and green where species can travel. The second is to create a low-environmental load for living and businesses. The third is to create a culture that values living with nature. The fourth is to establish a system to protect, nourish, and utilize nature.

3.2.3 Land Use, Green, and Transportation Master Plans of Nagoya City

Nagoya's "Green" 2050 Strategies, in order to enhance environmental sustainability, entail visions and strategies of urban structure, land use, transportation, and water/green and became the bases of plans such as the "Nagoya Green Master Plan 2020" (Nagoya City 2011a, March), the "Nagoya City Urban Master Plan" (Nagoya City 2011b, December), and the "Nagoya Transportation Urban Development Plan" (Nagoya City 2014, September).

3.2.3.1 Nagoya City Urban Master Plan

The "Nagoya City Master Plan," adopted in 2011, sets the basic city planning policies for 2020. It is composed of the city's vision, urban development policies (urban structure based on networked station areas, etc.), sectorial policies, strategic *Machizukuri* (three strategies and key areas), and the promotion of community-based *Machizukuri*.

One feature of the plan is its emphasis on the realization of urban structure based on networked station areas. The stations areas are defined as those within an 800 m radius from the railway stations. Various urban functions, such as commercial, business, housing, service, and cultural, should adequately be located in walkable station areas with attractive urban spaces, and the station areas are networked by public transit. In addition, land use zones such as the urban center area, urban area, and harbor/waterfront, environmental corridors of green and water, and station areas are designated. This urban structure is reflected in the land use policies. Assuming that the population of Nagoya City will increase until 2025, the plan seeks an increase of population, during both day and night, in the station areas. On the other hand, adequate measures should be considered to deal with areas outside station areas that have decreasing populations and increasing numbers of vacant properties.

In 2016, a group of experts in various fields, including the author, was conveyed to draft the Siting Optimization Plan for Nagoya City. Based on the city's urban master plans and other related plans, the lines of Community Amenity Advancement Districts and Residential Advancement Districts within Urbanization Promotion Area will be drawn. There are many elements to consider, but one important element would be the creation of a green infrastructure in the urban center, inner-city, and suburban areas with the help of these lines, since it is possible to designate areas within Urbanization Promotion Area to depopulate and de-intensity. The increase of vacant properties can help create a green infrastructure if planned, designed, and managed properly.

3.2.3.2 Nagoya Green Master Plan 2020

Nagoya Green Master Plan 2020, adopted in 2011, is a master plan based on the Urban Green Space Conservation Act to promote the appropriate conservation of green space and the promotion of greening. Herein, "green" includes plants such as trees and flowers, land such as forests, farmlands, rivers, and ponds, and the base to support the ecological system, such as soil and water. Basic policies and leading projects are presented to respond to the situation where quantity and quality of green is decreasing due to urban development. There are three leading projects. The first is to increase the citizens' concern about green through the expansion of citizen participation and the strengthening of collaboration with civic organizations. The second is to create a green water corridor through creating green streets, developing the waterfront environment, developing green blocks, utilization of local seeds, and promotion of environmentally conservative agriculture. The third is to retain as much existing green as possible through conservation of forests and farmlands with new ideas, by building a system of green conservation and management, and through the promotion of utilizing urban parks. The development of green blocks, which is deeply related with architecture and urban planning, involved intensively plant trees in public spaces and increasing high-quality green through a green zoning system and district plan system. The creation of green streets involves the creation of an aesthetically beautiful urban landscape and introduction of cool wind in the summer through growing street trees.

The green zoning system introduced in 2008 mandates a certain percentage of green coverage in new construction or building extensions in parcels larger than a certain size based on the provision of Urban Green Space Conservation Act. The amount of green coverage is checked through building confirmation and completion inspection.

The boundaries of parks and green land can be designated in urban planning in advance. In Nagoya City, approximately 2,800 ha are designated as parks and green lands in the official urban plan. Parks and green lands designated in urban planning are sequentially being developed through land acquisition and facility development. However, there are undeveloped areas that are not used as urban parks for a long period of time. The Long-term Undeveloped Parks and Green Lands are designated urban planning parks and green spaces that Nagoya City will develop in the future through the acquisition of private lands, but they remain undeveloped for a long period of time after the designation mainly due to the shortage of public money. There are 40 such parks and green lands in Nagoya City. Of the 1,150 ha designated as parks and green spaces, about a quarter are private lands that need to be purchased.

3.2.3.3 Nagoya Transportation Urban Development Plan

The "Nagoya New Transportation Strategy Promotion Plan" is the city's master plan for the field of transportation. There are six basic policies, including the realization of a transportation system that everyone can easily use, the development of a safe transportation environment, the realization of a low-carbon transportation system, the promotion of transportation eco-life, the creation of a walkable city, and the development of a transportation environment that supports the vitality of the city. In order to strategically implement these policies, the leading project, "Street Community Development," is proposed to change the street environment to be more human-oriented and to create a safe and vibrant urban area.

The "Nagoya Transportation Urban Development Plan" is the action plan for this "Street Community Development" and presents vision, policies, and contents of measures, measures for different areas, and a short, mid, and long-range roadmap for the urban center. The measures include clarification of the roles, reconsideration of the space distribution of arterial roads, and the creation of street space based on local characteristics, including the control of automobile entries to commercial areas and station areas, the expansion of the sidewalks along local streets, and the creation of vibrant blocks. Through this idea of transforming the design of streets, there will be opportunities to create green streets by planting trees and bushes and installing devices to retain and encourage permeation of rainwater.

3.3 The Issues of Urban Landscapes

The author introduced an innovative "green" urban planning scheme for Nagoya City, which is a central city of Ise-Bay Basin Area. However, there are no such "green" urban planning schemes in the municipalities around Nagoya City, urban development that destructs the natural environment and farmland still continues, and land coverage with un-permeable concrete and asphalt is still increasing. It is natural that each municipality promotes a certain level of urban development to sustain their economy. In order to control further urbanization and protect the natural environment and farmland, there is a need for the Ise-Bay Basin Area as a whole to pay for the cost of conservation. If there is a need for new urban development, the municipality should make an effort to minimize the environmental load of the development and also consider a way to mitigate the environmental impact by restoring ecological services in other areas. In such an undertaking, calculation of conservation cost plays an extremely important role.

Owing to the situation of limited public money to develop and maintain parks and green lands as facilities, it is important to shift the emphasis to increasing greenery in private lands, including low-story residential areas, multi-family residential areas, parking spaces, and temples and shrines. There are also opportunities to redesign the streets to be more human-oriented by expanding sidewalks and planting trees and bushes.

References

Ministry of Land, Infrastructure, Transport and Tourism (2003) Introduction of urban land use planning system in Japan. http://www.mlit.go.jp/common/000234477.pdf. Accessed 31 Mar 2016

Ministry of Land, Infrastructure, Transport and Tourism (2006) What are national land use plans? http://www.mlit.go.jp/english/2006/b_n_and_r_planning_bureau/02_useplans/. Accessed 31 Mar 2016

Ministry of Land, Infrastructure, Transport and Tourism (2014) Grand design of national spatial development towards 2050, Japan

Murayama A (2016) Land-use planning for depopulating and aging society in Japan. In: Yamagata Y, Maruyama H (eds) Urban resilience—a transformative approach. Springer (in press)

Nagoya City (2009a) Low carbon city 2050 Nagoya strategy

Nagoya City (2009b) Water circulation restoration 2050 Nagoya strategy

Nagoya City (2010) Biodiversity 2050 Nagoya strategies

Nagoya City (2011a) Nagoya green master plan 2020

Nagoya City (2011b) Nagoya urban master plan

Nagoya City (2014) Nagoya transportation urban development plan

Chapter 4
Urban Periphery Landscape: Dichotomization of Urban and Rural Dimensions

Toru Terada

Abstract This chapter focuses on the urban periphery landscape—urban fringe and suburban rural areas—in which both urban development and rural conservation should be considered. In Asian countries, urban periphery areas are often threatened by urban sprawl, land abandonment, or disordered land uses. In Japan specifically, the Area Division Scheme, a land use policy that divides urban and rural areas, was established in 1968 under the New City Planning Act. Even currently, however, a great deal of urban–rural disordered land mixture remains in urban periphery due to loose land use regulations. Farmlands are protected on some level through the easement of land taxes, but most have served to induce urban sprawl. Remaining forests, former coppiced woodlands, are attracting public attention as Satoyama (a term that combines village and forest), but conducting appropriate management is a challenge. Urban periphery faces a host of challenges and will need to be re-designed in an effort to respond to depopulation and city shrinkage.

Keywords Land use policy · Farmland and forest management · Urban sprawl · Land fragmentation · Satoyama

4.1 Introduction

This chapter focuses on the urban periphery landscape—urban fringe and suburban rural areas—in which both urban development and rural conservation should be considered. Land use control in these regions has to consider both urban and rural planning. In most Western countries, urban periphery is clearly divided into urban and rural areas by strict land use planning that maximizes the functions of each respective area. In Japan, however, land owners' rights take precedence over public land use regulation, resulting in "urban sprawl" and farmland appearing within

---

T. Terada (✉)
Department of Urban Engineering, School of Engineering,
The University of Tokyo, 7-3-1 Hongo, Bunkyo, Tokyo 113-8656, Japan
e-mail: terada@k.u-tokyo.ac.jp

H. Shimizu et al. (eds.), *Labor Forces and Landscape Management*,
DOI 10.1007/978-981-10-2278-4_4

urban areas (Hebbert 1994; Yokohari et al. 2000; Sorensen 2002). In this chapter, land use policy in urban periphery landscape is introduced from both urban and rural planning points of view; in addition, problems of farmland and forest management are discussed based on a review of existing literature.

4.2 Issues on Land Use Planning

4.2.1 Land Use Policy

The most important factor for land use control within the urban periphery landscape is the New City Planning Act, established in 1968 but still in force today. Based on the act, a prefectural government can determine an area, such as a concentrated-population area, to be a City Planning Area candidate. Within these areas, measures including land use control can be enforced. In the 1960s, the massive influx of rural migrants to cities resulted in control of urban expansion becoming the biggest concern in urban planning at that time. In response, the Area Division scheme was formulated in the new City Planning Act. The Area Division is a land use policy that divides a City Planning Area into two parts: Urbanization Promotion Area (UPA) and Urbanization Control Area (UCA) (MLIT 2003). Previously developed areas and those slated to be urbanized within 10 years are designated as UPA. In contrast, rural areas located outside of the city intended to be conserved (i.e., under regulated development) are designated as UCA. In addition, the Act on Establishment of Agricultural Promotion Regions (APR) was established in 1969 and states that prefectural government and local government can designate prime farmland in APR as Farmland for Agricultural Promotion (FAP). Once farmland is designated as FAP, development is strictly regulated (Fig. 4.1).

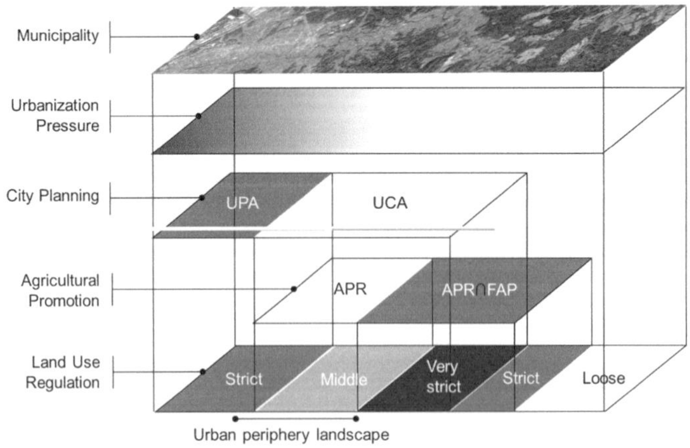

Fig. 4.1 Pattern diagram of land use regulation

The Area Division by City Planning Act appears to be functioning well through cooperation the designated APR areas; however, in reality, land use is not being ideally controlled in urban periphery areas. The reasons for this are likely due to the following three points: (1) optimistic prospects for future development in UPA, (2) loose land use regulation in UCA, and (3) emphasis of private property rights. Examples of these points are given in the next section.

4.2.2 Urban-Rural Land Use Mixture

The City Planning Act and Building Standards Act permit urban development if the plans for such meet the infrastructural standard (e.g., connecting to a road at least 4 meters wide, as well as having electricity and a water supply). During the rapid immigration from rural areas during the economic growth period, low-standard housing spread without well-ordered urban forms (e.g., the town brock system). In particular, in the inner edge of UPA—where development potential was overestimated—the area reveals a disordered landscape that is a combination of small houses, fragmented farmland, and forest.

Urban sprawl is often seen in the outer edges of UPA, as the City Planning Act makes an exception to permit development in UCA close to the city if emerging buildings would connect to a developed area of UPA (Article 34, item 11 and 12). This development permission is due to the following factors: (1) UCA is understood to be land reserved for future development, and (2) most farmland on the outer edges of UPA is not designated as FAP; rather, it is comprised of scattered small farmlands that are not suitable for large-scale and intensive farming (Fig. 4.2).

As discussed, although the ideal land use policy would outline a clear division of urban and rural areas, mixed and disordered landscapes have become the typical scenery of urban periphery, illustrating the problems with the current policy (Yokohari et al. 2000; Sorensen 2002). This is because individual land property rights are fundamentally guaranteed by the constitution (Article 29). As such, public intervention is not always prioritized and tends to be less strict, making it difficult to resolve urban and rural land mixture. These property rights should be evaluated more proactively to develop a unique land use system in which farm and wooded landscapes provide key ecological functions, visual amenities, and cultural services (Yokohari et al. 2010).

Management of farm and wooded urban periphery landscapes, however, is a challenging task, since the measures formulated for industrial farmland and forest management are not consistently applied. The following two sections discuss the challenges of farmland and forest management within an urban periphery landscape.

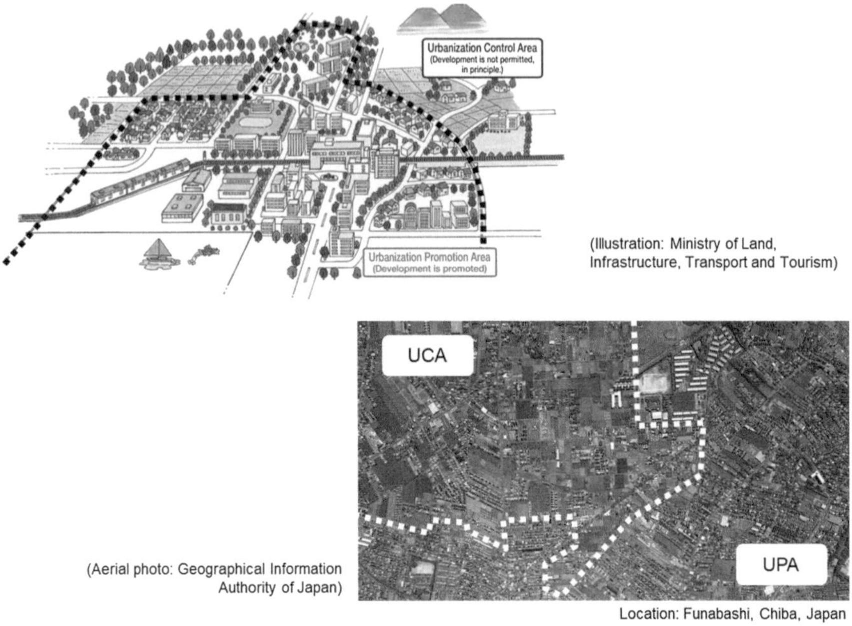

Fig. 4.2 Ideal (*top*) and reality (*bottom*) of the area division scheme

4.3 Issues of Farmland Management

4.3.1 Farmland in Urban Areas

Farmland in urban areas (UPA) is understood as land reserved for future urban development; therefore, land taxes are evaluated as developed land (i.e., land with building structures on it) even though it is currently farmland. Such farmland is called Farmland for Building Site (FBS). Farmers are typically unable to handle the tax burden on FBS and simply sell their land to developers; however, for those farmers who wish to continue their farming practice, even public policy cannot force land use change, as the constitution preferentially guarantees their property rights. In response, the Productive Green Land Act was established in 1974, which states that farmland in UPA can be designated as Productive Green (PG), the taxes for which are evaluated as farmland. While farmers can drastically reduce their land tax burden, particularly their real estate and inheritance taxes, by registering their farmland as PG, they have to maintain the land as farmland for the 30 years following the registration. Once farmland is designated as PG, it is basically impossible to eliminate this designation. In 1991, the Productive Green Land Act was revised and the amount of land area designated as PG increased since 1992 due to the elimination of the supplemental institution that moderated the tax burden on FBS (Fig. 4.3).

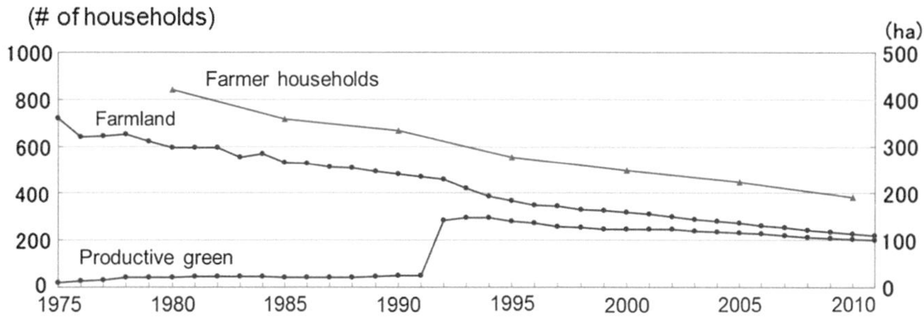

Fig. 4.3 Changes in the farmland, farmer households, and productive green in Setagaya Ward, Tokyo, 2011 (Setagaya Ward 2015)

There are two key challenges of farmland in UPA: (1) the possibility of a great number of PG dissolving around 2022 (i.e., the 30-year mark from the initial 1992 designations), and (2) maintenance of the farmland. The Productive Green Land Act determines that local government should purchase any dissolved farmland at the market price and use the land for public purposes; however, this may not happen due to government budget shortages, at which time it is highly likely that the land would be sold out in the market. Unless appropriate regulations or incentive measures are enacted, disordered and scattered development is likely to occur on dissolved farmland, resulting in disordered land use. Regarding the maintenance of PG farmland, appropriate agricultural production activities are not always carried out, since PG is often understood to be solely a tax-planning activity. For example, "mimic farming", such as planting chestnut trees on the farmland to make maintenance easier, is sometimes seen in PG. Additionally, farmland maintenance is difficult due to an aging farmer population. As such, new initiatives should be considered to maintain urban farmland, such as urban-resident support of agriculture (Namiki et al. 2006).

4.3.2 *Farmland in Peri-Urban Areas*

Taxes on farmland in peri-urban areas (UCA) are lighter than those enforced in UPA, but UCA faces urban sprawl and encroached farmland resulting from loose land use control. Farmland encroachment results in declining farmland areas, leading to easing of regulations for its development. This urban sprawl and abandonment of nearby farmland have been increasing since the revision of the City Planning Act in 2001, which promotes decentralization and permits prefectural government to enact ordinances that seek to ease regulations of urban development in UCA (Mitsui et al. 2009). In addition, farmers in UCA tend to stop their farming practices if there are alternative urban job opportunities that could bring in more household income (Hattori and Yamaji 1998). Compared to large paddies on river plains, paddy fields in small valleys also tend to be abandoned as they are targets of the governmental rice-reduction program.

The benefits of peri-urban farming include its proximity to cities as areas of consumption. Some farmers seek to meet the demand of urban residents by growing delicate leaf vegetables in greenhouses or running pick-your-own options for urban residents. Allotment gardens (or share plots) for urban residents is also increasing. Such strategic acts toward farmland maintenance and taking advantage of proximity to cities are essential for continuing successful farmland in UCA.

4.4 Issues on Forest Management

4.4.1 Satoyama

Forests in urban and peri-urban areas are often called Satoyama, a Japanese term coined from *sato* (village) and *yama* (forest). Satoyama is typical secondary forest in Japan, which has historically been used for collection of organic materials such as firewood, charcoal, leaves, and foodstuffs (e.g., mushrooms, wild vegetables). However, the relationship between people and Satoyama has changed, mainly due to the introduction of fossil fuels, reflecting the ecological condition of Satoyama. As Satoyama covers a wide variety of Japanese land areas, the socio-ecological interactions of people to this forest areas also varies among areas. As such, this overview is divided into three parts: Satoyama in (1) urban areas, (2) peri-urban areas, and (3) rural areas.

4.4.2 Satoyama in Urban Areas

Most cities in Japan have expanded and intermixed with formerly rural environments. This situation is typical in UPA. In this process, a great number of Satoyama were transformed into areas for buildings or urban infrastructure, with some isolated forest patches left within the urban fabric (Ichikawa et al. 2006). The key characteristic of such urban Satoyama (equivalent concept to the urban secondary forest discussed in Chap. 12) is its high value as "urban nature" derived from its multi-functionality of aesthetic and recreational value. For example, urban Satoyama can create biological value by offering patches for bird species migration (Hashimoto et al. 2005) and climate mitigation (e.g., creating cool air flow by combined canopies) (Hamada and Ohta 2010). Generally, the value of urban nature becomes increasingly higher as its surrounding areas become more and more urbanized. In order to enhance such ecological functions, urban Satoyama are protected from development by governmental land acquisition and land use regulation that includes compensation to land owners, the latter of which includes options such as Special Greens Conservation Areas, which are regulated as privately held Satoyama (MLIT 2014). However, reliance on government-led policies and programs are being eroded due to the loss of tax revenue caused by depopulation. In order to address this issue, citizen participation in management of urban Satoyama is increasingly encouraged through the Designated Manager System, which invites the private sector to take part in the maintenance of public green spaces (Tsuchiya et al. 2013).

4.4.3 Satoyama in Peri-Urban Areas

Challenges in Satoyama conservation can be divided into two aspects: (1) declining its area, the "quantitative" aspect, and (2) abandonment of management, the "qualitative" aspect. The former is more serious in urban areas, while the latter will become increasingly problematic as the Satoyama become located farther from cities. In peri-urban areas, both aspects should be considered simultaneously. Thanks to the designation as UCA, Satoyama in peri-urban areas are relatively difficult to become approved for development, comparing to those in UPA; however, land owners are increasingly not motivated to manage Satoyama by themselves due to the loss of economic viability. Consequently, peri-urban Satoyama in UCA tend to be abandoned, and the resulting overgrown vegetation can negatively affect the surrounding environment. Satoyama in UCA often physically and functionally interlink with other agricultural land uses, and such environmental unity can bring high ecological value (Katoh et al. 2009). In addition, the high recreational value of peri-urban Satoyama is derived from proximity to cities, so conservation measures should be aimed at attracting urban residents' involvement. Establishment of Forest Parks, Nature Observation Forests, and Citizen Forests, which opens private forests to urban residents, are the current options led by government. The private sector can also cooperate with the National Trust, and can contribute to substantive management by conducting volunteer work (Kobori and Primack 2003). Continuity of forest management, however, is problematic because most people who have engaged in Satoyama maintenance are currently aging. In addition, the measures tend to be applied to specific forests, typically those that stand to provide high recreational potential. To expand Satoyama management, substantial economical enhancement will be indispensable; for example, management as a resource for community-based wood energy might be an option (Terada et al. 2010).

4.4.4 Satoyama in Rural Areas

Most Satoyama in Japan are located in rural areas and have been abandoned due to the economic loss of trying to use them as fuelwood forests. Rural Satoyama stands used to have diverse ages as a result of frequent cutting and served as part of the mosaic landscape along with paddy fields, crop fields, grasslands, irrigation ponds, and other agro-related environments. Since the high economic growth period of the 1960s, however, such diversity has been lost due to the elimination of consistent Satoyama maintenance and the homogenization of the rural landscape (Fukamachi et al. 2001). Elimination of the maintenance also affected the level of degradation and the forests' physiologic health, accompanying forest diseases over a period of time. During the 1970s, red pine (*Pinus densiflora*) was seriously damaged by the nematode *Bursaphelenchus xylophilus*; currently, oaks (particularly the *Quercus* species) are being attacked by the insect *Platypus quercivorus*, especially in areas along the Sea of Japan (Kuroda et al. 2012). Conversely, some municipalities have started to reuse rural Satoyama as fuelwood forests for generating heat and electricity from woody biomass. This revitalization is promoted by the Feed-in Tariff policy that introduced renewable energies in 2012 (Nakata et al. 2005; METI 2012). Rural Satoyama are also highlighted as places for "green" tourism and as venues to hold art festivals (Klien 2010). However, revitalization of entire areas of rural Satoyama would be challenging due to the severe depopulation and rapid aging in rural Japan. Satoyama management should be integrated into the larger strategic plan for forest management that includes prospected operations for artificial and natural forest, with conversion to more naturalistic forests considered for the long-term vision (Fig. 4.4).

Fig. 4.4 Example of an abandoned Satoyama (*Photo* Author)

4.5 Conclusion

The urban periphery landscape can serve as a reflection of fundamental incoherence within the ideas under urban and rural principles. As long as such a dichotomy exists between the two, alternative solutions will not be possible. Furthermore, the urban periphery will be seriously affected by city shrinkage in the near future, factoring in depopulation due to inconvenient access. From these two aspects, management of the urban periphery landscape is a serious challenge; however, city shrinkage can also be a catalyst to orchestrate urban–rural mixture toward the creation of a sustainable urban–rural fabric. Related proposals are discussed in Chap. 14.

References

Fukamachi K, Oku H, Nakashizuka T (2001) The change of a satoyama landscape and its causality in Kamiseya, Kyoto. Landscape Ecol 16:703–717

Hamada S, Ohta T (2010) Seasonal variations in the cooling effect of urban green areas on surrounding urban areas. Urban Forest Urban Greening 9:15–24

Hashimoto H, Natsuhara Y, Morimoto Y (2005) A habitat model for Parus major minor using a logistic regression model for the urban area of Osaka, Japan. Landscape Urban Plan 70: 245–250

Hattori T, Yamaji E (1998) Factors to abandon cultivation by the condition of farm household in suburban area. J Assoc Rural Plan 16(4):325–333 (in Japanese with English abstract)

Hebbert M (1994) Sen-biki amidst Desakota: urban sprawl and urban planning in Japan. In: Schapira P, Masser I, Edgington D (eds.) Planning for cities and regions in Japan. Liverpool University Press, Liverpool, pp 70–91

Ichikawa K, Okubo N, Okubo S, Takeuchi K (2006) Transition of the satoyama landscape in the urban fringe of the Tokyo. Landscape Urban Plan 78:398–410

Katoh K, Sakai S, Takahashi T (2009) Factors maintaining species diversity in satoyama, a traditional agricultural landscape of Japan. Biol Conserv 142:1930–1936

Klien S (2010) Contemporary art and regional revitalisation: selected artworks in the Echigo-Tsumari Art Triennial 2000–6. Jpn Forum 22(3–4):513–543

Kobori H, Primack RB (2003) Participatory conservation approaches for satoyama, the traditional forest. J Hum Environ 32(4):307–311

Kuroda K, Osumi K, Oku H (2012) Reestablishing the health of secondary forests "Satoyama" endangered by Japanese oak wilt: a preliminary report. J Agric Extension Rural Dev 4(9): 192–198

METI (Ministry of Economy, Trade and Industry) (2012) Feed-in Tariff Scheme in Japan. http://www.meti.go.jp/english/policy/energy_environment/renewable/pdf/summary201207.pdf. Accessed 4 Aug 2015

Mitsui T, Sakuma Y, Akasaki K (2009) Current state of farmland diversion and relationship to changes in current usage of surrounding farmland in urbanization control area at the metropolitan fringe: a case study of Sakai city, Osaka prefecture. City Plan Rev 44(3):49–54 (in Japanese with English abstract)

MLIT (Ministry of Land, Infrastructure, Transport and Tourism) (2003) Introduction of urban land use planning system in Japan. http://www.mlit.go.jp/common/000234477.pdf. Accessed 1 Feb 2015

MLIT (Ministry of Land, Infrastructure, Transport and Tourism) (2014) City policies in Japan. http://www.mlit.go.jp/en/toshi/. Accessed 27 July 2015
Nakata T, Kudo K, Lamont A (2005) Design for renewable energy systems with application to rural areas in Japan. Energy Policy 33:209–219
Namiki R, Yokohari M, Hoshi T, Watanabe T, Amemiya M (2006) Agro-activities by urban residents on the farmlands of urban promotion area. J Rural Plan Assoc 25:269–274 (in Japanese with English abstract)
Setagaya Ward (2015) Land use of Setagaya 2011. http://www.city.setagaya.lg.jp/kurashi/102/118/329/d00123789_d/fil/P74-76.pdf. Accessed 1 Mar 2016
Sorensen A (2002) The making of urban Japan: Cities and planning from edo to the twenty-first century. Routledge Curzon, New York
Terada T, Yokohari M, Bolthouse J, Tanaka N (2010) "Refueling" satoyama woodland restoration in Japan: enhancing restoration practice and experiences through woodfuel utilization. Nat Cult 5(3):251–276
Tsuchiya K, Okuro T, Takeuchi K (2013) The combined effects of conservation policy and co-management alter the understory vegetation of urban woodlands: a case study in the Tama Hills area, Japan. Landscape Urban Plan 110:87–98
Yokohari M, Amati M, Bolthouse J, Kurita H (2010) Restoring urban fringe landscapes through urban agriculture: the Japanese experience. disP Plan Rev 46(181):51–59
Yokohari M, Takeuchi K, Watanabe T, Yokota S (2000) Beyond greenbelts and zoning: a new planning concept for the environment of Asian mega-cities. Landscape Urban Plan 47:159–171

Chapter 5
Agricultural Landscape: Farmland Abandonment and Direct Payments

Kazuaki Tsuchiya and Kazu Hagihara

Abstract This chapter shows the historical shift in Japanese land improvement projects and the borrowing and lending of farmland since World War II, referring to recent trends in agricultural land abandonment and extensification in Japan. This chapter also reviews causes of agricultural land abandonment, focusing on socio-economic causes and biophysical causes. In addition, comparative research on the direct payment system between Japan and the EU is reviewed, as is the history of direct payment schemes in agricultural and rural development policies in Japan. In a review of existing studies, this chapter identifies the characteristics of agricultural landscape management in Japan.

Keywords Land improvement · The borrowing and lending of farmland · Agricultural land abandonment · Direct payment system

5.1 Introduction

Japanese agricultural landscapes, which mainly comprise paddy fields, vegetable farmlands, orchards, and coppice forests, have experienced substantial changes in the previous century under rapid agricultural moderation, urbanization, and an

---

The original version of this chapter was revised: In Fig. 5.2 caption: rations was changed to ratios; In Fig. 5.5 caption: text was added (Some parts of this figure were altered). The erratum to this chapter is available at 10.1007/978-981-10-2278-4_26

K. Tsuchiya (✉)
Department of Ecosystem Studies, Graduate School of Agricultural and Life Sciences, The University of Tokyo, 1-1-1 Yayoi, Bunkyo, Tokyo 113-8657, Japan
e-mail: tcy@live.jp

K. Hagihara
Center for Community Co-design, The University of Shiga Prefecture, 2500, Hassaka-cho, Hikone 522-0057, Shiga, Japan
e-mail: hagihara.k@office.usp.ac.jp

H. Shimizu et al. (eds.), *Labor Forces and Landscape Management*,
DOI 10.1007/978-981-10-2278-4_5

aging population. Unlike other developing Asian countries, Japan is facing a severe issue related to agricultural land abandonment and extensification, particularly in remote mountainous areas. This chapter explains previous changes in the agricultural landscape of Japan, the consequences of agricultural land abandonment, and recent policy responses to the loss of environmental, social, economic, and cultural values in rural areas.

5.2 The Historical Shift in Japanese Land-Improvement Projects and the Borrowing and Lending of Farmland Since WW II

The history of land improvement in Japan can be divided into three main periods (Rural Development Planning Center 2003). The first period is regarded as one of increased food production during which, in 1949, Japan's Land Improvement Law changed the emphasis from "reclamation work" to "land improvement." To give actual examples, the Aichi Waterworks and the reclamation of Hachirogata were characteristic of Japan's mega-development during this period. The second period occurred when Japan's Land Improvement Law had reached a stage of concentration using allocated land and mechanization. The revision of this law in 1961 drove agricultural field improvement projects that consisted of land readjustment and agricultural water. In the third period, a revision of the law in 1972 drove an improvement in groundwork for agricultural productivity. In particular, the promotion of a system of land exchange for agricultural land use started during this period. As a result, the maintenance of living environments and structural improvements have hastened the development of rural areas. In this way, following a period of high economic growth in Japan, improvements in the groundwork for agricultural productivity were made. Now, in the present day, the focus is shifting to an improvement of the living level in rural areas.

On another point, historical change regarding the borrowing and lending of farmland is a very important issue (MAFF 2009). Prior to World War II, a landowner's farmland consisted mainly of agricultural land, and the difference between tenant farming and large landowners was obvious. The GHQ's push for democracy gave a great impetus to agricultural reform in Japan, and, as a result, the government redistributed farmland from large landowners. This allowed more tenants to farm their own land, and the "Agricultural Land Act" was established in Japan for the advancement of farmers and the regulation of the conservation of agricultural land. However, the established act did not increase the amount of available farmland. Therefore, the government promoted an increase in the borrowing of farmland estates (Shimizu 2007). In addition, the "Act on Establishment of Agricultural Promotion Regions" in 1969 implemented a comprehensive agriculture promotion policy in an attempt to meet the transformation envisioned in the revised "City Planning Act." The "Agricultural Land Act" and the "Act on Establishment of Agricultural Promotion Regions" were very important laws for the conservation of goof quality farmland.

The 2009, revision of the "Agricultural Land Act" constituted a significant change from prior and existing policies. The chief features of this revision are basically as follow: the borrowing and lending of farmland by private-sector companies and progress in strengthening the improvement of "Agricultural Promotion Areas" (Hashimoto 2009). In addition, the "Act for the Promotion of Development of Private Farmland for Community Use" in 1991 promoted the concept of non-farming households being actively involved in farmland. In conclusion, Japan's agricultural administration has aimed at achieving: (1) an increase in the amount of farmland in a variety of forms and (2) the promotion of agriculture by a diverse range of agencies.

5.3 Recent Trends in Agricultural Land Abandonment and Extensification in Japan

Agricultural land management in Japan has experienced both intensification and extensification in past decades under substantial changes in the social, economic, and environmental factors related to farming. The intensification of agricultural lands includes the development of greenhouse facilities and higher use of inputs. The extensification of agricultural lands is particularly visible in the increasing abundance of abandoned farmlands. While both intensification and extensification processes have crucial effects on multiple dimensions of current and future Japanese society and are related to each other, here we focus on reviewing the extents, causes, and consequences of agricultural land extensification, which may have great effects on future Japanese land uses under the prolonged aging, decreasing, and urbanizing trends of the Japanese population in the Twenty-First Century.

Several national-level indicators clearly illustrate substantial extensification trends in the Japanese agricultural sector. The extensification is most obvious in the increasing share of abandoned farmland. Figure 5.1 shows recent trends in the areas and ratio of abandoned farmland in Japan. The area of abandoned farmlands continued increasing in past decades, and the rate of change became even greater after the 1990s. Here, abandoned farmlands are defined as "previously cultivated lands with no farming records in the past year or more at the time of the surveyed period and with no clear signs of restarting cultivation in the near future shown by landowners." Japan's Ministry of Agriculture, Forestry, and Fisheries (MAFF 2012) indicated that nearly half of the abandoned farmlands are owned by "non-farming households," who possess between 500 and 1000 m^2 of cultivated and/or abandoned farmlands and whose agricultural income is less than 150,000 yen/year. This means that the abandonment of farmlands is particularly common in smallholding farm households in the non-agricultural sector and that they only use farmlands for subsistence or hobby purposes.

While the increases in abandoned farmlands can be observed elsewhere in Japan, the extent of the abandonment is diverse among regions and prefectures. Figure 5.2

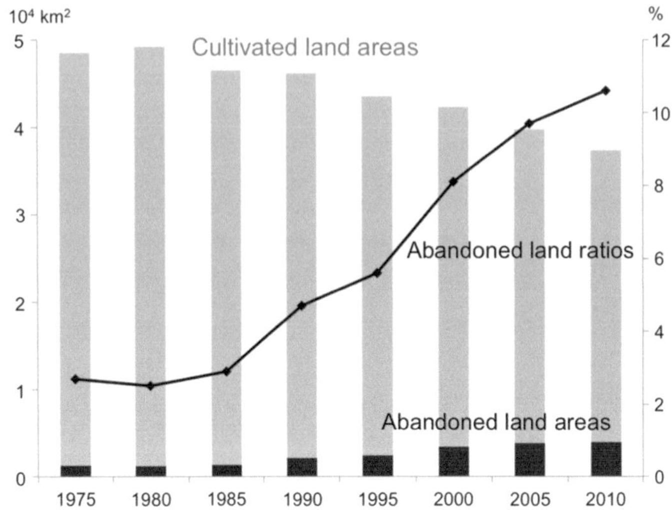

Fig. 5.1 Changes in cultivated and abandoned agricultural land areas (*left-side* bar) and abandoned land ratio (*right-side* bar) in Japan

shows the maps of abandoned farmlands in 47 prefectures of Japan. The abandoned farmland ratio compared to that of all cultivated lands per prefecture is highest at Nagasaki prefecture (25.95 %) and lowest at Hokkaido prefecture (1.84 %). The Kanto, Chugoku, and Shikoku regions have relatively higher levels of abandoned farmlands, while the Tohoku, Chubu, Kansai, and Kyusyu regions have lower levels. Figure 5.3 shows how the areas of abandoned farmlands have increased from 1990 to 2010. This map, together with Fig. 5.2, illustrates that the ratio of abandoned farmlands is currently lower in Tohoku region than it is in others; however, the region has also experienced steep increases in recent years. The 2011 Tohoku earthquake and tsunami significantly affected agricultural activities in this region, and the number of abandoned farmlands may have experienced a huge increase in those regions. The Hokkaido region, on the other hand, has a relatively lower level of agricultural land abandonment, and the increase in numbers is relatively lower. This is partly because the average size of farmlands in the Hokkaido region is larger than it is in other regions; thus, they have higher productivities, particularly in major crops such as rice.

5.3.1 Causes of Agricultural Land Abandonment

There are two major causes of agricultural land abandonment in Japan: socio-economic and biophysical reasons. Socio-economic reasons include labor resources, land ownership status, rent opportunity, agricultural infrastructure development, and product prices. Biophysical reasons include land conditions and

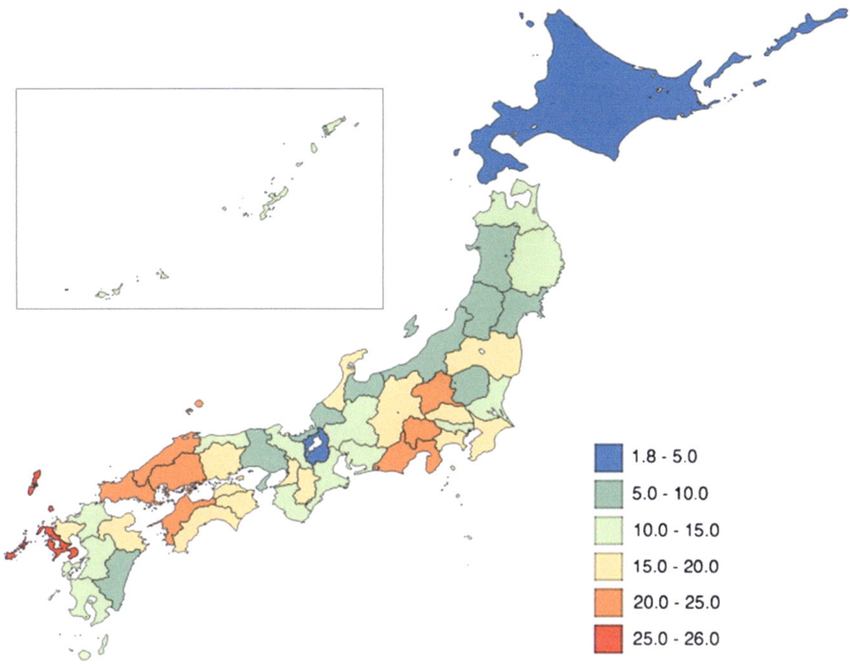

Fig. 5.2 Farmland abandonment ratios compared to the total of all cultivated lands in 47 prefectures of Japan

damages from wild animals. All causes in socio-economic and biophysical domains are not exclusive, and some of them are closely related to one another and have complex effects on farming activities. While major causes of abandonment vary across regions in Japan (Takayama and Nakatani 2011), common major drivers of abandonment across the nation can be summarized based on a review of previous studies. Here, we listed how those identified processes cause agricultural land abandonment in Japan together with empirical evidence from international and Japanese academic papers (Table 5.1). The use of agricultural census, questionnaires administered to farmers or local officials, field surveys, and GIS techniques are common methodological approaches that were used in said papers.

5.3.1.1 Socio-Economic Causes

Labor Resources. Lack of labor for continuing agricultural activities is one of the most frequently mentioned causes of agricultural abandonment in Japan. The observed parameters in this group include general shortages of household labor (Hattori and Yamaji 1998; Kurita et al. 2009), subsistence farming household ratios, the absence of a successor in farming activities (Yoshida et al. 2004), aged farmer ratio, size of machines (Senda 1998), number of household members (Sakurai et al. 2007), farmers

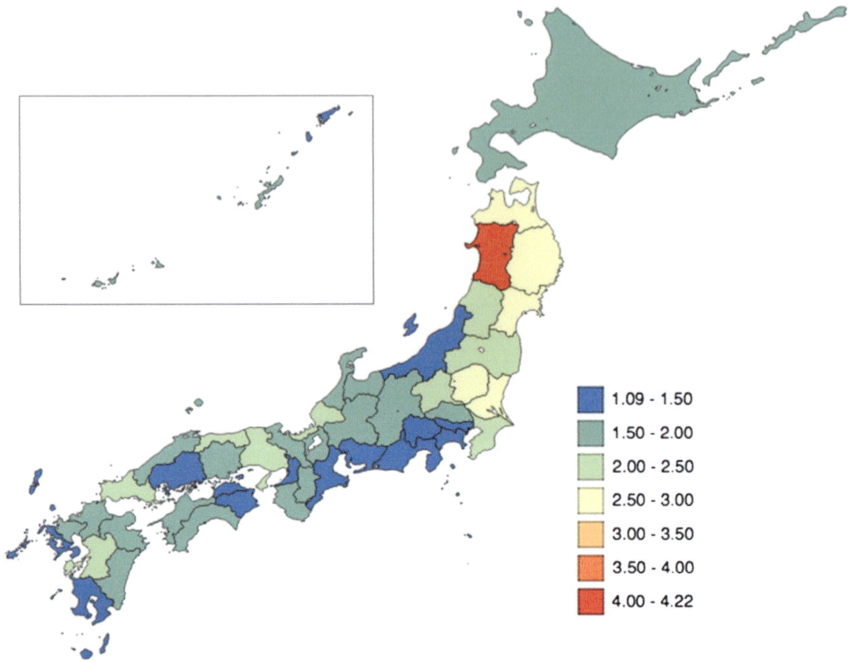

Fig. 5.3 Increase in abandoned farmland ratios in 2010 compared to 1990 in 47 prefectures of Japan

over 65 years of age (Takayama and Nakatani 2011), and the use of agricultural machinery (Matsui et al. 2014).

Land Ownership Status. This is another reason for agricultural abandonment by farming households and is generally represented by difficulties in maintaining agricultural activities (e.g., farmland is too small to earn income). The observed parameters in this group include distances from agricultural lands to houses and the number of small-holding owners in a village (Nomura et al. 2003), farmland size (Chung et al. 2004; Sakurai et al. 2007), land owners' speculative purposes (Hattori and Yamaji 1998), the number of subsistence farmers and large-holding farmers in a village (Takayama and Nakatani 2011), and increases in the share of non-farming owners in a village (Ishida 2011).

Rent Opportunity. Compared to the two above-mentioned household-level causes, rent opportunity is more closely related to social relationships among farmers within a village. The observed parameters in this group include the existence of large-scale farms within a village (Senda 1998), the number of farmers who expect to extend their farmlands (Ishida 2011), the ratio of rented farmland and large-scale farms (Takayama and Nakatani 2011; Matsui et al. 2014), and the conditions of the agricultural labor market, outsourcing, and social capital (Sakurai et al. 2007).

Table 5.1 Major causes of agricultural abandonment in Japan

	Factor	Indicator	Data source
Socio -economic	Labor resource	Lack of labor (Hattori and Yamaji 1998; Kurita et al. 2009)	Questionnaire to farmers
		Subsistence farm household ratio, successor (Yoshida et al. 2004)	Agricultural census
		Aged farmers' ratio, size of machine, successor (Senda 1998)	Agricultural census
		Aging of farmers (Ishida 2011)	Questionnaire to officials
		Age, successor, number of household members (Sakurai et al. 2007)	Questionnaire to farmers
		Farmers over 65 years old (Takayama and Nakatani 2011)	Agricultural census
		Lack of labor, machine use, aged farmers' ratio (Matsui et al. 2014)	Agricultural census
	Land ownership status	Living far away, small-holding owners (Nomura et al. 2003)	Field survey and ownership record
		Farmland size (Chung et al. 2004)	Field survey and GIS datasets
		Farmland size (Sakurai et al. 2007)	Questionnaire to farmers
		Speculative purposes, small-holding (Hattori and Yamaji 1998)	Questionnaire to farmers
		Subsistence farming, large-holding farmers (Takayama and Nakatani 2011)	Agricultural census
		Living far, increasing share of non-farming owners, speculative (Ishida 2011)	Questionnaire to officials
	Rent opportunity	Large-scale farm (Senda 1998)	Agricultural census
		Lack of farmers who expect to extend their farm (Ishida 2011)	Questionnaire to officials
		Rented farmland ratio, large-scale farm (Takayama and Nakatani 2011)	Agricultural census
		Agricultural labor market, outsourcing, social capital (Sakurai et al. 2007)	Questionnaire to farmers
		Large-scale farm, farmland rent (Matsui et al. 2014)	Agricultural census
	Agricultural infrastructure	Irregular shape, insufficient infrastructure, accessibility (Hattori and Yamaji 1998)	Questionnaire to farmers
		Distance to canals, spatially aggregated farmlands (Shima et al. 2004)	Field survey and GIS datasets
		Irrigated paddy fields (Sakurai et al. 2007)	Questionnaire to farmers
		Land readjustment (Ishida 2011)	Questionnaire to officials

(continued)

Table 5.1 (continued)

	Factor	Indicator	Data source
	Product prices and agricultural incomes	Lack of high-price crop choices (Ishida 2011)	Questionnaire to officials
		Involvement in farmers markets (Sakurai et al. 2007)	Questionnaire to farmers
		Growing product types, agricultural income (Matsui et al. 2014)	Agricultural census
Biophysical	Land condition	High elevation and steep slopes (Chung et al. 2004)	Field survey and GIS datasets
		Steep slopes (Shima et al. 2004)	Field survey and GIS datasets
		Hilly areas (Senda 1998)	Agricultural census
		Hilly areas (Takayama and Nakatani 2011)	Agricultural census
		Hilly areas (Ishida 2011)	Questionnaire to officials
	Damage by animals	Damage caused by animals (Ishida 2011)	Questionnaire to officials

Agricultural Infrastructure Development. This is another village-level reason for agricultural abandonment, along with the rent opportunity. The observed parameters in this group include irregular farmland shapes, insufficient infrastructure, and accessibility of farmlands (Hattori and Yamaji 1998), distance to canals from farmland and the coverage of spatially aggregated farmlands (Shima et al. 2004), the areas of irrigated paddy fields (Sakurai et al. 2007), and the history of land readjustment programs (Ishida 2011).

Product Prices. The last socio-economic reason concerns the prices of products and, more broadly, agricultural incomes. The observed parameters in this group include the lack of suitable crops grown on a farmland (Ishida 2011), the lack of involvement in farmers' markets (Sakurai et al. 2007), product types, and agricultural incomes (Matsui et al. 2014).

5.3.1.2 Biophysical Causes

Land Conditions. Hills and the steepness of farmland and surrounding areas are the most commonly mentioned causes of agricultural abandonment in mountainous areas of Japan. These areas are often referred to as *Chusankan-chiiki* (middle and mountainous areas) and cover more than 70 % of national land. The observed parameters in this group are the land conditions of farmlands, including high elevation and steep slopes (Chung et al. 2004; Shima et al. 2004), and the coverage of farmlands located in hilly areas (Senda 1998; Takayama and Nakatani 2011; Ishida 2011).

Damage from Wildlife. There are many reports in newspapers and other media that mention damage from wildlife (e.g., wild pigs) as a cause of agricultural abandonment; however, only a limited number of reviewed papers (e.g., Ishida 2011) actually evaluated this process.

5.3.2 Consequences of Agricultural Land Abandonment

Unlike studies on the causes of agricultural abandonment, there has been only a limited number of studies on its consequences. The most commonly mentioned effect is the negative effects on surrounding agricultural activities, including pests, illegal dumping, and the establishment of habitats for wildlife (e.g., wild pigs) that often damage agricultural products. These effects are measures by recognized effects on farmers (Yagi 2003), effects on the price of harvested rice (Yagi 2006), and reports on damages by wild pigs (Shimizu et al. 2013). The other major effect is biodiversity loss in and around agricultural ecosystems. This includes wetland vegetation (Ikegami et al. 2011), amphibian, fish, and bird species (Katayama et al. 2015), and threatened plant species that rely on continuous human disturbances on agricultural lands (Osawa et al. 2013). A unique example of the effects of agricultural abandonment is assessed by Fujimi et al. (2006), who evaluated how agricultural land abandonment affects the aesthetic values of terrace paddy fields by using preference and environmental economics methods. Table 5.2 shows the consequences of agricultural abandonment in Japan.

5.4 The History of Direct Payment Schemes in Agricultural and Rural Development Policies in Japan

This section provides an overview of the current state of research on the evaluation of farmland management policies after organizing their operations. As in other countries—including members of the EU—Japan enforces measures that can be classified as either hardware or software projects. As will be described later, the shift from hardware projects to software projects is underway in Japan. One software project is currently centered on direct payment, although subsidies are also available. Therefore, this section focuses on the direct payment system involved in farmland management.

Table 5.2 Consequences of agricultural abandonment in Japan

Type of effects		Indicator of effects	Indicator of agricultural abandonment
Effects on surrounding agricultural areas	Pest	Recognized pest impacts (Yagi 2003)	Abandoned ratio within 1 km grid
		Price of harvested rice (Yagi 2006)	Distance to abandoned land
	Damage by animals	Recognized animal impacts (Yagi 2003)	Abandoned ratio within 1 km grid
		Damage by wild pigs (Shimizu et al. 2013)	Minimum distance to abandoned land
	Illegal dumping	Recognized impacts of dumpings (Yagi 2003)	Abandoned ratio within 1 km grid
Biodiversity		Wetland vegetation (Ikegami et al. 2011)	Years since abandonment
		Amphibian, fish, and bird species (Katayama et al. 2015)	Various indicators (review)
		Threatened plant species (Osawa et al. 2013)	Abandoned ratio within 10 km grid
Aesthetic value		Preferences and economic valuation (Fujimi et al. 2006)	Abandoned ratio with a view

5.4.1 Support for Farmers Through the Direct Payment System

In contrast to supporting farmland management by improving the infrastructure—including the farmland itself—the direct payment system is based on asking people other than farmers to cover the expenses involved by paying considerable attention to the multiple functions of farmers and the benefits of agriculture. In this connection, expanding and improving the direct payment system was a turnaround in agricultural policy (Iiguni 2010). To be specific, it was a shift from "price support" to "direct payment" (direct transfer of income from the government to farmers). In particular, the Japanese-style direct payment system originates from the "Direct payment system in hilly and mountainous areas" that was introduced in 2000. Subsequently, Japan enforced the "Direct payment system for wheat, soybean, and others" and the "Support for eco-friendly farming activities" in 2007. As a result, Japan established three frameworks to support farmers: support for disadvantaged areas, support for compensating income, and support for conserving the environment. Based on the system adopted by the EU, these three frameworks of the direct payment system were modified to suit the regional situation in Japan and established as the Japanese-style direct payment system. The conceptual difference between the Japanese-style direct payment system and the direct payment system in the EU is that the former focuses on community activities, while the latter focuses on the behavior of each farmer (Ichida 2006). In this respect, her analysis concludes that "farming villages in EU countries including Germany ask each farmer to be totally responsible for conserving nature and the environment because of widespread dry field farming and stock raising that do not require joint activities (p. 11)."

5.4.2 *Review of Research on the Evaluation of Farmland Management Policies*

The above policies are examined with reference to the research results of major academic societies in the field of agricultural policy in Japan. We reviewed major publications including the collection of papers of the Agricultural Economics Society of Japan, the Journal of Rural Economics, the Japanese Journal of Farm Management, the Journal of Rural Planning Association, the Reports of the City Planning Institute of Japan, and the Journal of Environmental Information Science.

We organized the accumulated research results according to the following seven viewpoints.

(1) Experimental verification of the overall direct payment system.

As described above, the Japanese direct payment system originates from the "Direct payment system in hilly and mountainous areas" introduced in 2000. This research developed various theories by applying the ideas upon which a direct payment system is based to various situations, including ecosystem service, eco-friendly agriculture, and forest management. Specifically, it explored the future direction of the direct payment system, farmland management, and forest management by analyzing the direct payment systems that Shiga and Fukuoka prefectures enforced ahead of the central government.

(2) Discussion of policy change to the payment system in hilly and mountainous areas.

The "Direct payment system in hilly and mountainous areas" has been revised several times since its introduction, and this research evaluates the revisions as well as their effects. The "Direct payment system in hilly and mountainous areas" can be reviewed from the three viewpoints of *expansionism*, *prolonged vision*, and *strengthened inter-village cooperation*, and the three viewpoints have maintained agriculture in hilly and mountainous areas for a long period of time (Hayashi 2013). On the other hand, a previous study noted that the change in the payment system for hilly and mountainous areas discontinued 4,278 agreements, and it explored the necessary conditions for maintaining and developing a village agreement by examining the awareness and recognition of each farming family that participates in a village agreement (Tsushima and Katsura 2010).

(3) Research on organizing the operation of the direct payment system.

As mentioned earlier, the direct payment system in Japan can basically be divided into two parts: support for a village-wide approach and support for farming conducted by individual farming families. In the case of joint management conducted by all village members, an organization that receives direct payment needs to be established. However, joint management varies by region, and comparative studies on the actual situation of organization management have been conducted since joint management was introduced. For example, joint management should be classified in three ways: by village, by water system, and by project (Hoshino 2010) (Fig. 5.4).

(4) Discussion of the scale, factors for conclusion, and operating system (by a third-sector company or incorporation) of a village agreement.

Various themes were created under this subject to develop research on the organization discussed in (3), and further research was conducted to go into detail regarding what form an agreement should take and to explore the factors of concluding an agreement. For example, a previous study verified the characteristics of village agreements and the continuity of farmland management by classifying the relationships between village agreements and the five kinds of farmland locations: plateau, mountainous area, gentle mountain slope, hilly area, and lowland in the valley's bottom (Yonezawa and Takeuchi 2006).

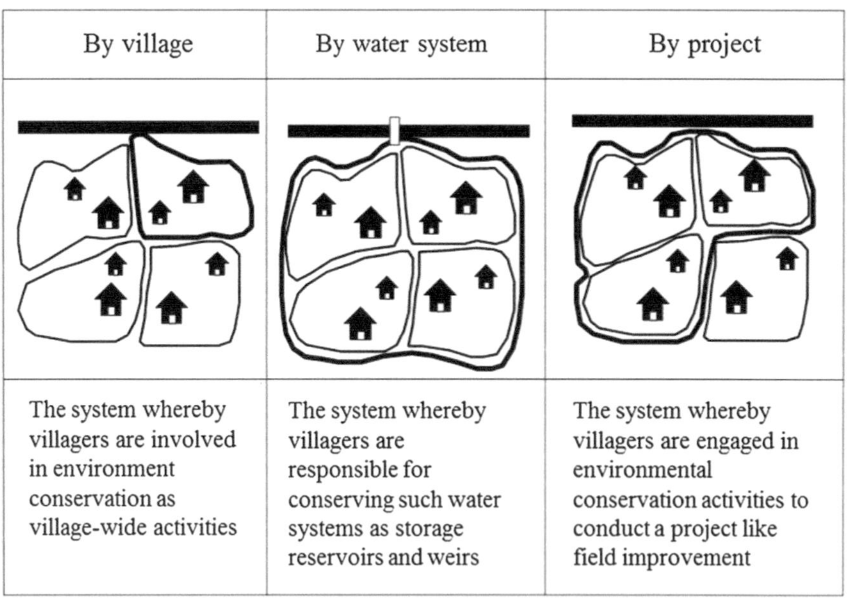

Fig. 5.4 Example of classifying target areas (*Source* Hoshino (2010)). (Some parts of this figure were altered)

(5) Discussion of the direct payment system in forest management.

The "Direct payment system in forest management and environment conservation" came into force in 2011, and many papers were published on the environmental payment system in forest management before and after its enforcement. In particular, some researchers analyzed the environment payment system from the viewpoint that it is critical to highlight forest improvement rather than forestry promotion (Ito 2011). On the other hand, the "forest environment tax" is clearly for taxation, but its effect is subject to intense discussion. Hence, many papers were written about what form should be taken by measures based on the "forest environment tax" as discussions of them grew serious.

(6) Verification of the effect and scenario of the direct payment system.

As the research results of the direct payment system accumulated, some researchers focused on verification of the effects of the policy. For example, the direct effect of the direct payment system was that it secured farmland conservation and multiple functions, and one scholar was motivated to conduct research that focused on the relationship between social capital and farming facilities (Endo 2011).

A previous study verified the viability of large-scale paddy field management in hilly and mountainous areas from the viewpoint of production infrastructure through verification of the normative model by agricultural region that takes the direct payment to slope areas into consideration (Yagi and Nagaki 2004). These kinds of efforts to construct a model are very important to the discussion about how to develop system operations that take rationality into consideration instead of assuming the direct payment system to be a uniform system throughout the country.

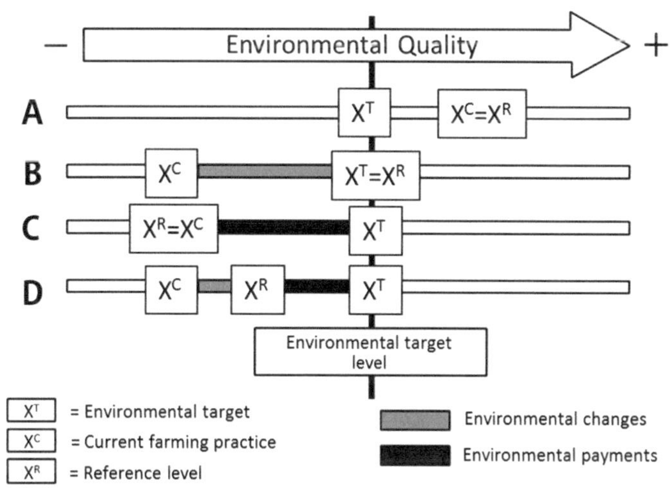

Fig. 5.5 The position relationships among the targets of environment improvement (X^T), RL (X^R), the existing farming method, and policy approach (*Source* OECD (2001)) (Some parts of this figure were altered)

(7) Comparative research on the direct payment system between Japan and the EU.

The Japanese direct payment system has been making progress in its own style by studying the operation of the direct payment system in the EU. Preceding studies clarified that the Japanese system is being operated in various organizational systems because joint management and farming systems vary by region. In particular, many researchers conducted studies on what approach is available for disadvantaged areas by investigating cases in Britain and Germany. One researcher emphasized the significance of analyzing subsidies and the specific conditions under which farmers are subsidized if they abide them (Shobayashi 2006). In particular, he emphasized the importance of setting a threshold for deciding whether farmers should take responsibility (Fig. 5.5). Some researchers narrowed down the operational unit from village to individual farming family to study the operation of the Japanese-style direct payment system. As a background to this trend, a previous study points out the problem that "neither areas without joint activities nor the approach conducted by an individual farmer was subject to support under the old measures (designed for projects to conserve farmland, water, and the environment)" (Nagasawa 2011).

References

Chung H, Yodogawa T, Yazawa M (2004) Space structural characteristic of agricultural settlement with an abandoned cultivation lands: case study in two municipalities of Hokkaido. J Rural Plan Assoc 23:259–264

Endo K (2011) A grasping of the rural social capital: observing the activation of communities discovered by introducing the direct payment system. Jpn J Farm Manag 49(3):91–96

Fujimi T, Watanabe M, Asano K (2006) Economic loss of landscape degradation in rice terraces caused from cultivation abandonment or land consolidation. Environ Sci 19(3):195–207

Hashimoto T (2009) "Syoyu" Kara "Riyou" Cyushin No Nouchiseido Eno Saikouchiku (Rebuild a farmland policy by shifting away from ownership to utilization). RITTPOU TO CYOSA, 1–9. http://www.sangiin.go.jp/japanese/annai/chousa/rippou_chousa/backnumber/2009pdf/20090501020.pdf. Accessed 14 April 2016

Hayashi K (2013) Changes in and effects of direct payments for hilly and mountainous areas on farming practices. J Rural Probl 49(2):304–309

Hattori T, Yamaji E (1998) Factors to abandon cultivation by the condition of farm household in suburban area. J Rural Plan Assoc 16(4):325–333

Hoshino S (2010) Measures for improving farmland, water-resource issues, and environmental conservation and regeneration of village communities (Feature: can a village revitalized? Measures for improving farmland, water-resource issues, and environmental conservation). Agric Econ 75(7):15–21

Ichida T (2006) Eu no kankyoshiharai to sonogenjyo -tokusyu kankyoshiharai wo kangaeru (Current state of environmental direct payment in the EU: considering a situation about environmental direct payment in Japan). AFC Forum 54(1):8–11

Iiguni Y (2010) Immerging issues and challenges of direct payment in Japan. Kochi Univ Rev Soc Sci 98:71–87

Ikegami Y, Nishihiro J, Washitani I (2011) Vegetation of abandoned rice fields in the head areas of small valley bottoms (*yatsu*) in the lake Kitaura watershed, Ibaraki. Jpn Jpn J Conserv Ecol 16 (1):1–15

Ishida K (2011) A study on factors and programs for abandoned cultivated land. Jpn J Farm Manag 49(1):99–104

Ito K (2011) Toward a Japanese style of direct payment for forest services appraisal of a grant system to promote regional activities for forest improvement. SATO Noriko For Econ 64 (3):12–16

Katayama N, Baba YG, Kusumoto Y, Tanaka K (2015) A review of post-war changes in rice farming and biodiversity in Japan. Agric Syst 132:73–84

Kurita H, Yokohari M, Yamamoto T (2009) The formation process of non-industrial agricultural land-uses in the Urban fringe area. J Jpn Inst Landscape Archit 72(5):727–730

Matsui T, Ugata T, Machimura T (2014) A development of factor analyzing and predicting model of abandoned agricultural land with machine learning algorithms. J Jpn Soc Civil Eng Ser G (Environ Res) 70(6):131–139

MAFF (2009) Nouchiseido no gaiyou (Delineation of the land tenure system). http://www.maff.go.jp/chushi/keiei/kaikaku/pdf/nouti-gaiyo.pdf. Accessed 14 April 2016

Ministry of Agriculture, Forestry, and Fisheries of Japan (2012) Report on results of 2010 world census of agriculture and forestry in Japan. http://www.e-stat.go.jp/SG1/estat/ListE.do?bid=000001037762&cycode=0. Accessed 14 April 2016

Nagasawa S (2011) Saishinno nousei nihongoban "kankyoshiharai" no sousetu: kankyohozengata nogyo cyokusetsushiharai taisaku (Current trend of farm policy: the foundation of environmental direct payment in Japan). Tuchidukuri to ekonogyo 43(4):70–74

Nomura W, Murakami A, Watanabe S (2003) A study on abandoned farmland in large-scale suburban agricultural area. J Rural Plan Assoc 22:91–96

OECD (2001) Agri-environmental policy measures: Overview of developments. OECD, Paris

Osawa T, Kohyama K, Mitsuhashi H (2013) Areas of increasing agricultural abandonment overlap the distribution of previously common, currently threatened plant species. PLoS ONE 8(11), e79978. http://dx.plos.org/10.1371/journal.pone.0079978. Accessed 14 April 2016

Rural Development Planning Center (2003) Nouson kankyoseibi no kagaku (Scientific understanding of rural development planning). Asakura Publishing, pp 37–45

Sakurai T, Shibahara M, Sakurai S (2007) Determinants of abandonment and lease of paddy field. Jpn J Rural Econ 2007:141–148

Senda T (1998) A quantitative analysis of abandoned cultivated land in Japan. Jpn J Farm Manag 36(1):57 62

Shima T, Ogawa S, Yoshisako H (2004) Classification and analysis of abandoned farmland in TUJYUN canal paddy field. J Rural Plan Assoc 23:271–276

Shimizu S, Mochizuki S, Yamamoto M (2013) The environmental factors of damage to rice paddies at the time of the expansion of wild boars distribution. Bull Int Assoc Landscape Ecol-Jpn 18(2):173–182

Shimizu T (2007) Nippon no nouchiseido to nouchiseisaku (Land tenure system and farmland policy in Japan). Mon Rev Agricult Forest Fish Finance 60(7), 346–356. https://www.nochuri.co.jp/report/pdf/n0707re1.pdf. Accessed 14 April 2016

Shobayashi M (2006) Policies for improving agri-environment in rural areas. J Jpn Soc Irrig Drainage Rural Eng 74(3):199–203

Takayama T, Nakatani T (2011) Preventive effect of the measure of direct payment in hilly and mountainous areas on abandonment of farmlands: evidence from the paddy and upland areas of Hokkaido. Agric Inf Res 20(1):19–25

Tsushima M, Katsura A (2010) Examining the factors determining the continuity of rural community agreement on direct payments in hilly and mountainous areas: a case of the Mitake District in Fukuchiyama City. Kyoto. J Rural Prob 46(2):232–238

Yagi H (2003) Clarification and prediction of the negative externalities on farm activities from non-farm land use. J Rural Plan Assoc 22:79–84

Yagi H (2006) An empirical application of the linear programming model for agricultural land use planning: through the valuation of negative externalities caused by farmland abandonment in marginal areas. J Rural Econ 78(1):12–21

Yagi H, Nagaki M (2004) Capability of large-scale rice farms on the production facilities in the mountainous area: normative model considering direct income support. J Rural Plan Assoc 23:169–174

Yoshida S, Sato T, Datai H (2004) Factor analysis for increasing abandoned cultivated land areas at Chugoku area: paying attention to differences among areas and relation among factors. J Rural Plan Assoc 23:277–282

Yonezawa K, Takeuchi K (2006) A classification of community agreements according to scale of the agreements and topographical location of farmlands for a direct payment measure for hilly and mountainous areas in Japan: a case study in former Tokamachi-shi, Niigata prefecture, Japan. J Rural Plan Assoc 25:497–502

Chapter 6
Plantation Forest Landscape: The Paradigm Shift in Forestry, a Belowground Ecosystem for Sustainability Land Use

Kazukiyo Yamamoto and Yasuhiro Hirano

Abstract Due to high demand for wood materials during and after World War II, expansive afforestation was promoted in Japan during the 1950s and 1960s. The area of plantation forests is currently 40 % of the total forest area in Japan. As economic growth in Japan continues, so has higher wood demand, but imported woods have been used due to their lower prices. Forestry has declined with increasing imported wood and decreasing population in the forest–rural area. Much of the forest area has been kept but the accumulation of forest volume has increased by three times during the past 50 years. Recently, to achieve the promise of the Kyoto Protocol, forest management has been accelerated, along with higher governmental subsidies. Soil degradation can occur after harvesting wood and even after re-planting the forest trees. Forest soils support an ecosystem that serves as carbon storage, nutrient cycling, biodiversity, and controlling water quality, but it can be degraded without sustainable forest management.

Keywords Depopulation · Ecosystem service · Expansive afforestation · Forestry · Soil degradation

---

K. Yamamoto (✉)
Graduate School of Bioagricultural Sciences, Nagoya University,
Furo-cho, Chikusa, Nagoya 464-8601, Japan
e-mail: kazukiyo@agr.nagoya-u.ac.jp

Y. Hirano
Earth and Environmental Sciences, Graduate School of Environmental Studies,
Nagoya University, Furo-cho, Chikusa, Nagoya 464-8601, Japan
e-mail: yhirano@nagoya-u.jp

H. Shimizu et al. (eds.), *Labor Forces and Landscape Management*,
DOI 10.1007/978-981-10-2278-4_6

6.1 Post-war Expansion Afforestation

During World War II, great amounts of wood were needed for war materials, and forests in great numbers were devastated through being cut down. Therefore, recovery of these devastated forests was a key challenge of forest administration in Japan. The "afforestation support projects" that were focused on the elimination of devastated areas (non-forested land of about 1.2 million ha in national forests and about 0.3 million ha in non-national forests) were incorporated into public projects in 1946 (Forestry Agency of Japan 2015). To ensure the execution of plantation on private forestlands, the government created an act named the "Special Measures for Reforestation Law" in 1950 (Ota 2013). This Forest Law was revised in 1951, and the "Forest Planning System" as well as a cutting permission system for privately owned forests was introduced to it (Nonoda 2008). As a result, planting in abandoned forests that were clear-cut mainly during wartime was concluded in 1956 (Forestry Agency of Japan 2014). Then, with the beginning of the high economic growth of the 1960s following the economic recovery of the 1950s, wood demand increased rapidly. Since domestic forest resources at the time were in a depleted state, wood prices skyrocketed as a result of supply shortage (Fig. 6.1). Therefore, in the forest law amended in 1962, the cutting permission system was abolished for ordinary forests with the exception of the protection forests (Forestry Agency of Japan 2014). Furthermore, in order to promote the supply of domestic timber corresponding to the strong wood demand, the "Forestry Basic Law," with the goal of expanding gross forestry production, was established in 1964 (Forestry Agency of Japan 2014). At the time, while the demand for softwood was great, the supply of domestic softwood had stagnated. In addition, because of the energy revolution during the high-growth period, fuelwood forests such as hardwood were no longer used. Therefore, "expansive afforestation" (afforestation in lumbered broad-leaf forests by fast-growing conifers) has been promoted on a nationwide scale. In addition, increasing wood demand and rising wood prices stimulate the afforestation willingness of forest owners (Nonoda 2008), and more than 300,000 ha of "expansive afforestation" was annually carried out until around 1970 (Fig. 6.1; Forestry Agency of Japan 2015). This "expansive afforestation" caused the plantation of 40 % of the forest area of Japan, and the age class structure of plantation for about 45 % (in 2012) of plantation was occupied by 41–55-year-old forests (Fig. 6.2; Forestry Agency of Japan 2014).

6.2 Forestry Decline Under High Economic Growth

In Japan, wood demand continued to expand during the period of high economic growth. However, the demand was covered by imported wood, which is entirely due to import liberalization in 1964, and the supply of domestic wood began to decrease

Year	Japanese cedar (yen/m³)	Hinoki cypress (yen/m³)
1955	4,478	5,046
1960	7,148	7,996
1965	9,380	10,645
1970	13,168	21,352
1975	19,726	35,894
1980	22,707	42,947
1985	15,156	30,991
1990	14,595	33,607
1995	11,730	27,607
2000	7,794	19,297
2005	3,628	11,988
2010	2,654	8,128
2011	2,838	8,427
2012	2,600	6,856
2013	2,465	6,493
2014	2,968	7,507

Fig. 6.1 Average log price of effective roadside volume for major plantation species in Japan (*Source* Forestry Agency of Japan 2015)

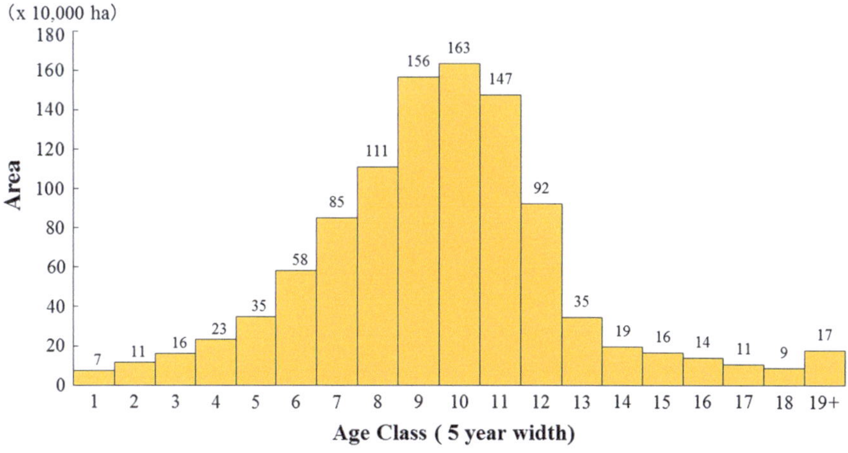

Fig. 6.2 Age class distribution of plantations in Japan (*Source* Forestry Agency of Japan 2014)

starting in the late 1960s due to competition with imported wood (Ando 1988; Sasaki 2011; Forestry Agency of Japan 2014). Problems such as aging and depopulation of the mountain village as well as increases in forest management costs (Tachibana 2013; Forestry Agency of Japan 2014) caused stagnant forestry activity. In part because of the decrease in wood demands due to the economic recession after the burst of Japan's economic bubble in 1991, wood prices declined for a long period of time, and forestry activities were more stagnant (Tanaka 2013; Forestry Agency of Japan 2014) (Fig. 6.1). On the other hand, since plantations by "expansive afforestation" were increasing growing stocks, the demand for requisite treatment of them also increased. In response, weeding, improvement cutting, and thinning were added to the list of the subsidy-eligible items of afforestation projects (Forestry Agency of Japan 2013). However, these subsidy projects interfered with the independence of forestry as an industry and functioned in opposition of the spirit of the "Forestry Basic Law" (Nakaoka 2015). Therefore, reinvestment afforestation and thinning became difficult, and issues arose as the abandonment of reforestation and management became more serious (Yabe 2002; Kajisa et al. 2011; Yoshida 2011; Nakamura 2012; Kakizawa 2013; Tachibana 2013). Traditionally, since forestry practices such as thinning have been done as part of forestry production activities in Japan, the government adopted a stance that forest improvement would progress together with forestry production activities and consequently the public would benefit (Forestry Agency of Japan 2014). Therefore, these situations began to be recognized as impairment of the public benefit function of forests (Tachibana 2013).

6.3 The Paradigm Shift in Forest Management

Greater emphasis came to be placed on public awareness of the multiple functional roles of forests in Japan. The government developed a fundamental reconstruction policy. The "Forestry Basic Act" was amended to the "Forest and Forestry Basic Act" in 2001, within which the "fulfillment of multiple functional roles of forests" was set as a target along with the "supply and use of forest products," and forest zoning for suitable and sustainable management was newly introduced (Yabe 2002; Fujie 2007; Kakizawa 2013; Forestry Agency of Japan 2014). In this act, the functions of forests in preventing global warming have also been highlighted. For the "first commitment period" (2008–2012) of the Kyoto Protocol adopted in 1997, Japan aimed to remove 3.8 % of its 1990 level of greenhouse gas (GHG) emissions through the promotion of forest sink activities, specifically by increasing the thinning area from an average of 350,000 ha per year to an average of 550,000 ha per year (Forestry Agency of Japan 2013). While converting the framework of forest policy, effective utilization of forest resources in plantations that have been grown as a result of "expansive afforestation" has become an issue. Therefore, political focus was devoted to the activation of forestry and the forest industry, and the "Forest and Forestry Revitalization Plan," aimed at the revitalization of the forestry industry, was designed in 2011 (Kojima 2013; Kakizawa 2013). Since about 50 %

of the area of plantations undergoing "expansive afforestation" had already exceed their standard cutting age (Fujie 2007; Forestry Agency of Japan 2014), there was a paradigm shift in that the plantations had reached the harvest stage from the silvicultural stage (Fujino 2013). However, even in the "Forest and Forestry Revitalization Plan," the mechanism to mobilize forest owners against the policies by economical induction, such as subsidies, has not changed (Kakizawa 2013). Therefore, for forestry in Japan to be reborn as an industry, it is important to cultivate human resources with management skills by breaking away from subsidy dependence.

6.4 Sustainable Land Use and Management of Forest Soils from Belowground Ecosystem Aspects

The cover ratio of forest area in Japan is about 67 % and is ranked third among advanced countries, indicating that Japan has one of the highest forest cover ratios in the world (Forestry Agency of Japan 2015). The higher ratio is due to mild, temperate climate conditions with higher levels of annual precipitation, such as 1500 mm, and a mean annual temperature of 15 °C. On the other hand, the Japanese islands are made up of a complex geology where four tectonic plates meet; therefore, there are frequent earthquakes and several volcanic mountains in Japan. The chemical properties of forest soils that generally regulate forest tree growth can vary with different surface geology (Imaya et al. 2005). Forest soils have gradually been matured not only by weathering subsurface rocks but also by the decomposition of plant litter by soil microorganisms. The five representative factors in soil formation are parent materials, climate, topography, time, and organisms (Jenny 1941). It appears that developments of 1 cm thickness in surface forest soil take more than 100 years.

Forest trees in Japan grow typically on steeply sloping mountainous areas. The slope conditions in forests can affect chemical and physical properties significantly in forest soils. Foresters have already learned that Japanese red pine (*Pinus densiflora*) should be planted at upper slopes, Japanese cypress (*Chamaecyparis obutsa*) at the middle, and Japanese cedar (*Cryptomeria japonica*) at lower parts of slopes. In general, the upper parts of slopes in forests have dry climatic conditions with lower water contents and forest soils that have lower concentrations of base cations (essential nutrients for trees, such as Ca and Mg). Both Japanese red pine and Japanese cypress can thrive in environments with poor nutrients and low water at the upper or middle parts of slopes, and they have been planted there conventionally. In contrast, Japanese cedar uses more water and is typically grown in conditions that are more fertile, such as the lower parts of slopes. Inorganic-nitrogen in forest soils is the most important essential nutrient in tree growth. Generally, nitrate-nitrogen can dominate at the lower parts of slopes, but ammonium-nitrogen does so at the upper parts (Tokuchi et al. 1999). The preference in inorganic-nitrogen uptake in roots differs among tree species and can affect tree height growth. A site index that expresses mean height in 40-year-old trees has been used to evaluate the soil productivity (Nishizawa et al. 1965). The index has been used to consider the management of the thinning and the final harvests and can vary regarding several factors such as soil types, topography, and climatic conditions (Yamane et al. 1990). Along the slope, the lower parts generally show higher index than the upper parts because of the differences in water and nutrient conditions.

In general, forestation from agricultural fields accelerates soil acidification because of nutrient uptake and the accumulated leaf litter of trees (Augusto et al. 2015). Soil acidification is one of the soil degradations in sustainable forests and leads to lower soil pH, leaching base cations such as Ca and Mg, and enhancing phytotoxic Al in soil solutions (Hirano et al. 2007). Forest soils have acid buffering capacities to buffer changing the soil pH. The capacities are different from the surface geology. High acid buffering capacities in the soil were seen in Hokkaido, northeastern Honshu, and Kyushu, where volcanogenic deposits were covered; soils with lower capacities were in western Honshu and Shikoku (Takahashi et al. 2001). Soils occupied by evergreen coniferous trees have acidified more than did those of deciduous trees (Augusto et al. 2015). Soil acidification in Norway spruce (*Picea abies*), which is one of the main tree plantation species in European countries, has been reported within 60 years of plantation (Priha and Smolander 1999). Of the two main plantation species in Japan, Japanese cypress showed similar accelerated soil acidification to Norway spruce, but Japanese cedar ameliorated and accumulated Ca in soils presumably because of the high Ca demands and the contents of its leaf litter (Baba et al. 2004). However, when planted in soils with lower base cation content, even Japanese cedar exhibits accelerated soil acidification (Tanikawa et al. 2014). When we plan to do reforestation after harvesting the existing plantation forests, we should note that soil degradation such as acidification has already occurred and determine which tree species are selected to protect further soil degradation. One suggestion is planting accompanying species to ameliorate soil acidification caused by accumulating base cations via leaf litter.

The decreasing birth rate and aging of the population of Japan have led to the decline of the forest industry. The total forest area has remained the same as it was 50 years ago, but the accumulation of wood volume in forests has increased threefold. One of the reasons for this is globalization that has caused the population to become concentrated in urban areas rather than rural regional districts, decreasing the demand for wood materials for energy and house construction. The decreasing population in the rural areas and the shortage of labor accounts has led to fewer relationships between human society and forest ecosystems. The lifestyle of the residents of rural areas has greatly changed as the landscape did. In the Satoyama area, bamboo had been planted for the purpose of eating new shoots and using the poles in the rice fields. However, the shortage of labor available to manage the bamboo area and new shoots has caused the bamboo to invade forests rapidly and has decreased their biodiversity (Umemura and Takenaka 2015). It is also obvious that the Sika deer (*Cervus nippon*) populations have increased rapidly. One of the reasons for this is the depopulation of the Satoyama area. Sika deer grazing has damaged not only the production of crop plants but also forest trees, leading to soil erosion because of scarce understory vegetation in surface forest soils. Moreover, the increased populations of bamboo vegetation and Sika deer in Japan can lead indirectly to changes in the ecosystem; for example, the degradation of plantation forests is linked with the aging population, and energy conversion to fossil fuels is linked with the increase of imported woods. Namely, degradation of the ecosystem's services occurs, particularly in rural areas of Japan, in concert with the progress of globalization (Hirano 2014).

Forest soils support the ecosystem services of forests concerning carbon storage, nutrient cycling, and biodiversity as basic services as well as controlling water quality and climatic conditions as adjustment services (Kosugi 2014). It should be noted that changes in the use of forest soils can lead to degradation of the ecosystem's services and, therefore, the sustainable management of land use in forests is crucial to maintaining them.

References

Ando Y (1988) Trend in timber import and emerging new face of trade under Yen's strength. J Jpn Soc Distributive Sci 1:109–119

Augusto L, De Schrijver A, Versterdal L, Smolander A, Prescott C, Ranger J (2015) Influences of evergreen gymnosperm and deciduous angiosperm tree species on the functioning of temperate and boreal forests. Biol Rev 90:444–466

Baba M, Kato M, Sugiura T, Kobayashi H (2004) Calcium accumulation alleviates soil acidification in Japanese cedar (*Cryptomeria japonica*) stands. Soil Sci Plant Nutr 50:403–411

Forestry Agency of Japan (2013) Annual report on forest and forestry in Japan FY2012, Tokyo

Forestry Agency of Japan (2014) Annual report on forest and forestry in Japan FY2013, Tokyo

Forestry Agency of Japan (2015) Annual report on trends in forests and forestry FY2014, Tokyo

Fujie T (2007) The turning point of forest management: direction of forest management practices under the "Renewed Forest and Forestry Basic Plan" (ad hoc meeting of the Japanese Forest Economic Society on the "Renewed Forest and Forestry Basic Plan"). For Econ 59:17–21 (In Japanese)

Fujino M (2013) Is the forest planning institution to function? Nougyou-to-Keizai 79:31–39 (In Japanese, tentative translation of the original Japanese title)

Hirano Y, Mizoguchi T, Brunner I (2007) Root parameters of forest trees as sensitive indicators of acidifying pollutants: a review of research of Japanese forest trees. J For Res 12:134–142

Hirano Y (2014) Clinical environmental studies on Kushida-river basin. In: Watanabe S, Nakatsuka A, Ow T (eds) Clinical environmental studies. The University of Nagoya Press, Nagoya, pp 168–179

Imaya A, Ohta S, Tanaka N, Inagaki Y (2005) General chemical properties of brown forest soils developed from different parent materials in the submontane zone of the Kanto and Chubu districts, Japan. Soil Sci Plant Nutr 51:873–884

Jenny H (1941) Factors of soil formation. McGraw-Hill, New York

Kajisa T, Yoshida S, Nagashima K, Murakami T, Mizoue N, Sasaki S, Kuwano Y, Saho K, Shimizu M, Miyazaki J, Fukuzato K, Oda M, Shimozono H (2011) Situation of erosion, landslide, and limiting factors of vegetation recovery on abandoned clear-cut sites in Kyushu region. J Jpn For Soc 93:288–293

Kakizawa H (2013) Trajectory and challenges of forest policy in Japan. Nougyou-to-keizai 79:5–14 (In Japanese, tentative translation of the original Japanese title)

Kojima T (2013) Overview of the Japanese forester system. Forest-consaru 131:4–13 (In Japanese, tentative translation of the original Japanese title)

Kosugi K (2014) Adjustment services. In: Ide Y, Okouchi I, Inoue M (eds) Forest science as a culture. Buneido, Tokyo, pp 97–105

Nakamura F (2012) Land conservation and ecosystem management: in anticipation of a regime shift. Sanrin 1537:2–11 (In Japanese, tentative translation of the original Japanese title)

Nakaoka S (2015) Underlying policies of the forestry basic act 50 years on. For Econ 68:1–8

Nishizawa M, Mashimo Y, Kawabata K (1965) Estimation methods of site index by quantification. Bull Gov For Exp Stat 176:1–54

Nonoda T (2008) Current status of forest policy and forest management. In: Onda Y (ed) Actual condition of the artificial forest devastation and water and sediment outflow. Tokyo: Iwanami Shoten, pp 170–183 (In Japanese, tentative translation of the original Japanese title)

Ota I (2013) Present status around small-scale forestry in Japan. In: IUFRO 3.08 and 6.08 joint conference, future direction of small-scale and community-based forestry proceedings, pp 11–21

Priha O, Smolander A (1999) Nitrogen transformations in soils under *Pinus sylvestris*, *Picea abies* and *Betula pendula* at two forest sites. Soil Biol Biochem 31:965–977

Sasaki T (2011) Japanese politics and forest: on the international year of forests. Shinrin Kagaku 63:40–49 (In Japanese)

Tachibana S (2013) Establishment and desired deployment direction of forest resources. Nougyou-to-keizai 79:15–21 (In Japanese, tentative translation of the original Japanese title)

Takahashi M, Sakata T, Ishizuka K (2001) Chemical characteristics and acid buffering capacity of surface soils in Japanese forests. Water Air Soil Pollut 130:727–732

Tanaka A (2013) Determining whether domestic timber is selling: Japan's true forestry problems. Nougyou-to-keizai 79:24–30 (In Japanese, tentative translation of the original Japanese title)

Tanikawa T, Sobue A, Hirano Y (2014) Acidification processes in soils with different acid buffering capacity in *Cryptomeria japonica* and *Chamaecyparis obtusa* forests over two decades. For Ecol Manage 334:284–292

Tokuchi N, Takeda H, Yoshida K, Iwatsubo G (1999) Topographical variation in a plant-soil system along a slope on Mt Ryuoh, Japan. Ecol Res 14:361–369

Umemura M, Takenaka C (2015) Changes in chemical characteristics of surface soils in hinoki cypress (*Chamaecyparis obtusa*) forests induced by the invasion of exotic Moso bamboo (*Phyllostachys pubescens*) in central Japan. Plant Species Biol 30:72–79. doi:10.1111/1442-1984.12038

Yabe M (2002) Aims of the forest and forestry basic law. Shinrin Kagaku 34:59–61 (In Japanese, tentative translation of the original Japanese title)

Yamane G, Usui G, Goshu K, Kikuchi K, Terazawa K (1990) Multivariate analysis of site factors on afforested *Larix leptolepis* Gordon. Bull Hokkaido For Res Inst 28:54–63

Yoshida S (2011) The background of the research for abandoned plantation clearcut site. J Jpn For Soc 93:277–279

Chapter 7
Inland Water Landscape: Structural and Functional Changes in the Ecosystem

Shigeya Nagayama, Takashi Tashiro and Jyun-ichi Kitamura

Abstract This chapter introduces the structural and ecologically functional changes of riverscapes, including waterbodies in agricultural landscapes, within the context of Japan's socioeconomic backgrounds. Riverscapes, which are characterized by hydro-geomorphic processes of rivers in each geomorphic division, dramatically change along the longitudinal axis of the river from mountain to sea. The longitudinal diversity of riverscapes and their components provide essential habitats for riverine organisms and support freshwater biodiversity. Old agricultural systems, constructed in former floodplains, also contribute to biodiversity by providing substitute habitats for floodplain-dependent organisms. Intensive alteration of freshwater habitats, however, began during the Meiji Era for rivers and the early Showa Era for agricultural lands within the context of socio-economic growth, due to innovations in civil engineering technology. This chapter also provides an overview of the present state of freshwater habitats in Japan.

Keywords Riverscape · Agricultural landscape · Aquatic habitat · Habitat alteration · Management

S. Nagayama (✉)
Aqua Restoration Research Center, Public Works Research Institute, Kawashimakasada, Kakamigahara, Gifu 501-6021, Japan
e-mail: s-naga77@pwri.go.jp

T. Tashiro
Disaster Mitigation Research Center, Nagoya University,
#401 Disaster Mitigation Research Building, 1 Furo-cho, Nagoya, Aichi 464-8601, Japan
e-mail: ttashiro@nagoya-u.jp

J. Kitamura
Mie Prefectural Museum, 3060 Isshiden-kouzubeta, Tsu, Mie 514-0061, Japan
e-mail: kitamura@bio.sci.toho-u.ac.jp

H. Shimizu et al. (eds.), *Labor Forces and Landscape Management*,
DOI 10.1007/978-981-10-2278-4_7

7.1 Introduction

The riverscape, consisting of the entire spatially riverine components, including river channels, floodplains, and floodplain waterbodies (Fausch et al. 2002), had intensively changed in Japan since the Meiji era (1868–1912), when modern civil engineering technology was innovated from European countries. Further alteration of the riverscape had been driven since the Showa 20s (1945–1955) for water use, flood control, gravel mining, and developments of agriculture, city, and hydropower, which were urgent needs during the period of economic growth. Especially in the Japanese archipelago, the mountains higher than 2000 m form a back bone which brings short and steep rivers with rapid runoffs, and human activities are concentrated in limited alluvial plains, which have caused frequent and serious flood damages.

The river infrastructures and works have directly and indirectly altered riverscapes. Primarily, river channelization, levee and dam constructions, and gravel mining (dredging) have directly affected the riverscapes. These massive river system works also changed the dynamic processes of sediment transportation and flow discharge, which are dominant factors forming riverscapes, and have indirectly altered riverscapes to some degree. Most of them have been installed and managed as part of the "river improvement plan," based on its master plan, the "fundamental river management policy." The master plan is established by the river administrator, such as the Ministry of Land, Infrastructure, Transportation, and Tourism (MLIT), or the prefectural governor, and it determines the design flood discharge by taking into consideration the conditions of flood damage frequency, the present status of water resource utilization, their development, and the fluvial environment in accordance with the River Law (Infrastructure Development Institute, Japan 1999). Practically, there are two types of river systems: class A (109 major rivers), which are deemed important for the economy and peoples' lives and are administrated by the MLIT, and class B (other relatively small rivers), which are administrated by the prefectural governor (Infrastructure Development Institute, Japan 1999).

Floodplains, the inundation areas during river floods, were converted to agricultural land (mainly paddy fields). The reclamation of floodplains extended not only into the broad lowland area on the coastal alluvial plains but also into narrow valley plains. Dense and broad canal networks were developed on the reclamation area. In addition, a number of small irrigation ponds were placed in the lowland plains and constructed in the contiguous area between the agricultural land and the hills. However, these waterbodies in the agricultural landscape were also altered during the modernization of agricultural land development, particularly since the Showa 20s (1945–1955).

In this chapter, we introduce the structural and ecologically functional changes of riverscapes, including waterbodies in agricultural landscapes, and their socioeconomic backgrounds.

7.2 Mountain Streams

Mountainous areas, including hills, occupy at least 70 % of the land in Japan, and a huge number of branched streams drain there. Although most of the mountain streams are designated as class B systems, their total length is overwhelmingly long in comparison to the large lowland alluvial rivers designated as class A systems (Infrastructure Development Institute, Japan 1999). Mountain streams are generally characterized by a narrow and steep channel with large riverbed materials, which often cause serious disasters with debris flows due to heavy rainfalls, landslides, or eruptions. Longitudinal changes in the stream gradient and disasters form great longitudinal variation in the riverscape (channel morphology) of mountain streams (Kani 1981; Montgomery and Buffington 1997) (Fig. 7.1).

"Cascades" and "step-pool reaches," with steep bed slopes and large boulders, form headwater small streams. "Pool-riffle reaches" with moderate bed slopes and pool-riffle sequences typically appear in the downstream segments of step-pool reaches. "Bedrock" reaches irregularly appear along the longitudinal axis; they often appear in a narrow area constricted by solid rock. The depositional zone, in which moderate-gradient channels occur with floodplains, is formed in the upper

Fig. 7.1 Pattern diagram of distribution of reach types along a longitudinal profile and photographs of step-pool, pool-riffle, and bedrock reaches in mountain streams

area of such narrow areas, resulting in the diversification of longitudinal channel distributions in mountain areas. Debris and landslide dams also affect the diversity of channel distributions (Grant and Swanson 1995).

A large depositional zone in the mountain area is recognized as a valley plain or basin, which is one of the alluvial plains. A "braided" reach appears there when the sediment production in the upper area is high. In contrast, when the sediment load is low, the riverbed is degraded, and a single-thread channel is formed. The canyon, which penetrates bedrock, is also one of the primary channel types in mountain areas.

Although natural riverscapes in mountain areas are still conserved relative to large lowland alluvial rivers, riverscapes have been considerably altered by dam construction and channel modification (Fig. 7.2). In Japan, construction of large dams, which are at least 15 m in height, began for water use development starting in the Meiji era (1868–1912). In the Taisho era (1912–1926), large dam construction was accelerated for hydropower development because of the increasing demand for electric power. Many multi-purpose large dams for water use, hydroelectric power, and flood control were constructed in the Showa era (1926–1989). Presently, there are approximately 2700 large dams in Japan (Yoshimura et al. 2005; Japan Commission on Large Dams 2009). Large dam construction brought a number of artificial lakes to mountain areas. This means the intercalation of lentic systems into lotic river systems.

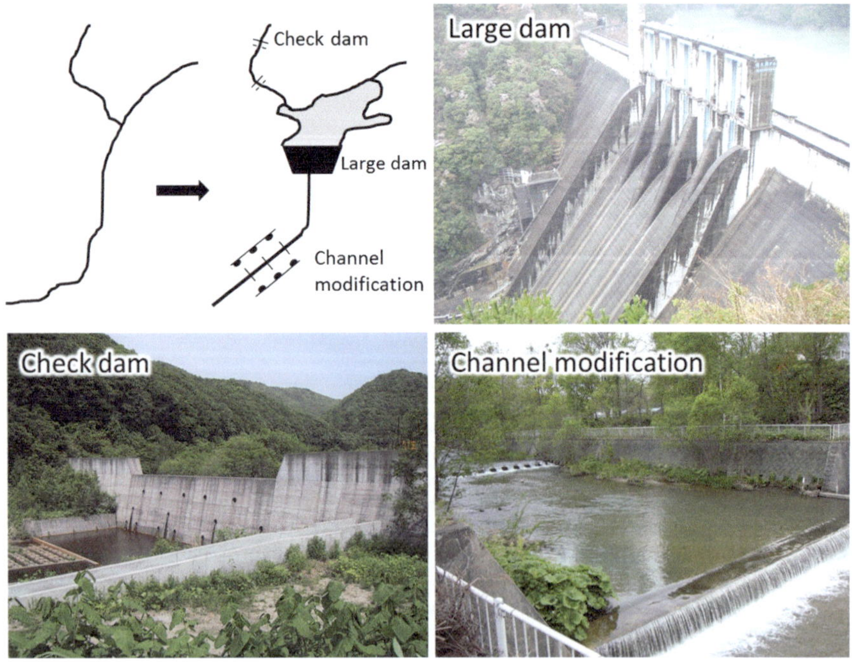

Fig. 7.2 Illustration of dam construction and channel modification and photographs of the same in mountain streams

Various phenomena attributed to the lentic environment, such as sedimentation, cold water discharge, turbidity, and eutrophication, have been observed with large dams. These phenomena and alteration of the flow regime have affected the habitats and distribution of riverine organisms in Japan (Yoshimura et al. 2005). Sedimentation in dams reduced sediment supply to the downstream segment of the dam, and altered the riverbed conditions to be coarse (i.e., "armoring") (Nishita et al. 2011). The suppression of large floods due to flow control by dams has caused a narrowing of the active channel and shrinking of the gravel bar in the Satsunai River, Hokkaido, northern Japan (Takahashi and Nakamura 2011). The intercalation of the lentic system allowed lentic species, including exotic species, to inhabit there (Mizuno et al. 1964; Kumazawa et al. 2012). In addition, prevention of the migration of aquatic organisms by large dams is a major concern. Local extinction and the possibility of anadromous salmon species in the upstream segments of small dams have been reported (Morita and Yamamoto 2002). Genetic differentiation of macroinvertebrates between the upstream and downstream segments of large dams has also been shown (Watanabe and Omura 2007).

Channel modification has directly altered riverscapes in mountain areas. In mountain streams with high stream power, bank and riverbed protection and small drops are typically incorporated in stream works to stabilize channel. Thereby, the original channel morphology and floodplain were often completely lost; instead, straightened channels characterized by homogeneous environmental conditions were constructed. In general, natural streams originally have a sequential structure of channel units, such as pools and riffles. These channel units satisfy various habitat requirements according to the species and life stages of fish (Wolter et al. 2016). Therefore, homogenization of channel units deteriorates fish assemblages.

Constriction of the stream channel via bank protection occasionally causes intensive riverbed degradation where the bedrock is soft (e.g., tuffaceous rock) (Ishiyama et al. 2009). Because the channel constriction increases the shear stress of the riverbed during floods, sediments that cover the bedrock become easily washed out. Once the soft bedrock is exposed, the riverbed is rapidly scoured and degraded.

Massive numbers of small dams (check dams), with heights of less than 15 m, have been constructed in Japanese mountain streams. There are two types of small dams in Japan: one is for preventing debris flow (*sabo* dams) and the other for conserving forests on mountain slopes by preventing riverbed degradation and landslides (*chisan* dams). The number of small dams has rapidly increased since 1960, when the "Soil Conservation and Flood Control Urgent Measures Act" was established by the government. Although its exact number throughout Japan is unknown, at least 35,000 check dams had been constructed as of 2004 in Hokkaido alone (Tamate and Hayajiri 2008), and approximately 85,000 *sabo* dams exist throughout Japan (Japan Sabo Association 2015). The construction of check dams changed riverscapes through the accumulation of sediment, the formation of wide active channels, and/or the establishment of riparian trees in the upper areas of the dams. The check dams also fragmented the stream channels, prevented migration of aquatic organisms, and caused local extinction of fish populations (Morita and Yamamoto 2002).

7.3 Alluvial Rivers

Most large alluvial rivers drain in the coastal alluvial plains that occupy approximately 15 % of the land in Japan, although alluvial rivers also appear in valley plains and basins in mountain areas. The large alluvial rivers designated as class A generally flow on the three geomorphic divisions of the coastal alluvial plain: fan, floodplain, and delta (Fig. 7.3). Because human populations and activities are concentrated there, a lot of flood disasters with serious damage to human lives and activities have occurred with severe inundations due to levee breaches.

Although the alluvial fan is a fluvial landform created over a long span of time, the riverscape is limited in an inundation area by floods along the present river channels. The riverscape primary consists of riparian forests, braided channels, bars, and isolated waterbodies. A braided channel consists of a network of small channels separated by islands (bars). Those riverscape components other than riparian forest are active and considerably vary with flood events. Therefore, bars are bare or covered by grasses.

The floodplain and delta divisions originally had broad inundation areas (floodplains) in their riverscapes. In the floodplain division, the riverscape primarily

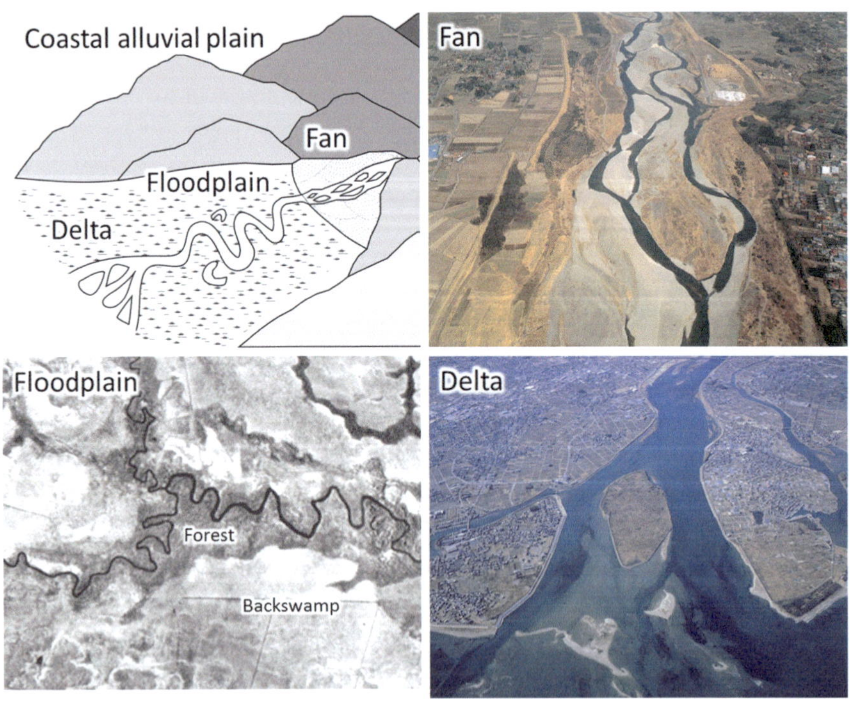

Fig. 7.3 Illustration of a typical coastal alluvial plain and photographs of the riverscape in each geomorphic division

consists of a single-thread meandering channel, point bars, riparian forests on the natural levees, backswamps, and oxbow lakes. These components of the riverscape are relatively stable in comparison to those of the alluvial fan, but vary over time. Channel shortcuts occasionally occur, and this creates an oxbow lake. The estimated persistence of oxbow lakes was 600–1300 years in the Ishikari lowland, western Hokkaido (Ishii and Hori 2016). There are many natural levees and backswamps associated with old channels, which are alternately distributed in the cross-sectional profile. In deltas, river channels are branched toward the sea, and six types of deltas are identified based on their geomorphic characteristics (Suzuki 1998). The characteristics of deltas are influenced by various factors such as the coastal current of the sea, the way it faces toward the sea (bay or outer sea), and the sediment supply from the watershed.

The riverscapes in coastal alluvial plains have been directly altered by levee construction and channel modification (Fig. 7.4). The direct alteration was particularly intensive in the floodplain and delta divisions. River channels were consolidated, straightened, and channelized for navigation and flood control. This channel modification negatively affected aquatic organisms, such as fish and macroinvertebrates, through the loss of habitat diversity characterized by channel

Fig. 7.4 Present riverscape of the lowland Nagara and Kiso Rivers in the Nobi Plain, central Japan

meandering (Nagayama et al. 2008; Nakano et al. 2008; Nakano and Nakamura 2008). Long and continuous levees were constructed to exceed the "high water level" (HWL) during the peak stage of designed floods and to prevent the agricultural lands and urban areas from inundating. The Tokai Heavy Rainfall in 2000 is recognized as one of the severest flood events, which includes not only exceeding HWLs (with a return period of 30 years for class B river systems and 200 years for class A river systems, in those days) but also causing levee breaches in some of the class B rivers. After this Tokai flood, a series of river works were undertaken using the framework of the "Special Recovering Project against Serious Disaster" and successive river improvement plans in accordance with the master plan. In addition to these structural measures against floods, non-structural measures have been established, such as the flood hazard map for understanding potential flood risks, radar observations of precipitations, "XRAIN" for issuing alerts of real-time heavy rainfalls, and evacuation information in order to reduce exposure to inundation risks (Thomas et al. 2015).

The levee construction resulted in the thorough loss of the broad inundation area (floodplain) through the isolation of the floodplain from the river channel. A river project in the Nobi plain, central Japan, during Meiji 20–45 (1887–1912) is particularly famous (Fig. 7.5); three large lowland rivers (the Kiso, Nagara, and Ibi Rivers), which had connected with each other through many branched channels, were isolated and channelized under the guidance of a Dutch engineer, Johannis de Rijke. The floodplain, which was protected by levees and no longer inundated, was extensively converted into paddy fields and city, resulting in the loss of the

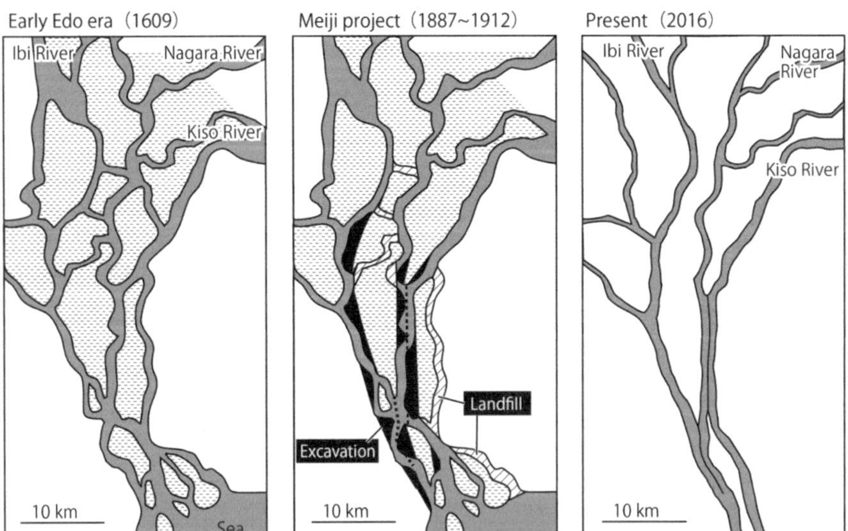

Fig. 7.5 Primary river channels of the Kiso, Nagara, and Ibi Rivers before, during, and after the river project in the Meiji era

backswamp and oxbow lake (Fig. 7.4). Although floodplain alteration would affect the populations of riverine organisms, agricultural land (paddy fields) would provide substitute habitats for many floodplain-dependent organisms and created a new agroecosystem (see below).

Moreover, river works have indirectly altered riverscapes in the coastal alluvial plains. Expansion of the riparian forest along the river channel is one of the most noticeable indirect changes of riverscapes in Japan (Fig. 7.6). One reason for the forest expansion is riverbed degradation in main channels. Riverbed degradation is probably associated with the reduction of sediment transportation due to dam construction and gravel mining and with the intensification of hydraulic power due to channel consolidation and narrowing. Riverbed degradation causes a decrease in the inundation (disturbance) frequency of inter-levee floodplains and lowers groundwater levels. The inter-levee floodplain was gradually terrestrialized with the riverbed degradation and was ultimately covered by trees (Negishi et al. 2008; Nagayama et al. 2015).

In the alluvial fan, the expansion of riparian forest causes the reduction of bare or grassed gravel bars and the shrinking of active channels (Takahashi and Nakamura 2011), resulting in the deterioration of populations of gravel bar-dependent organisms, including grasses, insects, and birds (Kuramoto et al. 1992; Takeuchi and Fujita 1998; Yabuhara et al. 2015).

Although natural meandering rivers in the floodplain division generally have a single-thread channel with point bars, air photos taken in the early Showa era (around 1945) show that active braided channels separated by sand bars occurred in the inter-levee riverscape of some Japanese large rivers (Fig. 7.6). This channel morphology that differs from the general pattern might be associated with the massive amount of sediment production from the mountain area during the time spanning from the Edo era to the early Showa era due to the intensive deforestation in the upper mountain areas in the context of a population increase. However, the

Fig. 7.6 Expansion of riparian forest in the inter-levee floodplain of the Kiso River (Nagayama et al. 2015)

expansion of riparian forest following riverbed degradation has also occurred in the floodplain division since around the 1980s, and the inter-levee riverscape has considerably changed. There are waterbodies isolated from and connected to the main river channel in the forested riverscape, which often function as habitats for floodplain-dependent aquatic organisms, including threatened freshwater mussels and endangered bitterling fish (Sagawa et al. 2011; Negishi et al. 2012; Kume et al. 2014). However, habitat degradation due to the reduction of the inundation frequency in the waterbodies attributed to the riverbed degradation is a concern (Negishi et al. 2008, 2012).

Riverbed degradation has also greatly changed the riverbed condition and bathymetry feature in some large lowland rivers. In the Kiso River, the firm clay layer, the deposition in a former floodplain (uppermost alluvium), was exposed in the riverbed surface, resulting from the washout of the sandy gravel that covered the clay layer (Takaoka et al. 2014). Some cases resulted in intensive local riverbed scouring due to the loss of the clay layer followed by the scouring of the lower sand layers (Fukuoka et al. 2004; Takaoka et al. 2014), which is a considerable issue in river management (Kurihara et al. 2013). In the Kiso River, some intensive local scouring occurred and in one case reached the basal gravel layer from the last glacial period (Kurihara et al. 2013; Takaoka et al. 2014).

7.4 Agricultural Canals and Ponds

Although paddy fields have been developed since the Yayoi era (3 B.C.–3 A.D.) in Japan, reclamation of broad floodplains in coastal alluvial plains might have started in the Edo era in the context of the advancement of civil engineering technology, social stability, and population expansion. Further and intensive reclamation has been driven since the Meiji era, when modern engineering technology was innovated from European countries and continuous levees were constructed along the large lowland rivers to protect paddy fields and residences. The backswamp of the floodplains were converted into paddy fields, and natural levees were used as residences. The agricultural landscape has a few types of waterbodies: paddy fields, irrigation/drainage canals, irrigation ponds, and rivers (Fig. 7.7). In general, ponds and rivers are permanent waterbodies throughout the year, and the others often dry up during non-irrigation periods (autumn and winter). These waterbodies constitute a dense aquatic network in the agricultural landscape and contribute to the maintenance of a unique ecosystem (agroecosystem) that includes originally floodplain-dependent organisms (Washitani 2007). This ecological function of the aquatic network had probably been maintained until at least the year Showa 24 (1949), when the “Land Improvement Act” established by the government drove modern agricultural land development.

The substitute function of floodplain habitats could be ensured by the synchronization of the irrigation period for rice cultivation (spring and summer) and important life stages of floodplain-dependent organisms. Irrigated paddy fields and

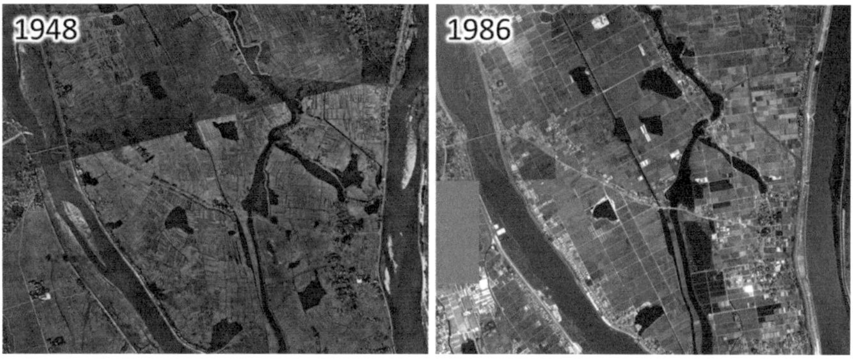

Fig. 7.7 Change of agricultural landscape in the Nobi Plain before and after the enforcement of the "Land Improvement Act" in the year Showa 24 (1949)

canals create a similar environmental condition to backswamps and provide aquatic organisms with spawning, growth, and refuge habitats (Saitoh et al. 1988; Iwata 2006). In particular, this synchronization is essential for those organisms that evolutionarily acquired the behavior of spawning in inundated areas during the pluvial period (spring to summer) of the Asian Monsoon climate system.

In addition, connectivity among the waterbodies in the agricultural landscape would also be an essential factor of the substitute function. Until the Meiji era, dual-purpose canals for both irrigation and drainage were generally used. The difference in height between paddy fields and dual-purpose canals was low so that paddy fields could be easily irrigated from the canal by simple manipulation, such as a small weir and mound made of a wood plate and soil. The difference in height between the canal and the river was also low because drainage into the river was generally dependent on the natural ground slope at that time. Ponds created in the plain also connected the canals smoothly. The small differences in height among the waterbodies involved a high connectivity among aquatic habitats and probably encouraged high biological diversity (Ishiyama et al. 2014). In addition, irrigation water was also used for daily life and fire extinction at that time, and thus the water flowed throughout the year in many of the canals, probably enhancing the ecological function of the aquatic network. Furthermore, most dual-purpose canals were unlined, sinuous channels with natural bed materials such as sand and gravel, providing local habitats suitable for aquatic organisms (Akiyama and Maruyama 2010; Nagayama et al. 2012). Therefore, many of the floodplain-dependent species might have flourished in the aquatic networks of the past agricultural landscape.

The agricultural landscapes have been rapidly altered since the "Land Improvement Act" was established in the year Showa 24 (1949) (Fig. 7.7). Until then, paddy fields were wet throughout the year because natural backswamps were used as paddy fields without advanced land improvement. After 1949, however, the wet paddy field was reformed to the dry one, which was heightened and drained well. In the reformation, former paddy fields, which had irregular and small dimensions, were consolidated and gridironed. Canals were straightened and separated according to their function (irrigation and drainage). The drainage canals and small rivers were deepened to enhance water drainage from paddy fields, and vertical drops were also constructed. Some drainage canals directly connecting to large rivers were extended to drain the water into the downstream part of the river with a lower altitude. In some regions, drainage water was forcibly discharged to the river by pump. The necessity of irrigation water for daily life and fire extinction gradually decreased over time. Thus, many paddy fields and canals in lowland plains come to experience drought during the non-irrigation period. Most ponds in the plain were entirely or partly landfilled during the reformation.

Although the reformation of paddy fields increased rice productivity due to the improvement of environments for cultivation and farm works, it led to drastic changes in the agroecosystem, including the aquatic network. The periodic drought of the paddy fields and canals narrow the available habitat for sedentary organisms, such as mussels. Reduction of the connectivity among waterbodies due to the construction of vertical drops constricts the distribution of migrant organisms, such as fish. Straightened, concrete-lined canals degrade the local habitat quality for all aquatic organisms. Although agricultural ponds presently function as habitats for native and threatened species (Kitamura 2008; Uchida and Inoue 2010), many of them are losing their ecological function due to habitat degradation, such as homogenization of the shoreline habitat with concrete-lined banks, decrease in the quality of water and bottom sediments, and invasion (or intentional/accidental introductions) of exotic species. Thus, available habitats for floodplain-dependent organisms have become extremely limited in agricultural landscapes (Negishi et al. 2014).

References

Akiyama YB, Maruyama T (2010) Habitat characteristics influencing distribution of the freshwater mussel *Pronodularia japanensis* and potential impact on the Tokyo bitterling, *Tanakia tanago*. Zoolog Sci 27:912–916

Fausch KD, Torgersen CE, Baxter CV, Li HW (2002) Landscapes to riverscapes: bridging the gap between research and conservation of stream fishes. Bioscience 52:483–498

Fukuoka S, Ikeda T, Tamura H, Toyoda H, Shigematsu R (2004) Cause of the six scouring portions and evaluation of correcting the main channel in the lower Tone River. Adv River Eng 10:119–124 (In Japanese with English abstract)

Grant GE, Swanson FJ (1995) Morphology and processes of valley floors in mountain streams, Western Cascades, Oregon. In: Costa JE, Miller AJ, Potter KW, Wilcock PR (eds) Natural and anthropogenic influences in fluvial geomorphology. Geophysical monograph series, vol 89. American Geophysical Union, Washington DC, pp 83–101

Infrastructure Development Institute, Japan (1999) The River Law with commentary by article: legal framework for river and water management in Japan supervised by River Bureau, Ministry of Construction Japan. IDI water Series No. 4

Ishii Y, Hori K (2016) Formation and infilling of oxbow lakes in the Ishikari lowland, northern Japan. Quatern Int 397:136–146

Ishiyama N, Akasaka T, Nakamura F (2014) Mobility-dependent response of aquatic animal species richness to a wetland network in an agricultural landscape. Aquat Sci 76:437–449

Ishiyama N, Watanabe K, Nagayama S, Nakamura F, Kenmotsu H, Takahashi H et al (2009) The influence of bedrock outcrop on stream fish habitat, and the assessment of restoration in a bedrock stream channel. Ecol Civ Eng 12:57–66 (In Japanese with English abstract)
Iwata A (2006) Significance of paddy field agriculture for the existence of the kissing loach (*Leptobotia curta*). Jpn J Conserv Ecol 11:133–141 (In Japanese with English summary)
Japan Commission on Large Dams (2009) Dams in Japan: past, present and future. CRC Press, Netherlands
Japan Sabo Association (2015) Sabo Binran. Tokyo, Japan (In Japanese)
Kani T (1981) Stream classification in "ecology of torrent-inhabiting insects" (1944): an abridged translation. Physiol Ecol Jpn 18:113–118
Kitamura J (2008) Bitterling fishes (Cyprinidae: Acheilognathinae): current threats and conservation. Jpn J Ichthyol 55:139–144 (In Japanese)
Kumazawa K, Osugi T, Nishita M, Azami K, Kamata K, Okitu J et al (2012) Invasive exotic fish control by application of encircling net and water level drawdown operation. Ecol Civ Eng 15:171–185 (In Japanese with English abstract)
Kume M, Negishi JN, Sagawa S, Miyashita T, Aoki S, Ohmori T et al (2014) Winter fish community structures across floodplain backwaters in a drought year. Limnology 15:109–115
Kuramoto N, Takenaka A, Washitani I, Inoue K (1992) A conservation biology of Aster kantoensis growing along the Tama River. J Jpn Inst Landscape Architects 55:199–204 (In Japanese with English summary)
Kurihara T, Asano K, Kikuchi H, Takahashi S, Kuroda N (2013) Estimation of generating factors for local scouring at Kisogawa River. Adv River Eng 19: 165–170 (In Japanese with English summary)
Mizuno N, Nagoshi M, Mori S (1964) Fishes of the Sarutani Reservoir in Nara Prefecture, Japan I: an outline of their abundance in the reservoir and its tributary. Jpn J Ecol 14:4–9 (In Japanese with English summary)
Montgomery DR, Buffington JM (1997) Channel-reach morphology in mountain drainage basins. Geol Soc Am Bull 109:596–611
Morita K, Yamamoto S (2002) Effects of habitat fragmentation by damming on the persistence of stream-dwelling charr populations. Conserv Biol 16:1318–1323
Nagayama S, Harada M, Kayaba Y (2015) Can floodplains be recovered by flood-channel excavation? An example from Japanese lowland rivers. Ecol Civ Eng 17:67–77 (In Japanese with English abstract)
Nagayama S, Kawaguchi Y, Nakano D, Nakamura F (2008) Methods for and fish responses to channel remeandering and large wood structure placement in the Shibetsu River Restoration Project in northern Japan. Landscape Ecol Eng 4:69–74
Nagayama S, Negishi JN, Kume M, Sagawa S, Tsukahara K, Miwa Y et al (2012) Habitat use by fish according to seasons and life stages in small perennial agricultural canals. Ecol Civ Eng 15:147–160 (In Japanese with English abstract)
Nakano D, Nagayama S, Kawaguchi Y, Nakamura F (2008) River restoration for macroinvertebrate communities in lowland rivers: insights from restorations of the Shibetsu River, north Japan. Landscape Ecol Eng 4:63–68
Nakano D, Nakamura F (2008) The significance of meandering channel morphology on the diversity and abundance of macroinvertebrates in a lowland river in Japan. Aquat Conserv Mar Freshw Ecosyst 18:780–798
Negishi JN, Kayaba Y, Sagawa S (2008) Ecological consequences of changing riverscape: terrestrialization of floodplain and freshwater mussels. Civ Eng J 50:38–41 (In Japanese)
Negishi JN, Sagawa S, Kayaba Y, Sanada S, Kume M, Miyashita T (2012) Mussel responses to flood pulse frequency: the importance of local habitat. Freshw Biol 57:1500–1511
Negishi JN, Tamaoki H, Watanabe N, Nagayama S, Kume M, Kayaba Y et al (2014) Imperiled freshwater mussels in drainage channels associated with rare agricultural landscape and diverse fish communities. Limnology 15:237–247
Nishita M, Azami K, Ishizawa N, Kumazawa K, Nakazawa S (2011) Changes in physical habitat and benthic animals in the Miharu Dam downstream river in response to river regulation and

sediment replenishment after initial impoundment: the result of 11 years monitoring. Ecol Civ Eng 14:63–74 (In Japanese)

Sagawa S, Kayaba Y, Kume M, Mori S (2011) Comprehension of floodplain in the Kiso River and its restoration project. Civ Eng J 53:6–9 (In Japanese)

Saitoh K, Katano O, Koizumi A (1988) Movement and spawning of several freshwater fishes in temporary waters around paddy fields. Jpn J Ecol 38:35–47 (In Japanese with English summary)

Suzuki T (1998) Introduction to map reading for civil engineers, vol 2: Plains. Kokon Shoin, Tokyo

Takahashi M, Nakamura F (2011) Impacts of dam-regulated flows on channel morphology and riparian vegetation: a longitudinal analysis of Satsunai River, Japan. Landscape Ecol Eng 7:65–77

Takaoka H, Nagayama S, Kayaba Y (2014) Investigation of river-bed morphology and generating process of local scouring in the Kiso River. J Jpn Soc Civ Eng Ser B1 (Hydraul Eng) 70: I_1015–I_1020 (In Japanese with English abstract)

Takeuchi M, Fujita H (1998) Habitat status of the grasshopper, *Eusphingonotus japonicas* (Saussure), in Kanagawa Prefecture, Japan. Jpn J Appl Entomol Zool 42:197–200 (In Japanese with English summary)

Tamate T, Hayajiri M (2008) The relationship between the number of main dams and the coastal catch of masu salmon (*Oncorhynchus masou*) in Hokkaido: implications for river ecosystem conservation. Water Sci 52:72–84 (In Japanese)

Thomas M, Obana M, Tsujimoto T (2015) Adaptation of resilience against disaster: case study of 2000 Tokai flood and 2011 flood in Shonai River, Japan. Nat Sci 7:32–41. http://dx.doi.org/10.4236/ns.2015.71004

Uchida Y, Inoue M (2010) Fish species richness in spring-fed ponds: effects of habitat size versus isolation in temporally variable environments. Freshw Biol 55:983–994

Washitani I (2007) Restoration of biologically-diverse floodplain wetlands including paddy fields. Glob Environ Res 11:135–140

Watanabe K, Omura T (2007) Relationship between reservoir size and genetic differentiation of the stream caddisfly Stenopsyche Marmorata. Biol Conserv 136:203–211

Wolter C, Buijse AD, Parasiewicz P (2016) Temporal and spatial patterns of fish response to hydromorphological processes. River Res Appl 32:190–201

Yabuhara Y, Yamaura Y, Akasaka T, Nakamura F (2015) Predicting long-term changes in riparian bird communities in floodplain landscapes. River Res Appl 31:109–119

Yoshimura C, Omura T, Furumai H, Tockner K (2005) Present state of rivers and streams in Japan. River Res Appl 21:93–112

Chapter 8
Coastal Area Landscape: Environmental Changes and the Characteristics of Labor Activities

Hiromi Yamashita and Taito Yasufuku

Abstract In this section, we look at the definition of a coastal area, environmental changes observed in coastal areas in recent years, and the characteristics of labor activities in coastal areas, as well as the challenges of calculating them. Taking labor accounts in coastal areas is not as straight forward as one might hope, and there are still many obstacles and barriers in understanding coastal management activities as a whole, institutionally, geographically, periodically, and mentally. This exercise on calculating labor to maintain and restore coastal biodiversities hopefully provides a small starting point in linking numerous activities together.

Keywords Labor accounts · Labor activities in coastal areas

8.1 Introduction

Coastal regions are becoming an increasingly important part of our lives, particularly in island states such as Japan. The combined length of the world's coastline is approximately one million kilometers, of which 30,000 km are in Japan (equivalent to 85 % of the Earth's circumference with Japan having the 6th longest coastline in the world due to its intricate nature). Coastal areas have served as a source of "livelihood, enjoyment, and inspiration for humans throughout much of their history" (Walker 1994, p. 103). For island nations, coasts have also served as defense frontiers for military purposes and as ports for international trade. In the course of history, much of the coastline has been modified and developed to suit human purposes (Palmer 1998; PEMSEA 2003). Walker explains, "the most conspicuous

H. Yamashita (✉)
Ritsumeikan Asia Pacific University (APU), 1-1 Jyumonji-baru, Beppu, Oita 874-8577, Japan
e-mail: hiromiya@apu.ac.jp

T. Yasufuku
Graduate School of Environmental Studies, Nagoya University,
Furo-cho, Chikusa-ku, Nagoya, Aichi 464-8601, Japan
e-mail: taitoyasufuku@gmail.com

H. Shimizu et al. (eds.), *Labor Forces and Landscape Management*,
DOI 10.1007/978-981-10-2278-4_8

boundary on Earth is that which joins land and sea… It is a zone whose resources have been exploited heavily and whose physical form has been modified drastically" (Walker 1994, p. 103).

In this section, we would like to look into the definition of coastal areas, environmental changes observed in coastal areas in recent years, and the characteristics of labor activities in coastal areas, as well as the challenges of calculating them.

8.2 Definition of a Coastal Area and the Boundaries for This Study

According to the Japanese Association for Coastal Zone Studies, a coastal zone is defined as "a narrow strip-shaped space along the shore including shallow areas of the ocean with lands attached to them" (Japanese Association for Coastal Zone Studies 2000, p. 1). The Ramsar Convention of Wetlands defines coastal wetlands as including: permanent shallow marine waters (less than 6 m deep at low tide); marine sub-tidal aquatic beds (underwater vegetation, including kelp and seagrass beds, and tropical marine meadows); coral reefs; shores; saline or brackish water; intertidal areas (e.g. mud, sand or salt flats, and marshes); lagoons (brackish or coastal freshwater); and estuarine waters. What is clear is that when we use the term coastal area or coastal zone, or even coastal wetlands area, it includes areas that are semi-permanently or permanently underwater. A coastal neighborhood landscape complex unit looks like Fig. 8.1. In this study, among these coastal neighborhood landscape complexes, we will be especially focusing on coastal wetlands, including tidal flats, seaweed beds, coral reefs, and beach ecosystems, since they provide homes for coastal biodiversities and have been closely linked to human activities for centuries, both through protection and exploitation activities.

8.3 Environmental Changes Observed in Coastal Areas

The importance of coastal areas for human survival is more than apparent. Coastal regions harbor many of the Earth's most diverse, complex, and productive ecosystems (UNESCO 1997).

Many early human settlements are found by the seashore together with shellfish fossils. Coastal wetlands such as tidal flats, seaweed beds, coral reefs, mangroves, and beaches provided varieties of ecological services such as water purification, juvenile fish nursery provision, high tide barriers, recreation places, and food (e.g. Lee et al. 1998, Fujimae higata o mamorukai 2013). For example, seaweed beds purify seawater and being regularly cut and managed by people encourages its health and sustainability (Nomura 1998). Nojima (1996) also notes that seaweed beds provide nurseries for benthos and juvenile fish (see also Fig. 8.2).

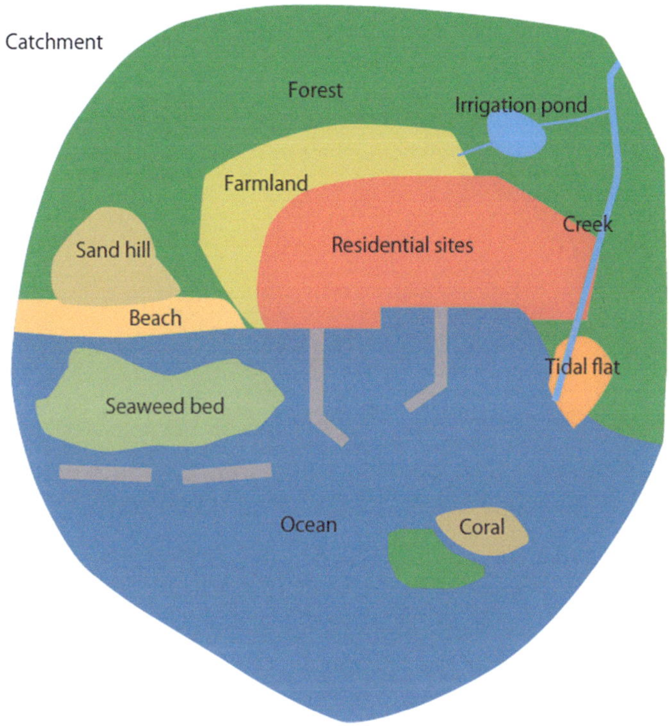

Fig. 8.1 Model of a coastal neighborhood landscape complex unit

In many countries, more than half of the population lives in a coastal zone, a percentage that is increasing. Japan has more than 4000 harbors, one for every 8 km of shoreline; 51 % of its shoreline is bordered by artificial structures. Other examples include England with 38 % and Belgium with 85 % (Walker 1994). Although the ecological importance of wetlands and tidal flats has been widely communicated in recent years (e.g. Smardon 2009), they are still under great pressure from urban and coastal development projects in Japan and abroad. In Japan, between the 1940s and 1980s nearly 40 % of the natural tidal flats were lost through reclamation, and currently it is said to be 50 % or more (e.g. Baba et al. 2003). Ise Bay also had many shallow areas of water (tidal flats) in 1922, but most of them were lost due to land reclamation, and one of the few small areas left is the Fujimae tidal flat Ramsar site. Many of the city developments in coastal areas have utilized coastal wetlands, including tidal flats, to expand available land to use for living, transportation, and infrastructure.

When we look at the data from the baseline research for natural environmental conservation conducted by the Ministry of Environment of Japan, this trend of coastal wetland loss is clear (see Table 8.1). The extent of coastlines has expanded over the years, yet the number of natural or semi-natural coastlines is shrinking.

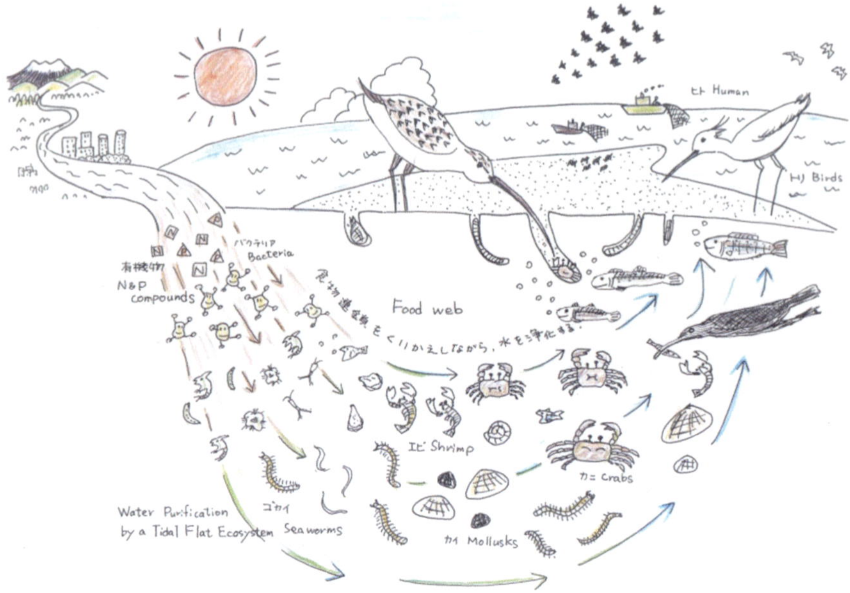

Fig. 8.2 The food chain and water purification mechanism of living creatures on tidal flats (Fujimae higata o mamorukai 2013)

This means artificial coastlines have been created on natural tidal flats, seaweed beds, and in coral reef areas, and they were lost together with their rich biodiversity and ecological functions.

In addition to the loss of coastal wetlands, land-based activities have also influenced the deterioration in the health of the coastal environment. Influence from land-based activities include untreated discharges from factories, homes, illegal waste dumping along the watercourse, and building numerous dams preventing the natural supply of soil to coastal wetlands (see river section for details). Early coastal engineering using concrete structures to prevent the imbalance of soil erosion of coastal wetlands might itself have contributed to the destruction of natural ecosystems in many cases. Table 8.2 summarizes various possible reasons for coastal environment degradation, from both anthropogenic and natural perspectives.

The idea of promoting "Satoumi" [literally, village and the sea] has been introduced to encourage balanced relationships between human activities and biodiversities (Yanagi 2006, 2010). According to the Ministry of Environment (2009), "Satoumi" is defined as a coastal area that includes nature and human beings, as well as an area in which both high biological productivity and biodiversity are expected. Within the "Satoumi" concept, it is expected that "the material circulation function is appropriately maintained, when integrated and comprehensive management of the land and coastal area is performed, and when the rich and

Table 8.1 Changes in the Japanese coastline (Biodiversity Center of Japan 1980, 1994, 1998)

			1994–1980	1998–1994	1998–1980
Coastline (km)	1980	32,170.2	608.6 (102 %)	20.2 (100 %)	628.8 (102 %)
	1994	32,778.8			
	1998	32,799.0			
Natural coastline (km)	1980	18,967.1	−861.5 (95 %)	−691.7 (96 %)	−1553.2 (92 %)
	1994	18,105.6			
	1998	17,413.9			
Seminatural coastline (km)	1980	4340.3	127.3 (103 %)	−214.9 (95 %)	−87.6 (98 %)
	1994	4467.6			
	1998	4252.7			
Artificial coastline (km)	1980	8598.9	1342.8 (116 %)	879.8 (109 %)	2222.6 (126 %)
	1994	9941.7			
	1998	10,821.5			
Estuary (km)	1980	263.7	0.2 (100 %)	46.8 (118 %)	47.0 (118 %)
	1994	264.0			
	1998	310.7			
Seaweed bed (ha)	1980	207,615.0	−6403.0 (97 %)	−58,752.2 (71 %)	−65,155.2 (69 %)
	1994	201,212.0			
	1998	142,459.8			
Tidal flat (ha)	1980	53,856.0	−2413.0 (96 %)	−2062.7 (96 %)	−4475.7 (92 %)
	1994	51,443.0			
	1998	49,380.3			
Coral reef (ha)	1980	36,153.8	−1511.7 (96 %)	−1111.0 (97 %)	−2622.7 (93 %)
	1994	34,642.1			
	1998	33,531.1			

(Natural coastline) The coast holds its natural state without being altered artificially; (Semi-natural coastline) The coast holds its natural state in the intertidal zone, although part of the coast is formed of artificial structures, such as roads, embankments, tetra pods; (Artificial coastline) The coast has been significantly artificially modified by civil engineering work, including port development, landfill, reclamation by drainage, dredging; (Estuary) The boundary of the land is downstream of the "river area" in accordance with the provisions of the River Law

diversified ecosystem and natural environment are conserved" (Ministry of Environment 2009, p. 3, see Fig. 8.3). Unfortunately, many of the current Japanese coastal areas have lost their maintained material circulation functions, and various attempts are being made to bring coasts back to a healthy state.

8.4 Overview of Labor Activities in Coastal Areas

There are various types of activities to maintain costal biodiversities that need human labors, and these have been often carried out by people who have lived nearby or utilized a particular coastal environment regularly. For example, "moku-tori", harvesting seaweed beds, used to be conducted by fishermen to sell or use the seaweed as fertilizers for their farmland. The labor provided benefit not only to the fishermen but also contributed to managing the productive density of

Table 8.2 Anthropogenic and natural cause of the coastal environmental changes in Japan

	Tidal flat (including mangroves)	Seaweed bed	Coral reef	Beach
Anthropogenic cause	• Land reclamation or land reclamation by drainage • Reduction of soil supplies due to construction of dams up river, or changes in river flows with river engineering • Extensive harvesting of resources • Selling certain species for commercial reasons (e.g. coral reefs, shellfish) • Lack of people to conduct biodiversity maintenance work (e.g. cutting seaweed beds regularly) • Coastal water pollution (e.g. oil tankers) • Water pollution from land-based activities (e.g. mines, factories, homes, agricultural pesticides) • Sudden soil discharges from the river after rain, due to lack of forest soil management activities • Increase of alien species			
Biological/natural cause	• Changes in ocean environment, due to high waves, high tides, tsunami, floods • Changes in water quality including nutrient content • Possible effects of global warming including rising acidity in coastal water • Occurrence of red tides and blue tides (these occurrences are said to be linked to the anthropogenic reason of too many nutrients being discharged) • Increase of one particular species which overtakes a coastal ecosystem			

seaweed beds. Since each village has boundaries for obtaining coastal resources, villagers themselves decided their own unique rules to use their coastal recourses sustainably. The management activities also involved a village as a whole. Because of decreasing usage of various coastal resources in their day-to-day lives, as well as degradation of the coastal environment, management work increasingly built up to keep the coastal environment as healthy as it was in the past (a summary of these activities is in Tables 8.3, 8.4 and Fig. 8.4).

Most of the activities that look after coastal areas have been conducted by local fishermen associations and small NGOs/NPOs at each coastal community. For example, for seaweed beds, regional councils lead training for smaller organizations and sometimes provide funding to implement national protection strategies. For drifting garbage, the Ministry of Environment conducted a research project to determine models of garbage reduction in 2007, and gathered data from 11 coasts to identify the types and quantity of coastal garbage. The follow-up research in 2009 included 10 coastal areas (including several from previous studies). After the Ministry of Environment issued the Bill for Promoting Dispositions of Coastal Driftage in 2009 [Kaigan hyocyakubutsu syori sokushin ho], local governments were encouraged to engage with local volunteer bodies to work on this issue.

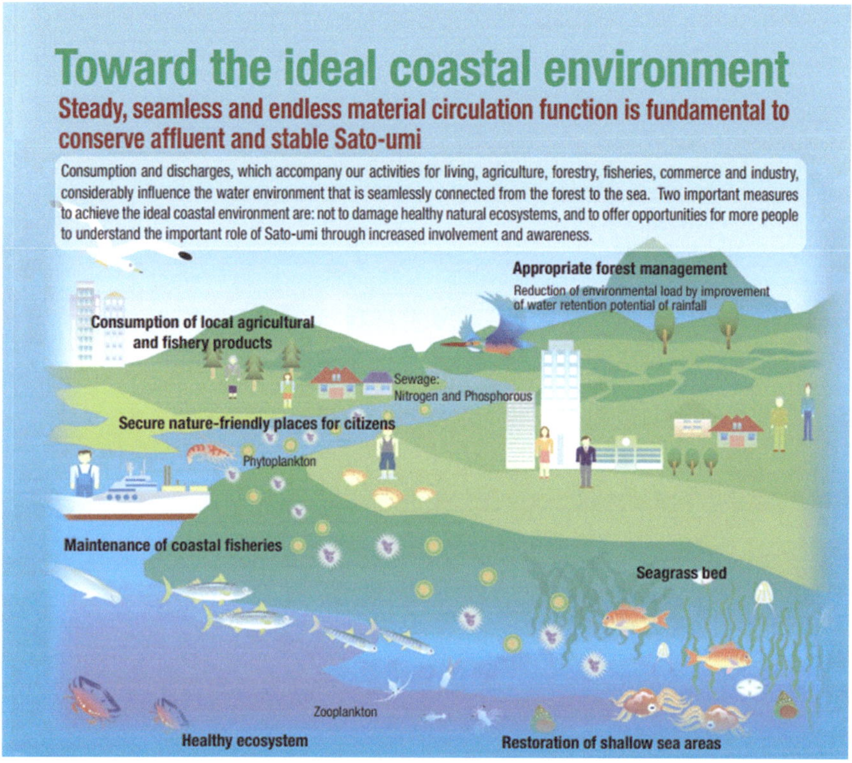

Fig. 8.3 Satoumi diagram to show the maintained and balanced material circulation functions of the sea (Ministry of Environment 2009, p. 3)

The various tasks conducted to look after coastal biodiversities are not necessarily traditional activities. They are tasks that became necessary due to degradation of coastal environments across Japan. For this study, we will look at the existing tasks conducted between 2014 and 2016; however, the amount and types of labor tasks might increase over the coming years. There are a few pieces of research look at labor accounts in forest environments. For example, Terada (2010) investigated the labor count in "Satoyama" (literally village + mountain) environments. "Satoyama" is an environment where people utilize the environmental ecological services of their mountains, as well as look after them to enhance their biodiversities). Tanizaki et al. (2005) studied conditions of citizen management activities and labor accounts in "Satoyama" environments. However, so far, there has not been a labor count study that looks at coastal areas; this study hopes to provide baseline data for future research activities, to understand management costs and possible strategies for conserving coastal areas.

Table 8.3 Various labor activities conducted to maintain or restore coastal environments

	Tidal flat (including mangroves)	Seaweed bed	Coral reef	Beach
Management for its environment	Soil management; removal of living creatures that invade the surface of tidal flats, picking up garbage	Planting and distributing seeds of eelgrass; harvesting them regularly	Removing starfish when they are overtaking the environment; re-introducing/planting reefs	Monitoring numbers of people utilizing the area; protecting coastal vegetation
Management for its resources	Releasing juvenile fish or shells; removing living creatures that eat juvenile fish or shellfish; setting up protection areas; observing implementations of fishery rules	Observing implementation of fishery rules	Observation of illegal fishing or illegal collection of corals	Observation of illegal collection of shellfish
Common activities across ecosystems	Picking up garbage; removing invasive species; preventing and monitoring sudden flow of soils after heavy rain; environmental monitoring and research activities; conducting environmental education activities; creating leaflets			

In Japan, fewer and fewer people are living in rural coastal areas. Also, after the Eastern Japan Earthquake and Tsunami in 2011, with nearly 18,500 people killed or still missing, living by the sea became less attractive to many people. The populations of fishermen and fisherwomen are aging too. At the same time, the task of looking after coastal environments or even restoring them seems to be escalating. The positive human inputs to nature needed for the future might have to come from those who are not necessarily living by the coast.

8.5 Summary

In this section, we have looked at the definition of a coastal area, environmental changes observed in coastal areas in recent years, and the characteristics of labor activities in coastal areas, as well as the challenge of calculating their number.

As discussed in this section, taking labor accounts in coastal areas is not as straight forward as one might hope. Although the discussion on "Integrated Costal Management (ICM)" or "Integrated Coastal Zone Management (ICZM)" (Ramsar Convention on Wetlands 2010) has been happening for decades, there are still many

Table 8.4 Average activity account and activity area for conservation activities, in the manual for environmental and ecological conservation activities (Fisheries Agency, Ministry of Agriculture, Fisheries, and Forestry 2009)

Resources	Activities	Grants from national and local governments (million yen)	Average activity area (ha)	Average activity amount (person a day)	Labor accounts (person a day)
Seaweed beds	Planting mother seaweeds	100	5	40	0.08
	Growing seaweed seeds	100	8	30	0.038
	Seaweed seedlings	100	0.4	30	0.75
	Planting seaweed	150	0.1	40	4
	Removing creatures that feed on seaweed (sea urchins)	150	4	40	0.1
	Removing creatures that feed on seaweed (fish)	250	60	100	0.017
	Deciding on protection areas	150	0.3	40	1.3
	Management of the concentration of sea urchins	250	20	100	0.05
	Provision of nutrient salt	250	8	40	0.05
	Cleaning of rock surfaces	350	9	100	0.11
	Planting trees in its bioregion	100	0.4	70	1.8
	Removing floating or deposited things	150	5	100	0.2
	Other specially admitted activities	100–200	–	–	–
Tidal flats Shallow waters	Prevention of soil or mud movement	150	0.5	40	0.8
	Adding soil	200	1	50	0.5
	Digging (tidal flats)	150	50	50	0.01
	Digging (shallow area)	150	50	75	0.015
	Removal of dead creatures	150	25	50	0.02
	Removal of creatures that weaken the ecosystem functions (gastropods)	200	80	130	0.016
	Removal of creatures that weaken the ecosystem functions (fish)	200	215	50	0.0023
	Removal of creatures that weaken the ecosystem functions (fushiashirui)	200	2	60	0.3
	Removal of creatures that weaken the ecosystem functions (others)	150	15	40	0.027

(continued)

Table 8.4 (continued)

Resources	Activities	Grants from national and local governments (million yen)	Average activity area (ha)	Average activity amount (person a day)	Labor accounts (person a day)
	Deciding on protection areas	250	140	90	0.0064
	Encouragement of juvenile shells to stay on the seabed	150	1	50	0.5
	Management of the concentration of juvenile shells	150	30	75	0.025
	Re-planting living creatures that help functions of the ecosystem	250	30	110	0.037
	Planting trees in its bioregion	100	0.4	70	1.8
	Removing floating or deposited things	250	2	130	0.65
	Other specially admitted activities	100–200	–	–	–
Reed beds	Cutting and culling reeds	150	1	100	1
	Re-planting reeds	150	6.5	90	0.14
	Management of competing plants	250	6	100	0.17
	Setting up the conservation hedge	100	0.2	60	3.0
	Deciding on conservation areas	100	0.3	30	1.0
	Removing floating or deposited things	250	15	150	0.1
	Other specially admitted activities	100–200	–	–	–
Coral reefs	Growing coral seeds	200	0.007	60	85.7
	Re-planting corals	100	0.005	20	40
	Removal of creatures eating corals	250	40	80	0.02
	Deciding on conservation areas	100	0.005	30	60
	Removing floating or deposited things	100	10	50	0.05
	Other specially admitted activities	100–200	–	–	–

Note "Other specially admitted activities" means conservation activities that are regarded as necessary by the local councils

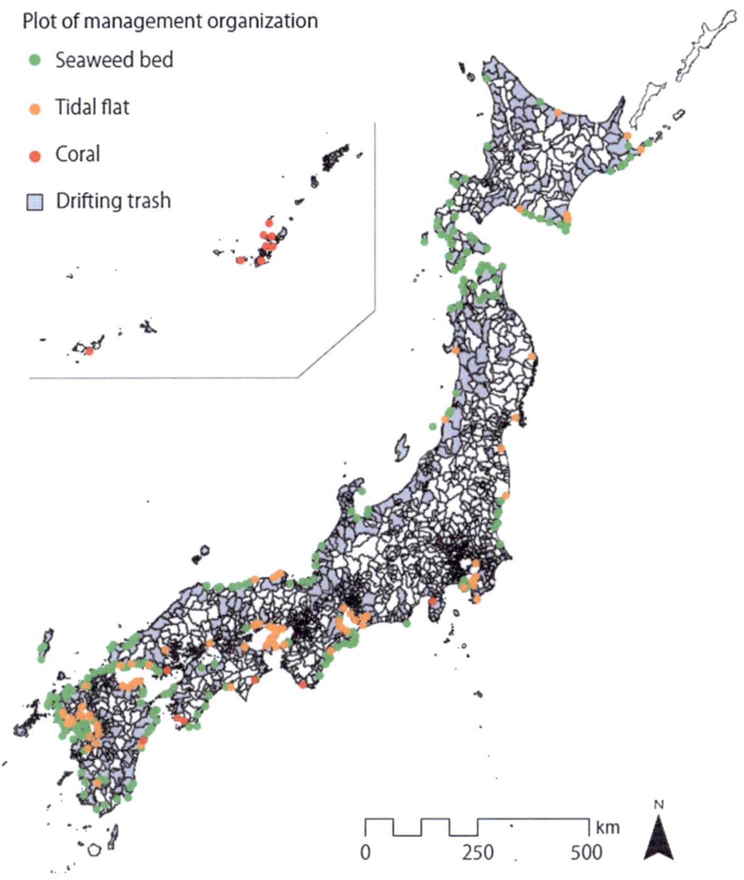

Fig. 8.4 Plot of coastal management conducted by local organizations and citizens in Japan (apart from fishermen associations)

obstacles and barriers to fully understanding coastal management activities as a whole, institutionally, geographically, periodically, and mentally. This exercise of counting labor to maintain and restore coastal biodiversities hopefully provides a starting point in linking numerous activities. In Chap. 16, the case study of Shima city in Mie Prefecture is presented to investigate the types and amount of labors currently existing, to ask ourselves how we would maintain these activities in the future.

References

Baba E, Kawarada H, Nishijima W, Okada M, Suito H (2003) Waves and tidal flat ecosystems. Springer, London

Biodiversity Center of Japan, Ministry of Environment (1980, 1994, 1998) Basic research on environmental conservation baseline research (Shizen kankyo hozen kiso cyosa). http://www.biodic.go.jp/trialSystem/shpddl.html. Accessed 16 June 2016

Fisheries Agency, Ministry of Agriculture, Fisheries, and Forestry (2009) The guideline of environment and ecosystem conservation activities. Fisheries Agency, Ministry of Agriculture, Fisheries, and Forestry, Tokyo

Fujimae higata o mamorukai (2013) Food chains and water purification mechanisms of living creatures on tidal flats. http://fujimae.org/. Accessed 16 June 2016

Japanese Association for Coastal Zone Studies (2000) Proposal for sustainable usage and environmental protection of coastal zones. Japanese Association for Coastal Zone Studies, Tokyo

Lee J, Nishijima W, Mukai T, Takimoto K, Seiki T, Hiraoka K, Okada M (1998) The organic matter purification function of natural and artificial tidal flats and its role in the purification of Hiroshima Bay (Shizen oyobi jinko higata no yukibutsu jyokano no teiryoka to Hiroshima wan no jyoka ni hatasu yakuwari). J Jpn Soc Water Environ (Mizukankyo gakkaishi) 21(3):149–156

Ministry of Environment (2009) Satoumi: a coastal sea area in which there is a harmonious coexistence of nature and human beings. Ministory of Environment, Tokyo

Nomura M, Sasaki M, Chiba N, Xu K, Sudo K (1998) Role of seaweed for water purification in inner bays (Naiwan no suishitsu jyoka niokeru kaiso no yakuwari). Article Collect Jpn Assoc Coast Zone Stud (Nihon engan gakkai ronbunshu) 10:125–136

Palmer J (1998) Environmental education in the 21st century: theory, practice, progress and promise. Routledge, London

PEMSEA (2003) Sustainable development strategy for the seas of East Asia: regional implementation of the world summit on sustainable development requirements for the coasts and oceans. Partnerships in Environmental Management for the Seas of East Asia (PEMSEA), Quezon

Ramsar Convention on Wetlands (2010) Coastal management: wetland issues in integrated coastal zone management, handbooks for the wise use of wetlands 4th edn., No.12. Ramsar Convention Secretariat. Gland

Smardon RC (2009) Sustaining the world's wetlands: setting policy and resolving conflicts. Springer, New York

Tanizaki S, Kaga H, Shimomura Y, Masuda N (2005) Study on estimation of management efficiency in preserving satoyama conducted by local residents (Shimin sankagata no satoyamakanri niokeru sagyo koritsu nikansuru kenkyu). Landscape Res [Landscape Kenkyu] 68(5):623–626

Terada T, Yokohari M, Bolthouse J, Matsumoto R (2010) A citizen-based maintenance scheme for conservation of suburban satoyamas based on the Forest Working Plan (Toshi kinko deno shinrin segyo keikaku nimotozuku shimin niyoru satoyama kanri katsudo no jittai). J Rural Plann Assoc (Noson keikaku gakkaishi) 29:179–184

UNESCO (1997) Oceans, environment and sustainable development: UNESCO implementing Agenda 21. Bureau for the Coordination of Environmental Programmes. UNESCO, Paris

Walker HJ (1994) Coastlines and artificial structures. In: Eblen RA, Arun WR (eds) The encyclopedia of the environment. Houghton Mifflin Company, New York

Nojima T (1996) Biodiversity and stabilization mechanisms in seagrass communities (Kaiso moba gunsyu no tayosei to tanteika kiko). Jpn J Ecol (Nihon seitaigakkaishi) 46(3):327–337

Yanagi T (2006) Theory of satoumi (satoumi ron). Koseisha Kouseikaku, Tokyo

Yanagi T (2010) Theory of creating satoumi (satoumi sosei ron). Koseisha Kouseikaku, Tokyo

Part III
Case Studies

Chapter 9
Methodology

Hiroyuki Shimizu, Chika Takatori and Nobuko Kawaguchi

Abstract In this chapter, the methodology for adapting landscape management labor accounts in Japan is developed. As a first step, definition of landscape elements, landscape units, and neighborhood landscape complex units will be set and the relationships among them will be clarified. Secondly, a methodology to calculate landscape labor accounts is shown. The classification of landscape units in Japan is also described, as are several definitions of landscape labor accounts. Lastly, methodologies toward connectivity of small neighborhood scales to the large regional scales is proposed.

Keywords Landscape management · Labor force · Definitions of landscape labor accounts

9.1 Application to Japanese Case Studies

The third part of this book aims to verify the effectiveness of the concept of landscape labor accounts through several case studies, and to show the possibility of research developments in different scales. In this chapter, the methodology for adapting landscape management labor accounts to each case study will be developed.

In this study, to gather data, the sites of the case studies were selected in the Chubu metropolitan area according to the seven Japanese basic landscape types classified in Chap. 1. The basic data for the landscape management labor accounts was collected through interviews of residents and workers, or through observation

H. Shimizu · C. Takatori (✉) · N. Kawaguchi
Graduate School of Environmental Studies, Nagoya University,
Furo-cho, Chikusa-ku, Nagoya, Aichi 464-8601, Japan
e-mail: takatori@corot.nuac.nagoya-u.ac.jp

H. Shimizu
e-mail: shimkankyou@msn.com

N. Kawaguchi
e-mail: nobqou@gmail.com

H. Shimizu et al. (eds.), *Labor Forces and Landscape Management*,
DOI 10.1007/978-981-10-2278-4_9

of landscape management activities at the sites. Also digitized maps and aerial photographs were used as calculation bases for GIS analysis.

As a first step in our analysis, several definitions of landscape labor accounts are shown, and their meanings and visions for the utilization as indicators are considered. As a second step, details of case studies in landscape units and complex unit levels are examined, and useful results are extracted. Then, as a third step, methodologies of the connectivity of small neighborhood scales to the large regional scales are proposed. And, then finally, the future situation, especially given the shortage of landscape management labor, is examined through the comparison of the populations of today and in the future.

9.2 Definition of Landscape Element, Landscape Unit and Neighborhood Landscape Complex Unit

9.2.1 Relation Between Landscape Element, Landscape Unit and Neighborhood Landscape Complex Unit

"Landscape elements" are basic and minimum land cover components for the calculation of labor accounts, such as "secondary forests", "kitchen gardens", "trees", "grasses" and so on. "Landscape units" are basic land use units such as "building lots", "parks", "paddy fields" and so on. If a landscape unit has a homogeneous characteristic like a paddy field, cultivated land, or artificial forest, etc., it could have the same meaning as a landscape element. If a landscape unit has a non-homogeneous characteristic like a park or a detached house site, a landscape element becomes a part of the landscape unit. A "neighborhood landscape complex unit" is a landscape complex unit with various landscape units which acts as a geographical, social or cognitive unit in a neighborhood scale (Fig. 9.1).

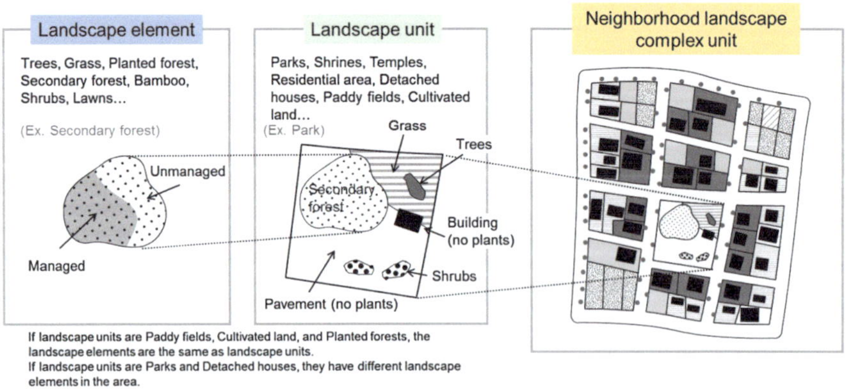

Fig. 9.1 Definition of landscape element, landscape unit and neighborhood landscape complex unit

9.2.2 Three Aspects of Neighborhood Landscape Complex Units

The concept of a neighborhood landscape complex unit is introduced in order to define the basic landscape area, which is regarded as an integrated unit for the management of local complex landscapes. There are some aspects used to determine what a neighborhood landscape complex unit is.

First is a geographical aspect. A geographical unit such as a watershed might be considered as an appropriate management unit of local complex landscapes because of its biological entity. Second is a social aspect such as an administrative boundary (such as an elementary school district or town) or a management boundary (such as a neighborhood association). As local landscape management is usually carried out by hand or by the initiative of a local municipality or local community, it makes sense to use a local boundary as a landscape complex unit. Third is a cognitive aspect. It is important to know which area is recognized as "the place we manage" by residents. In some special cases, these three different boundaries might overlap completely, but they might not overlap in normal cases (Fig. 9.2).

There also might be some gaps among the boundaries of these three aspects, and these gaps might cause management troubles (Fig. 9.3). In Japan, units of geographical and social aspects usually overlap, especially in the lowland agricultural fields, because the water management of paddy fields is the most important issue in the landscape management of local communities. Recently, however, vacant gaps or overlapping of social management areas can be seen even in the same geographical unit. This phenomenon can be seen especially in urbanized areas because of the collapse of old agricultural communities. This makes effective and comprehensive landscape management difficult.

On the other hand, in a mountainous area, the geographical boundary is important to evaluate its ecological values. However, the cognitive boundary by residents is weakened because of the aging and population outflow, and the cognitive blank by residents, i.e. gaps between cognitive and geographical units, is

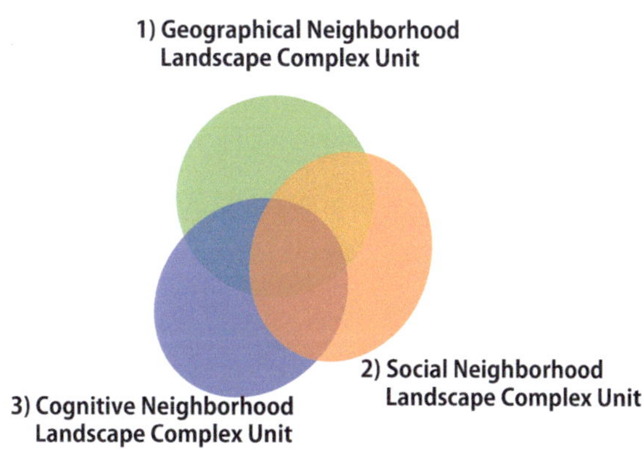

Fig. 9.2 Three aspects of a neighborhood landscape complex unit

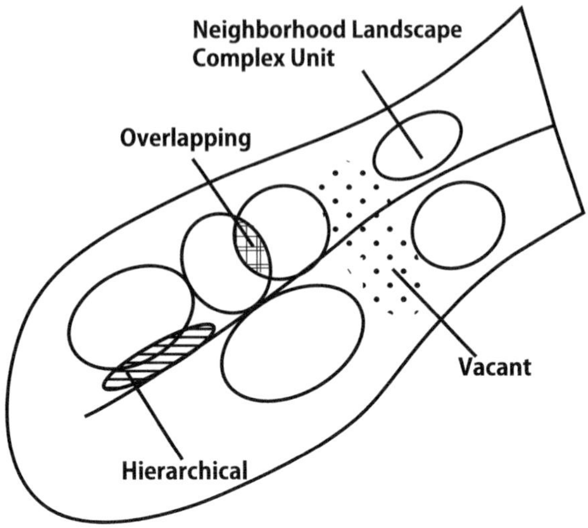

Fig. 9.3 Vacant, hierarchical, and overlapping areas among neighborhood landscape complex units

expanding. Therefore it is important to determine the place where the gaps of geographical and cognitive units exist, in other words, where appropriate ecological preservation is needed and the risk of abandonment is increasing.

In this book, the importance of the social aspect is emphasized, and neighborhood landscape complex units are used as a base of the research and consideration. But in the integration of landscape management labor accounts in large scales, i.e. regional scales, the lack or shortage of management labor caused by the cognitive gaps will be considered.

9.3 Calculation Methodology of Landscape Labor Accounts

9.3.1 Classification of Landscape Units in Japan

In this section, landscape units are classified by using the available datasets in Japan. The Datasets of *Vegetation survey, Natural Environmental Information GIS* (Ministry of the Environment of Japan 2005) (scale: 1/25,000) are used for the large classification of landscape units. The Vegetation Survey classifies the vegetation naturalness according to the extent of human intervention from the plant sociology aspect. And it can be said to have a strong relation with landscape management. Thus, in this research, vegetation naturalness will be selected as a major classification index.

However, there are some lacking areas in the *Vegetation Survey* (1/25,000), especially in mountainous areas. The data of the *Vegetation Survey* (scale: 1/50,000) is used as a complement of the area. Moreover, in the *Vegetation Survey*, "built-up areas" includes various land use types such as commercial, industrial, and residential areas, religious and public facilities, parks, and so on, which are assumed to have the variability of landscape elements. This kind of variety might occur in the diversity of labor accounts. Thus, the "built-up areas" are classified into more detailed landscape unit types by using the data of the *Basic Survey of City Planning* (Nagoya City 2012). "Water Surface" is also divided into river, irrigation channel, and pond by using the Basic Survey of City Planning data, excluding "Sea Surface". Table 9.1 shows the result of classification of landscape units based on the *Basic Survey of City Planning* (Nagoya City 2012).

In order to calculate labor accounts of each landscape unit, collection of the detailed area data of landscape elements which are included in each landscape unit is needed. Thus, the detailed data of landscape elements (land cover) are created from aerial photos in the focal sites (Nagoya City, Kushida Watershed, and Shima City) on ArcGIS software. As in Nagoya City, fine land cover data arranged by municipality (*Green Status Survey* (Nagoya City 2010)) are utilized as this purpose. Figure 9.4 shows the classification of landscape elements, landscape units, and neighborhood landscape complex units in this study.

Figure 9.5 shows the result map of land use units arranged in 1 km meshes in Chubu metropolitan area which includes eight prefectures (Aichi, Shizuoka, Gifu, Mie, Nagano, Toyama, Ishikawa, and Fukui prefectures).

9.3.2 *Definition of Indicators for Labor Accounts*

According to the target or intention of the usage of landscape management labor accounts, several different indicators, which are defined in this section, must be developed. Firstly the following variables are defined.

A_j Area size ((a), one are = 100 m^2) of a targeted element (j)
B_{ik} Average working hours per one day of a person (i) of a labor content (k)
C_{ik} Annual working days of a person (i) of a labor content (k)
S_l Total area size (a) of a targeted landscape unit (l)
A_l Total landscape element size (a) in a targeted landscape unit (l)
S Total area size of a targeted landscape complex unit
A Total landscape element size of a targeted landscape complex unit
A_i Total landscape element size of a targeted person (i)
P Population of a targeted landscape complex unit

(a) Personal annual labor account (hour) of a person (i) of a targeted element (j) of a labor content (k) in a landscape unit: CLAijkl

The calculation of the element level is the base of all types of labor accounts. There are many labor contents for the annual management of the element. For example, the management of paddy fields requires plowing, mowing, fertilizing and so on, which are called labor contents. Firstly, the personal annual labor account of a

Table 9.1 Classification of landscape units

	Landscape units	Landscape units	
	Basic vegetation survey by ministry of the environment (scale: 1/25,000)	Basic survey of city planning (land use) (scale: 1/2500)	
Built-up areas Developed lands	Built-up areas Residential district of high coverage of green, parks, cemeteries, etc. Golf courses/lawns	Private building lots	Commercial lots
			Industrial lots
			Multiple dwelling houses
			Detached houses
			Parking lots and other paved surfaces
			Shrines and Temples
		Public sites	Public facilities
			Parks
			Roads
			Railroads
Farmlands	Paddy fields	Paddy fields	
	Other crop fields	Cultivated lands	
	Tea plantations	Tea plantations	
	Orchards	Orchards	
		Arboretums	
	Abandoned farmlands	Abandoned farmlands	
Secondary grasslands Vacant lands	Vacant lands	Vacant lands, abandoned lands and grasslands	
	Secondary grasslands (low grassland)		
	Secondary grasslands (tall grassland and Shrubs)		
Forests	Tree plantings	Tree plantings	
	Artificial forests	Artificial forests	
	Bamboo forests	Bamboo forests	
	Secondary forests (*Quercus variabilis* or *Quercus serrata*)	Secondary forests (*Quercus variabilis* or *Quercus serrata*)	
	Secondary forests	Secondary forests	
	Secondary forests (almost natural forest)	Secondary forests (almost natural forest)	
	Natural forests	Natural forests	
Water surfaces	Water Surfaces	Rivers	
		Irrigation channels	Paved irrigation channels
			Unpaved irrigation channels
		Pond	
Coasts	Tidal flats	Tidal flats	
	Seaweed forests	Seaweed forests	

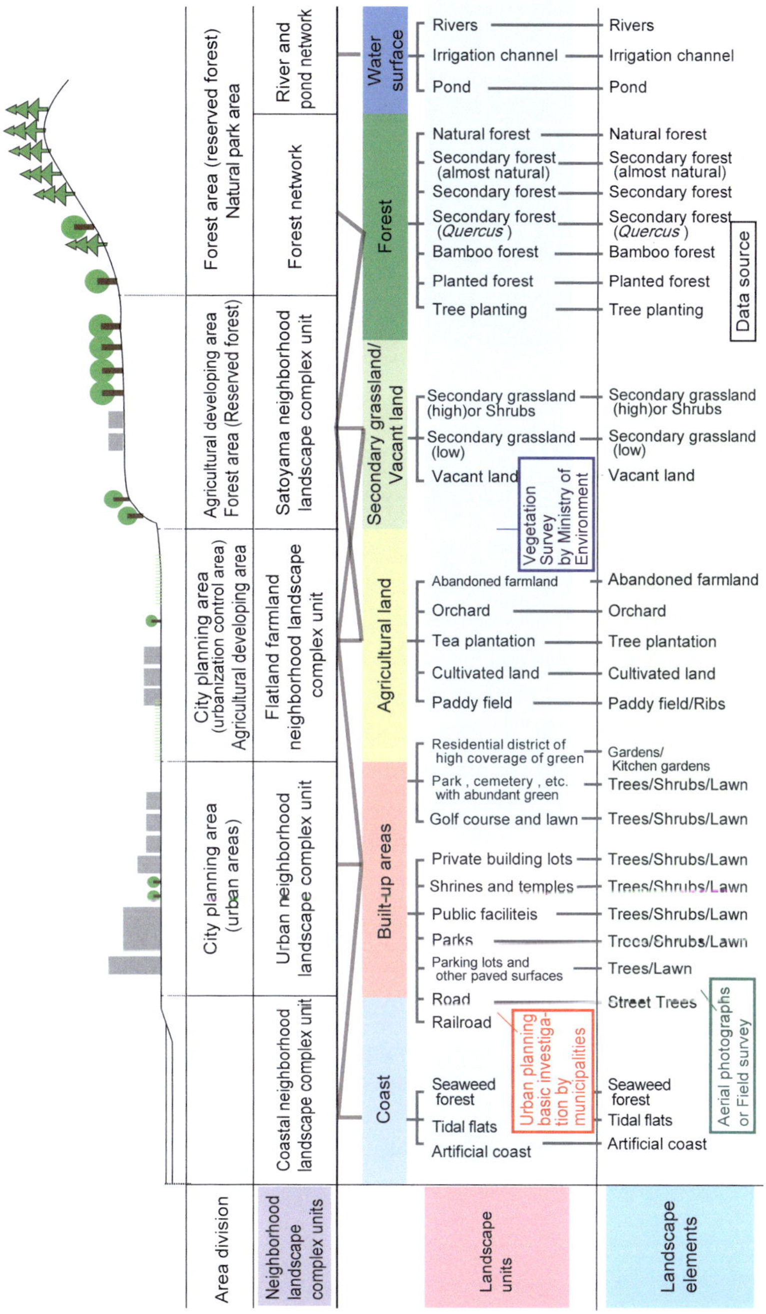

Fig. 9.4 Classification of landscape elements, landscape units and neighborhood landscape complex units in this study

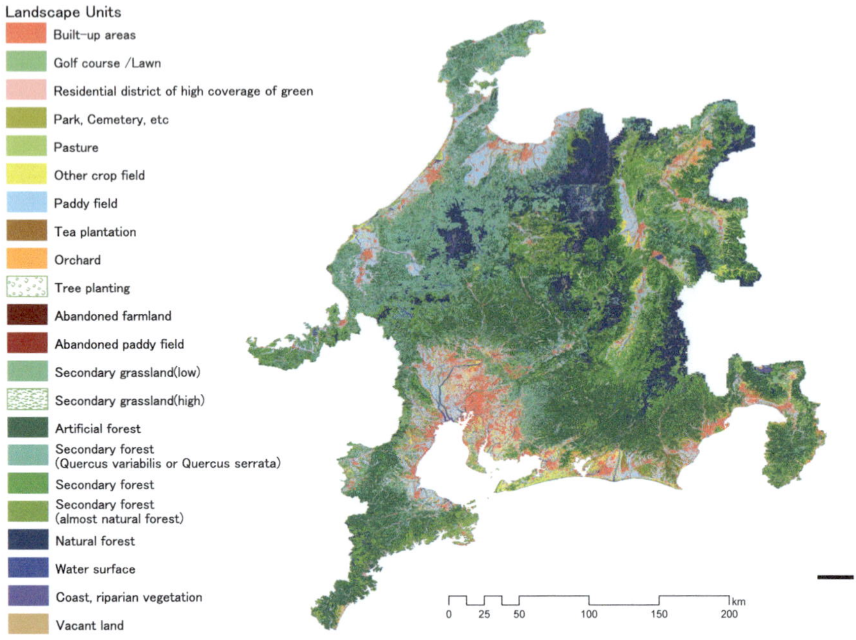

Fig. 9.5 Classification of landscape units in Chubu metropolitan area

person (i) of a targeted element (j) of a labor content (k) in a landscape unit (l) is calculated as follows:

$$CLA_{ijkl}(hr) = B_{ijkl} \times C_{ijkl}. \tag{9.1}$$

If the detailed work log is available, then this indicator can be calculated directly as the total value of the relevant working hour in the work log.

(b) Total annual labor account (hour) of a targeted element (j) of a labor content (k) in a landscape unit: $TCLA_{jkl}$

There are some cases in which only one person does all the labor of the targeted labor content of the targeted element. But in many cases, multiple workers take part in the labor of the targeted labor content of the targeted element. Thus the integration of multiple workers in one indicator is needed for the total annual labor account of a targeted element (j) of a labor content (k) in a landscape unit (l) as follows:

$$TCLA_{jkl}(hr) = \sum_{i=1}^{x} \left(B_{ijkl} \times C_{ijkl}\right) \tag{9.2}$$

(c) Total annual labor account (hour) of all labor contents in a targeted element (j) in a landscape unit (l): TLA_{jl}

Then the total annual labor account of all labor contents in a targeted element (j) in a landscape unit (l) is calculated as follows:

$$TLA_{jl}(hr) = \sum_{k=1}^{x} TCLA_{jkl}. \tag{9.3}$$

(d) Total annual labor density (hour per one are) of a targeted element (j) in a landscape unit (l): TLD_{jl}

The calculation of annual total amount of labor in an element level can show how much labor is needed in the management of the targeted element. But this indicator depends on the area size of the targeted element. As an independent factor, the definition of labor density is needed. Labor density can show the intensity of labor of a targeted element. Total annual labor density of a targeted element (j) in a landscape unit is calculated as follows:

$$TLD_{jl}(hr) = TLA_{jl}/A_l. \tag{9.4}$$

This indicator is very useful to recognize what kinds of labor content are most or least dominant by the management of a targeted landscape unit. For instance, in Japanese gardens, management of pine trees is the most elaborate and requires high labor density. Such a fact will be clarified by using this indicator.

(e) Total annual labor account (h) of a targeted unit (l): TLA_l

In cases which have many elements inside a unit, the total annual labor accounts of a targeted unit (l) must be calculated as follows:

$$TLA_l(hr) = \sum_{j=1}^{x} TLA_{jl}. \tag{9.5}$$

(f) Gross labor density (hour per one are) of a targeted unit (l): GLD_l

Also the labor density of a targeted unit (l) can be calculated. However, there are two types of labor density. One is gross labor density, which is defined as follows:

$$GLD_l(hr/a) = TLA_l/S_l. \tag{9.6}$$

This indicator can show how strong the labor density is for the management of the total area including occupied areas with no vegetation. For instance, this can be well applied to the calculation of the gross labor density of an urban park.

(g) Net labor density (hour per one are) of a targeted unit (l): NLD_l

In some cases, clarification of a net labor density of total element areas in a landscape unit (l) will be needed. In such a case, net labor density is defined as follows:

$$NLD_l(hr/a) = TLA_l/A_l. \tag{9.7}$$

(h) Total annual labor account (h) of a targeted complex unit: TLA

This is one of the important results of our research, which can show the base numerical value to consider how much labor is needed for the appropriate management of the local neighborhood landscape complex.

This value is calculated as follows:

$$TLA(hr) = \sum_{l=1}^{x} TLA_l. \tag{9.8}$$

(i) Gross labor density (hour per one are) of a targeted complex unit: GLD

This can be calculated as follows,

$$GLD(hr/a) = TLA/S. \tag{9.9}$$

The procedures mentioned above are used to integrate element landscape management labor to a complex unit. But the integration of a personal labor account, in other words how much labor can be offered by one person for the management of a landscape, is also important. The amount of annual labor performed by one person has a limit, and it depends on the working styles. For the examination of this factor the following two indicators are provided.

(j) Total personal annual labor account (h) of a targeted person (i): $TPLA_i$

This can be calculated as follows:

$$TPLA_i(hr) = \sum_{j=1}^{x}\sum_{k=1}^{y}\sum_{l=1}^{z} CLA_{ijkl}. \tag{9.10}$$

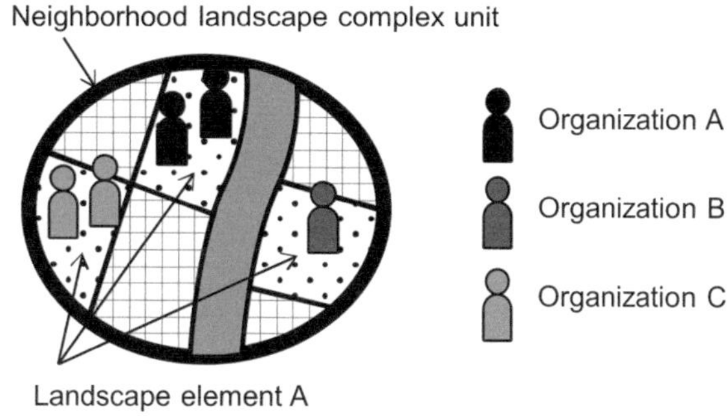

Fig. 9.6 Image of landscape management of landscape units by various stakeholders

(k) Personal labor density (hour per one are) of a targeted person (i): PLD_i

$$PLD_l(hr/a) = TLA_i/A_i. \tag{9.11}$$

(l) Social labor intensity: SLI

This can be calculated as follows:

$$SLI(hr) = TLA/P. \tag{9.12}$$

This is an indicator which can show the landscape labor intensity, or burden, which is borne by one population in the targeted landscape complex unit.

In this book the working style types are examined by using the above two indicators. Figure 9.6 shows the image of landscape management of landscape units by various stakeholders in a neighborhood landscape complex unit. And in Table 9.2, the definition of all indicators for labor accounts and labor density are shown.

9.3.3 Methodology of Case Studies on Neighborhood Landscape Complex Units

Some landscape units are managed by various management stakeholders or workers. For example, an agricultural landscape neighborhood complex unit is managed by professional farmers (individual farmers and corporate farmers), self-support farmers and collaborative works, "Deai", by all participants in a community. In this part of the book, the following steps will be conducted in each case study, considering such types of working complexity.

First, population, geographic and social characteristics, problems, and issues in each neighborhood landscape complex unit will be mentioned as an introduction with regard to landscape labor accounts and density. The types of landscape units

Table 9.2 Indicators for labor accounts and labor density

	Labor content level	Landscape element level	Landscape unit level	Neighborhood landscape complex unit level
Annual labor account (hour)	Total annual labor account (hour) of a targeted element (j) of a labor content (k) in a landscape unit (l): $TCLA_{jkl}$	Total annual labor account (hour) of all labor contents in a targeted element (j) in a landscape unit (l): TLA_{jl}	Total annual labor account (hour) of a targeted unit (l): TLA_l	Total annual labor account (hour) of a targeted complex unit: TLA
Annual labor density (h/a)	–	Total annual labor density of a targeted element (j) in a landscape unit (l): TLD_{jl}	Gross labor density of a targeted unit (l): GLD_l Net labor density of a targeted unit (l): NLD_l	Gross labor density of a targeted complex unit: GLD
Personal labor account (hour)	Personal annual labor account (hour) of a person (i) of a targeted element (j) of a labor content (k) in a landscape unit (l): CLA_{ijkl}	Personal annual labor account (hour) of a person (i) of a targeted element (j) in a landscape unit (l): CLA_{ijl}	Total personal annual labor account (hour) of a targeted person (i): $TPLA_i$	
Personal labor density (h/a)	–	Personal labor density of a targeted person (i) of a targeted element (j) in a landscape unit (l): PLD_{ijl}	Personal labor density of a targeted person (i): PLD_i	
Social labor intensity	–			Social labor intensity of a targeted complex unit: SLI
Basic variables	A_j Area size (a = 100 m^2) of a targeted element (j) B_{ik} Average working hours per one day of a worker (i)of a labor content (k) C_i Annual working days of a worker (i) of a labor content (k) S_l Total area size (a) of a targeted landscape unit (l) A_l Total landscape element size (a) in a targeted landscape unit (l) S Total area size of a targeted landscape complex unit A Total landscape element size of a targeted landscape complex unit A_i Total landscape element size of a targeted person Landscape element ratio of a targeted complex unit: LER LER = A/S (Nearly equal to the green coverage ratio) P Population of a targeted landscape complex unit			

and landscape management stakeholders in each neighborhood landscape complex unit will be also grasped.

Second, labor density and labor accounts in each landscape unit will be calculated. The matrixes of labor contents, annual labor schedules, and number of people who conduct each labor content, and management area and time of stakeholders in each landscape unit will be created by conducting field surveys. According to the datasets, annual labor accounts and annual labor densities in each landscape unit will be grasped. These calculations will be also made for each worker and the differences and similarities of labor density among the workers and the reasons of differences will be clarified. The level of landscape management in each landscape unit will be also discussed and compared to the average data of labor density from the national census, if it is available. These arranged datasets will be useful for the evaluation of the current situation and the discussion to build future perspectives, what kind of landscape management in a targeted neighborhood landscape complex unit will be appropriate to sustain its environment, and how much and what kinds of labor forces will be needed for them.

Third, working style types in each neighborhood landscape complex unit will be examined. The labor time and labor density of each working style type, i.e. total personal annual labor account of a targeted person and personal labor density of a targeted person, such as a professional individual farmer or a self-support farmer, in each landscape unit will be calculated and the characteristics will be graphed. It could be assumed that some people work by combining working style types. For example, some professional farmers also do self-support management of kitchen gardens and do collaborative work to clean the temples and shrines in the local community in the village. And some urban residents manage the gardens in each residential place and also manage the forest in the park as volunteers. These detailed lifestyle datasets will be obtained by hearing from the viewpoints of management stakeholders. These datasets will be useful to consider what kind of and how many people in each working style type will contribute to manage the future sustainable environment in each neighborhood landscape complex unit.

Finally, if possible, future landscape management perspectives in each neighborhood landscape complex unit will be discussed by considering labor density and labor accounts, working style types and estimation of future population.

9.3.4 Methodology of Integration of Landscape Labor Accounts in Larger Scales

There are three stages to integrate landscape management labor accounts in larger scales, i.e. from a neighborhood scale to a regional scale.

First the neighborhood scale will be focused on. The summary of landscape management labor accounts of the case studies will be examined from two viewpoints. One is the gross labor density (GDL) of a targeted neighborhood landscape complex unit. The amount and types of labor forces to maintain each neighborhood landscape complex unit will be clarified and compared to each other in a bird's-eye point of view. The second is the mixture of working styles. The typical working style types are classified by the relation between the total personal annual labor account of a targeted person ($TPLA_i$), which means the annual amount of personal labor, and personal labor density of a targeted person (PLD_i), which means the efficiency of labor. According to these indexes, appropriate combination of working style types, which will suit the different social and geographical conditions of landscape complex units, will be discussed.

Second is the integration into the city scale. Three steps will be conducted. In the first step, landscape units and elements will be classified over all the city areas. The target of the case study is Nagoya city. The land use data of the *Basic Survey of City Planning* (Nagoya City 2012) and the green coverage data of the *Green Status Survey* (Nagoya City 2010) are used to create a GIS dataset on landscape elements and units. Both types of data are in the scale of 1/2500. The former is used to create landscape units, i.e. parks, residential sites and roads, and the latter is used to create landscape elements, i.e. green zones and spots as trees and shrubs. By using these two types of data, landscape elements and units will be classified into a city scale. In the second step, total annual labor accounts of all labor contents in a targeted element in a landscape unit (TLA_{jl}) will be collected to sum up the whole city area. The value of total annual labor density of a targeted element (TLD_{jl}) of each classified landscape unit will be determined first. As the reference value of TLD_{jl}, the average values of TLD_{jl} calculated in the case studies in neighborhood scales are used. In the next step, by multiplying the area of the targeted element, the TLD_{jl} and TLA_{jl} of each type of unit over the whole city area are calculated. The obtained values of TLA_{jl} are used to determine the city scale landscape labor accounts, which can also compare the difference between public and private landscape labor accounts. In the third step, the summed up values of labor account (TLA_{jl}) divided by S (the total area of city area), which we call the Gross Labor Density of a targeted city area (GLD), will be obtained. The value of GLD will be used to evaluate the typical labor density of built-up areas in the regional scale.

Finally, the total labor accounts in the regional scale will be calculated. The focal site is Chubu metropolitan area, which includes eight prefectures, Shizuoka, Aichi, Gifu, Mie, Nagano, Toyama, Ishikawa, and Fukui. Two steps will be conducted. First, the total area of each landscape unit will be calculated in 1 km mesh by using the *Vegetation Survey* (Ministry of the Environment of Japan 2005) on the ArcGIS software (Figs. 9.4 and 9.5). After that, the total annual labor account of a targeted unit (TLA_l) will be calculated by multiplying the labor density of each landscape unit to the area of each landscape unit in each 1 km mesh. Secondly, the current population or working population of each 1 km mesh will be compared to each TLA_l and the amount of management density per person invested in each mesh will be discussed. This will enable us to discuss the differences of labor forces among coastal, urban, rural, agricultural, and mountainous areas. Finally, the future population in 2050 in each mesh will be estimated due to the data set of the *National Land Numerical Information, Population Projection mesh Data* (National land

Information Division, National Spatial Planning and Regional Policy Bureau, Ministry of Land, Infrastructure, Transportation and Tourism (MLIT) of Japan 2014) and compared to the current labor accounts of each mesh.

References

Ministry of the Environment of Japan (2005) Vegetation survey, natural environmental information GIS, 1:25,000 and 1:50,000. http://www.biodic.go.jp/trialSystem/EN/shpddl.html. Accessed 1 June 2014

Nagoya City (2010) Green status survey. Licensed by Nagoya city

Nagoya City (2012) Basic survey of city planning. Licensed by Nagoya city

National Land Information Division, National Spatial Planning and Regional Policy Bureau, Ministry of Land, Infrastructure, Transportation and Tourism (MLIT) of Japan (2014) National land numerical information, population projection mesh data. http://nlftp.mlit.go.jp/ksj/old/meta/mesh.html (Japanese). Accessed 1 June 2015

Chapter 10
Case Studies in a Variety of Urban Greenspaces: Nagoya City

Nobuko Kawaguchi

Abstract This Chapter illustrates the management of a variety of the urban greenspaces in Nagoya City, Aichi Prefecture. In recent years, maintenance costs for public greenspaces have been reduced due to it being done by public hands. In contrast, urban greenspace maintenance costs are growing, which results in the need for more landscape management labor. This chapter considers the landscape management labor accounts of administrative officer and volunteer groups who participate in the management of urban public greenspaces. In the case of private hands, in the urban center and its periphery residential areas, the amount of vacant spaces is increasing. Aging land owners are finding it hard to maintain their rich greenspaces due to the burden of care. The sustainable management of a variety of private greenspace sites is also considered in this chapter. The concept of landscape management labor accounts across different greenspace types is expected to contribute to creating a basis for the evaluation and combination of different greenspaces from the same point of view.

Keywords Urban green · Urban land use · Aging society · Environmental preservation activity · Japanese garden

10.1 Introduction

In this chapter, the landscape labor accounts of urban green spaces are considered. Urban green spaces are divided into two categories, i.e. green spaces managed mainly by public entities and private green spaces. In European countries, urban green spaces consist mostly of public green spaces like urban parks and gardens. The concepts of public parks were introduced in Japan in the latter half of the 19th century, and many public parks were developed. However, the number of public

N. Kawaguchi (✉)
Graduate School of Environmental Studies, Nagoya University,
Furo-cho, Chikusa-ku, Nagoya, Aichi 464-8601, Japan
e-mail: nobqou@gmail.com

H. Shimizu et al. (eds.), *Labor Forces and Landscape Management*,
DOI 10.1007/978-981-10-2278-4_10

parks in cities is relatively fewer than that of European countries. For instance, Nagoya city has public parks with a total area of 1,450 ha, which is 4.4 % of the total city area, and it corresponds to 6.7 m^2 per person. Even though only a small amount of the area of Nagoya city is green, management of public parks is in a crisis because of the matured economy and the aging society. Much of the public budget must be allocated to the health care of older people. Furthermore, the budget for green space maintenance has decreased year by year, even though the meaning of urban green spaces is increasing more and more from the viewpoint of mitigating urban heat islands and maintaining urban ecosystem services.

In these situations, conservation of urban green spaces depends much on private green spaces. These private green areas, however, have continuously decreased over the decades. Figure 10.1 shows the change of the unsealed area ratio and green coverage ratio of Nagoya city between 1990 and 2010 (Kawaguchi et al. 2014).

In hilly areas of the east part of Nagoya city, big public parks and private secondary forests are spreading widely. In the southwest part agricultural fields are distributed. These core green spaces create the green belt of Nagoya city, which is not developed but is regarded as very important, with the combination of river networks surrounding the city. The amount of the biomass of public and private secondary forests is growing, but this growth is due mainly to the result of the

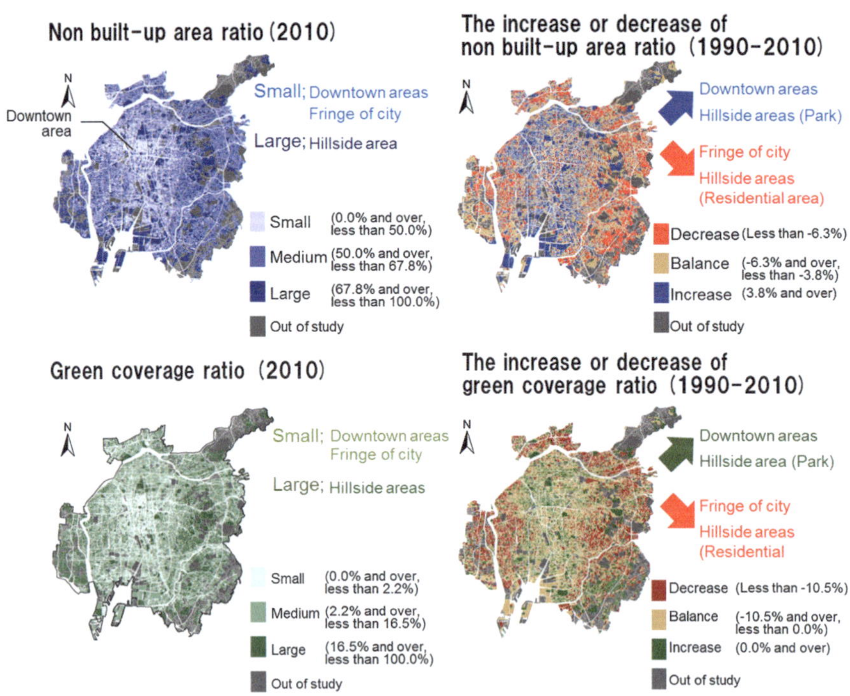

Fig. 10.1 Change of unsealed area ratio and green covering ratio of Nagoya between 1990 and 2010 (Kawaguchi et al. 2014)

management shortage, and has resulted in an increase in the number and amount of fallen trees and leaves.

On the contrary, in the urban center and peripheral residential areas, the amount of vacant spaces is increasing because of the hollowing out of urban spaces. The increase in vacant spaces, however, does not result in an increase in the green coverage ratio. Most vacant spaces, which are privately owned, are used as sealed spaces such as parking. Private owners have no incentives to keep their vacant places green, because greening creates no economic benefits and increases the burden of care. In other words, economic principle and aging tend to push the society to minimize the cost of green management and the labor forces needed to maintain green spaces.

Nagoya city has developed a new green master plan in which the green coverage ratio shall increase to 30 % of the total area (Nagoya City 2011). However, there is no clear, concrete road map to reach the goal. A new public policy will be needed to keep and create green spaces for the future that the new green master plan calls for.

The new public policy needs, however, an understanding of society. To incubate such an understanding, landscape management labor accounts, which can make visible to stakeholders all the burdens involved in the management of green spaces, will play a very important roll.

Figure 10.2 shows the amount of green areas categorized according to the land use types of Nagoya in 2005. It is obvious that not only public parks but also green spaces in private land uses must be considered, because they account for most of the urban green areas. For this reason, labor accounts will be considered in the two aspects, i.e. public green spaces, and private green spaces.

Figure 10.3 shows the types and issues of landscape elements and units in urban areas. Firstly, the green spaces are divided into three categories, i.e. large-scale green spaces, small-scale green spaces and urban farmlands. Large-scale green

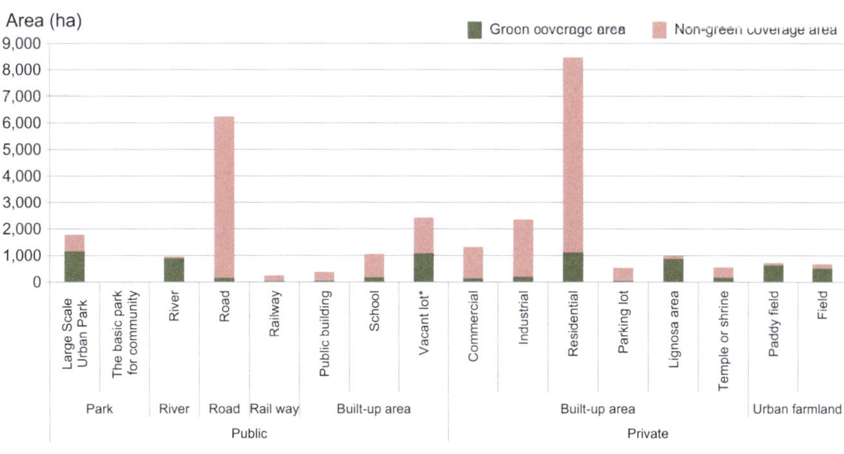

Fig. 10.2 Amount of green areas according to land use types of Nagoya in 2005

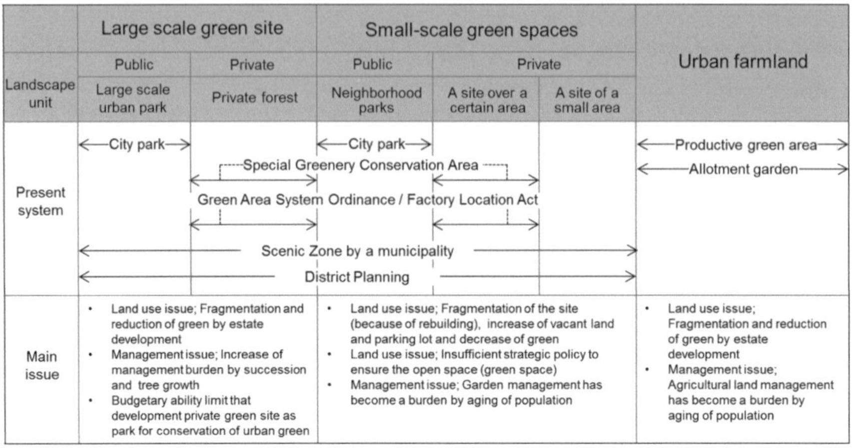

Fig. 10.3 Types and issues of landscape elements and units in urban areas

spaces are divided into two categories, i.e. public parks and private forests. Small-scale green spaces are divided into three categories, i.e. basic parks for the community, small private green sites and private green sites beyond certain area sizes, which are designated in various kinds of acts and ordinances related to green conservation. Public parks including basic parks for the community are managed publically as urban parks. Large-scale private forests are to be managed by the Green Space Conservation Area, but this system does not work well because of the protection of individual rights. Large-scale private forests and private green sites beyond certain area sizes are regulated by the systems of the Special Greenery Conservation Area, Green Area System Ordinance or Factory Location Act. The large-scale green spaces and private green spaces can be designated as Scenic Zones by a corresponding municipality. Urban farmlands are categorized into two types, i.e. Productive Green Spaces designated by the Productive Green Space Act and the other. Both in the productive green spaces and the other, citizen's allotment gardens are increasing recently. As shown above, there are many kinds of regulations and designations for the maintenance of green spaces in urban areas, but the practical efficiency has been limited, and there are not enough controls to stop the reduction of urban green spaces. One reason is that the connectivity of legal systems and measures between these different kinds of green spaces are not well considered.

The concept of landscape management labor accounts across different green space types is expected to contribute to make a basis for the evaluation and combination of different green spaces in the same point of view.

10.1.1 Hierarchy of Scales to Consider Urban Land Management

Figure 10.4 shows the hierarchy of scales to consider urban land management. As shown in Chap. 9, landscape management labor accounts in urban areas are considered in three levels. A site unit, which is owned by one or more related owners, corresponds to the "landscape unit" level. The second level is a "landscape management complex unit" level, which corresponds to one or several urban blocks managed by a neighborhood association, or an elementary school district or a residential council district, which is managed by an upper complex of neighborhood associations. The third is the municipality level. Public facilities such as parks, roads, rivers, public building spaces and vacant lands are managed by municipalities. In some cases, a wide scale voluntary association takes part in the management of public parks beyond district management complex unit levels.

In the following case studies, as targeted areas for the consideration of landscape labor accounts, several districts in Nagoya city are selected. These case studies are described based on the paper by Kawaguchi et al. (2016).

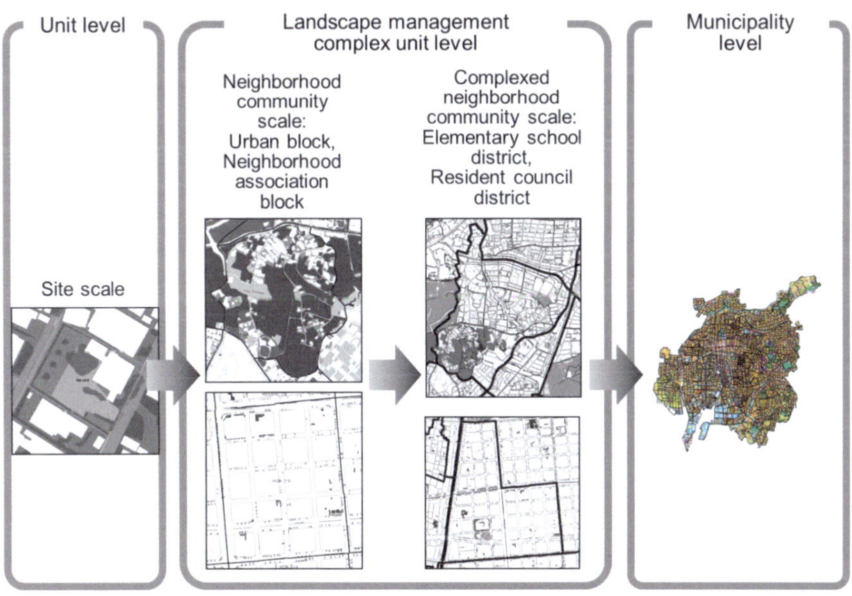

Fig. 10.4 Hierarchy of scales to consider urban land management

10.2 Public Parks, Gardens and Roadside Trees

10.2.1 Parks of Big Green Sites—Higashiyama Park

10.2.1.1 Overview

Figure 10.5 shows the spatial distribution of each type of park in Nagoya city.

Urban parks can be classified into two categories according to the characteristics of the purpose of use. The first category is called a large scale urban park (Toshi Kikan Koen in Japanese) because it has the role of being a large scale urban park in city planning policy that forms the overall perspective of an urban area. The second category is called a neighborhood park (Juku Kikan Koen in Japanese) because it has the criterion of being a comparatively small park determined by the range of life actions of the citizens.

In the east part of Nagoya city, hilly districts, where big green sites such as public parks, private forests and guardian forests of temples and shrines are agglomerated, are spread widely. These forests consist mainly of secondary forests with a highly progressed transition. In this section, *Kurashi No-Mori*, which is a zone within Higashiyama Park, and is a favorite park of the citizens and consists of secondary forests in a large scale, will be targeted as a case study for a public park.

Higashiyama Park is a public park which is publically owned and managed. The amount of GLD_l (Gross labor density of a targeted unit (l)) of the *Kurashi no mori*

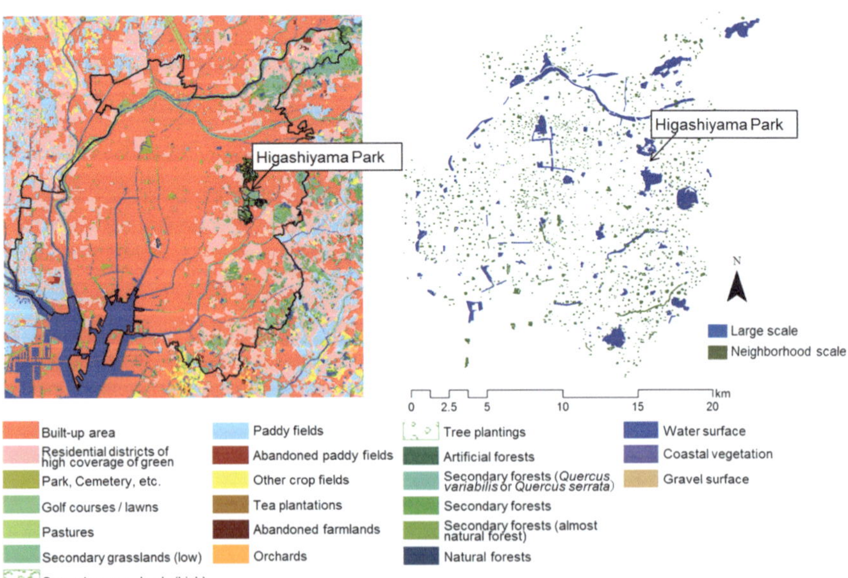

Fig. 10.5 Spatial distribution of each type of park in Nagoya city

zone (a part of Higashiyama park: the targeted area) is very low (0.97 h/a). Most labor is involved in the mowing of grasslands, logging of trees and removal of fallen trees along the roadsides. Most parts of the secondary forest remain unmanaged. Looking at this situation, some citizens have begun to manage the forests on their own. Now wide areas of the forests, bamboo shrubs and grasslands are managed by the voluntary group named "*Nagoya Higashiyama Moridukuri no Kai*", the *Nagoya Higashiyama Forest Management Association*, and the municipality of Nagoya has agreed to collaboratively manage the public park. Between the municipality of Nagoya and voluntary groups who engage in the preservation and regeneration of natural environments, a collaborative network named the "*Nagoya Moridukuri Partnership Network*", or the *Association of Nagoya Forest Management Partnership Network*, has been established for the management not only of Higashiyama Park, but also other public parks. The city of Nagoya gives subsidies to the certificated organizations under the Green Town Planning Ordinance. 29 groups were organized as of April, 2013. *Nagoya Higashiyama Forest Management Association* belongs to this partnership. Members, divided into eleven subgroups, are living in and around Nagoya city and manage forests, bamboo shrubs, farmlands, and conduct research on the natural environment.

In this section, the area where this group conducts most of its activities, *Kurashi No-Mori*, has been chosen as the location for investigation of its landscape management labor accounts.

10.2.1.2 Landscape Units and Organization

Table 10.1 shows the area sizes of landscape elements. Table 10.2 shows their management bodies. Figure 10.6 shows the spatial distribution of the elements, and Fig. 10.7 shows their managed areas. The total area size of a targeted landscape unit is about 5.4 ha and total landscape element size in a targeted landscape unit is about 5.0 ha. The volunteer group manages mainly grasslands, fields, orchards and the forests on their fringes. The municipality manages mainly the logging of trees and removal of fallen trees along roadsides. In other words, the management level is high on the fringes of forests and low inside the forests. The amount of the ratio of the managed area occupied by the total landscape elements area is very low. In this study, because it does not contain all of the labor contents, it does not mean that people do not enter in the unmanaged area or patrol inside the park or that tree maintenance around the park is done by full-time staff, or the cultivation of farmland is done by voluntary groups.

10.2.1.3 Landscape Labor Accounts and Densities

In Table 10.3, landscape labor accounts and labor densities of all areas of *Kurashi no Mori* in 2014 are shown. In Fig. 10.8, landscape labor accounts (TLA_l) and labor densities (TLD_{jl}) of the volunteer group in 2014 are shown. Labor accounts (TLA_l)

Table 10.1 Area sizes of landscape elements of Kurashi no Mori

Landscape element		A_l (a)	S_l (a)	
Secondary forest	Tree	4,088	4,088	4,173
	Shrubs	85	85	
	Bamboo forest	87	87	87
Vacant land	Shrubs	0	0	682
	Grassland	493	493	
	Tree	98	98	
	Bare field	0	91	
Farmland (Orchard)	Orchard	13	13	26
	Shrubs	8	8	
	Grassland	5	5	
Farmland (for zoological garden)	Eucalyptus	67	67	70
	Bare field	0	3	
Farmland (Other crop field)	Other crop field	36	36	42
	Grassland	3	3	
	Bare field	0	3	
Farmland (Paddy field)	Paddy field	4	4	5
	Bare field	0	2	
Water Surface (pond)	Tree	6	6	28
	Water Surface	0	23	
Stream	Grassland	10	10	10
Sidewalk (Unpaved)	Unpaved surface	0	154	154
Road (Paved)	Paved surface	0	92	93
	Road tree	1	1	
Parking lot	Tree	1	1	21
	Shrubs	1	1	
	Paved surface	0	19	
Total area (A, S)		5,004	5,390	

are high in bamboo forests and other secondary forests. In comparison to the low TLD_{jl} of other secondary forests which is 0.17 h/a, the labor density of bamboo forests is high at 9.45 h/a. The reasons are believed to be as follows;

(1) The working area is limited in the other secondary forests, but the management working in the bamboo forests is done in the entire area, even though the area size of bamboo forests is very small compared to the area size of the other secondary forests.
(2) Bamboo forests need intensive care for thinning inside and prevention from erosion to the surrounding forests, because of the rapid growth and expanding of the growing area.

The interest of the volunteer group in the management of bamboo forests was found to be high through interviews. It was also clarified that the labor accounts

Table 10.2 Management organization of Kurashi no Mori

Organization	Labor contents	Managed area size (a) (2014)
Public agency Department of Higashiyama park management (2014)	Mowing or lawn mowing along the sidewalk	314
Voluntary group Nagoya Higashiyama Moridukuri no Kai (Mowing and tree pruning of undergrowth and grassland along the main sidewalk) (2014)	Secondary forest management (Mowing of undergrowth, tree logging of secondary forest and mowing of orchard crop field)	159
	Bamboo forest management	
	Grassland management (Mowing of grassland and undergrowth along the sidewalk and pond)	
Unmanaged (Except for management contents of tree logging, cleaning and removal (weakening/withering) tree by Department of Higashiyama park management.) (2014)	–	4,918
Total area		5,390

given to the secondary forest were high in totality but were difficult to manage equally because of the shortage of feasible labor forces.

Figure 10.9 shows the average of the annual number of weakened trees, withered trees and removed trees throughout Higashiyama Park from 2007 to 2014. However, we were unable to obtain the data on the number of weakened trees, withered trees and removed trees limited to the area of the *Kurashi No-mori* zone. Therefore, the results of such data are not included in Table 10.2. The major causes of weakening and withering of trees are pests, typhoons or aging. The objectives of the maintenance work are accident prevention and health management of trees (i.e. for people to pass safely and to prevent the impact of pests). As the bar graph indicates, there are many removal works for weakening or withering trees in urban secondary forests. Moreover, the amount of standard deviations is also large. Therefore, the administrator cannot predict the quantity of removal works.

10.2.1.4 Personal Labor Accounts and Densities

In Fig. 10.10, the personal labor accounts and personal labor densities are shown.

This chart represents three people who belong to a public agency who mow grasslands, log trees and clean sidewalks in *Kurashi No-mori*. The amount of their TPLAi and PLDi are high. The working style is professional. The voluntary group works three hours/person each time. This activity is carried out about three times a month.

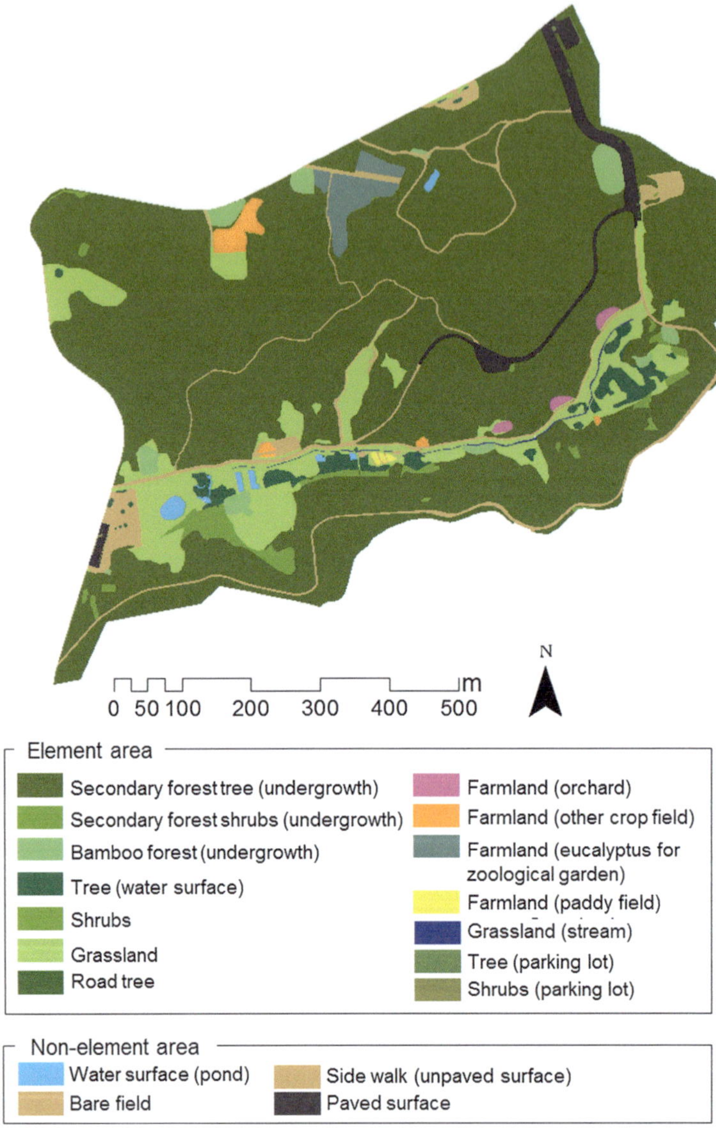

Fig. 10.6 Spatial distribution of landscape elements (Higashiyama park, *Kurashi No-Mori*)

The activities of the volunteer group are regular meetings held twice a month, subgroup meetings once a month and the activities of each subgroup on its local spot.

Anyone can participate in the management of the landscape elements, and the number of participant changes in each event, but is normally around twenty people. This volunteer group also has corporate members, and a company can participate as

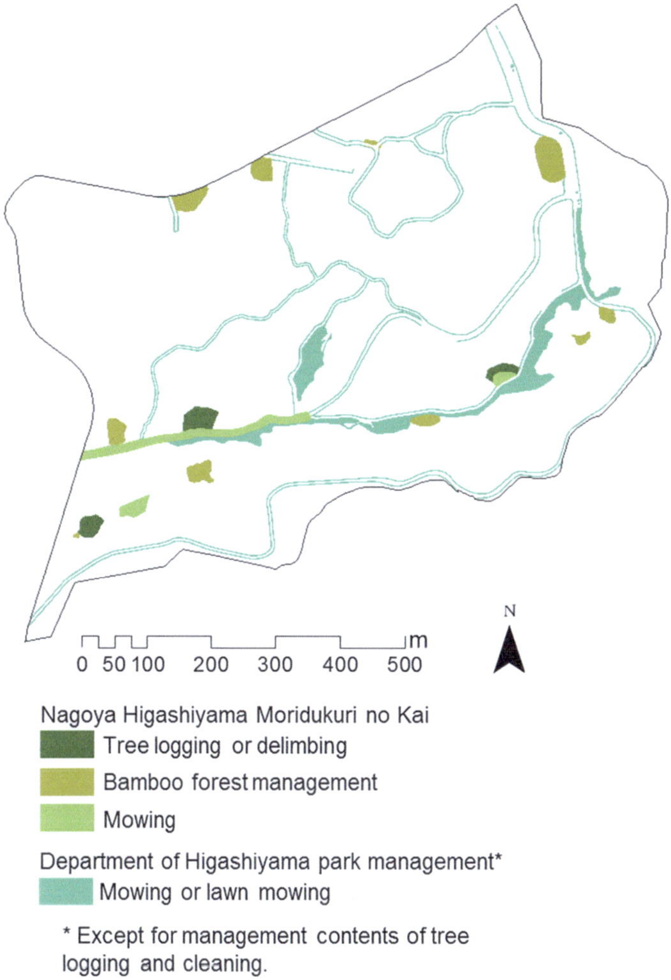

Fig. 10.7 Spatial distribution of managed area (public entities/voluntary group, 2014)

an activity of its corporate social responsibility. In such a case in the past, the participants reached 89 persons. This kind of system will act to get more attention of citizens and to collect more participants among the people who do not have a strong will to conserve nature.

10.2.2 Parks for Neighborhood Scale

In this subsection, the landscape labor accounts of parks for neighborhood scale are calculated. The utilized GIS data of landscape unit and element are provided by

Table 10.3 Landscape labor accounts and labor densities of all areas of Kurashi no Mori in 2014

Organization	Landscape element	Labor content	A_l (a)	S_l (a)		Annual work number of times	Number of people per time		Total number of people[a]	TLA_l (h)	TLD_{jl} (h/a)	NLD_l (h/a)	GLD_l (h/a)	Management area ratio (%)
							Mean	Stdev						
Public agency	Undergrowth and grassland along the sidewalk	Undergrowth and grassland along the sidewalk	314	314	314	53	7	–	3	2,541	8.09	1.04	0.97	9
	Trees and shrubs	Trees and shrubs	–	–	–	20	3	–		420				
	Sidewalk	Sidewalk	–	–	–	13	3	–		273				
Voluntary gro	Grassland of vacant land	Grassland of vacant land	37	37	168	4	26.5	9.6	106	444	11.86			
	Bamboo fores	Bamboo forest management	78	88		20	15.5	5.8	310	833	10.64			
	Undergrowth and grassland of secondary forest and orchard field	Secondary forest management (Mowing of undergrowth, tree logging of secondary forest and mowing of orchard, other crop field)	43	43		10	27.5	23.4	275	699	16.28			
Unmanaged	–	–	4,532	4,908	4,908	–	–	–	–	0	0			
Total			5,004		5,390	–	–	–	–	5,209	–	1.04	0.97	9.0

[a]Total number of people

Department of Higashiyama park management: The number of full-time management staff

Nagoya Higashiyama Moridukuri no Kai: The summary of participant numbers (Not considering the case of overlapping participation)

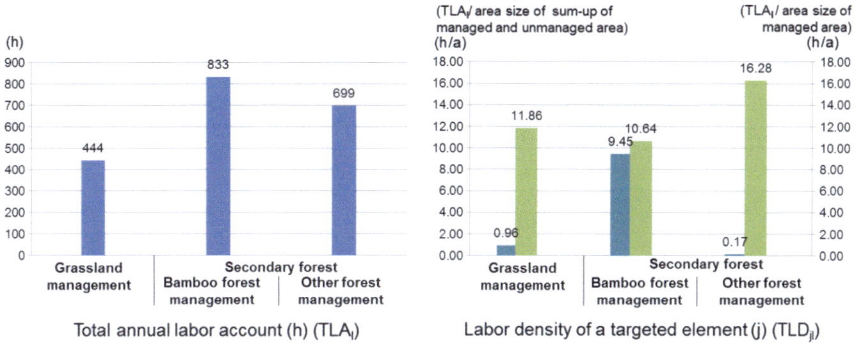

Fig. 10.8 Landscape labor accounts (TLA_l) and labor densities (TLD_{jl}) of the volunteer group in 2014

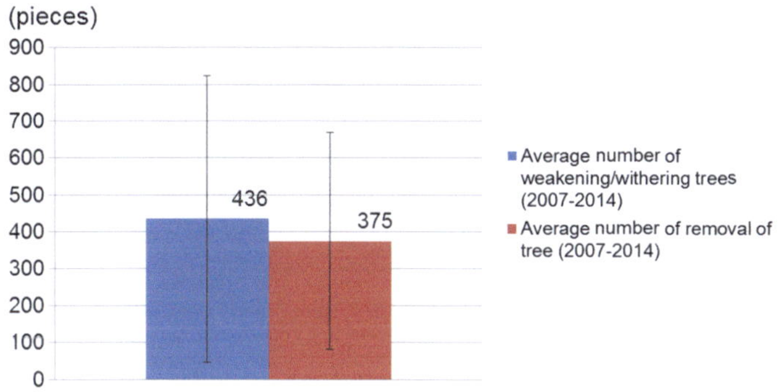

Fig. 10.9 Average annual number of weakened trees, withered trees and removed trees in 2007–2014

Nagoya city. The landscape labor accounts are calculated based on the record book of maintenance of the neighborhood park that was provided by Nagoya city in 2014. This record book described management (i.e. contents and amounts of management labor) of the parks that are located in three wards in Nagoya city (Naka ward, Nakagawa ward and Tenpaku ward). The landscape labor accounts are calculated based on the data and the *guidelines for cost estimating for public works* (Nagoya City 2015).

Table 10.4 shows the total area size of landscape unit and landscape element of neighborhood parks of each ward in the entire Nagoya city. Figure 10.11 shows the calculated results of the landscape labor accounts (TLA_l) for each item of landscape elements of targeted neighborhood parks. As the results indicate, the highest value of landscape labor accounts of each landscape element is that for the mowing of grasslands. Most of the grasslands are mowed by machine (Fig. 10.12).

Figure 10.13 shows the calculated results of the landscape labor accounts of trees. The contents of maintenance of trees can be classified in several categories

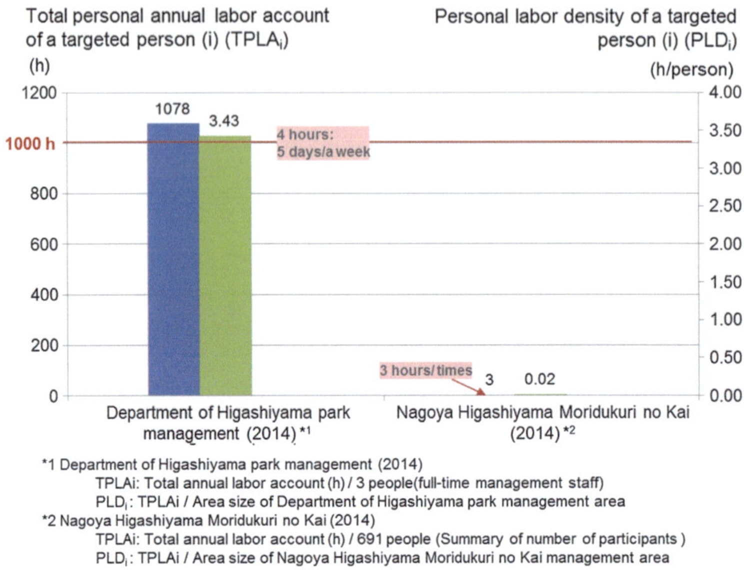

Fig. 10.10 Personal labor accounts and personal labor densities in 2014

like Crown Thinning, Crown Reduction, Crown Raising, Removing of weakened trees and "Raking up fallen leaves" according to the characteristics of its purposes. The usual purposes of maintenance of trees are aesthetics, safety and health. For example, the purpose of "Crown Thinning" is the pruning work to keep the trees to a certain size, and the purpose of "Removing weakened trees" is the removing work to cope with a serious safety hazard for weakened trees falling. As the results indicate, the amount of landscape labor accounts of "Crown Thinning" and "Crown Reduction" occupies most of the total amounts. Especially, tall trees need the use of a tall mobile vehicle, which requires additional workers to operate the machine.

Table 10.5 shows the calculated result of total landscape labor accounts and densities in Nagoya city. The total landscape labor account is 410,696 h, and Net labor density (NLD) is 16.43 h/a and Gross labor density (GLD) is 7.41 h/a.

10.2.3 *Roadside Trees*

In this subsection, the landscape labor accounts of roadside trees are calculated. The utilized GIS data of landscape unit and element are provided by Nagoya city. The landscape labor accounts are calculated based on the record book of management of the roadside trees provided by Nagoya city in 2014. This record book has described maintenance of the road, i.e. contents and amount of management labor which are located in eight wards within Nagoya city (Naka ward, Nakagawa ward, Tenpaku ward, Chikusa ward, Nishi ward, Nakamura ward, Higashi ward and Kita ward).

Table 10.4 Total area size of landscape unit and landscape element of neighborhood parks of each ward in the entire Nagoya city

Ward	S (a)	S (study area) (a)	A (a)	LER (%)	A_l (Each type of elements) (a)			A (study area) (a)	LER (study area)
					Trees, shrubs, bamboo forests	Grasslands and lawns	Paddy fields, other crop fields, orchards		
Naka ward[a]	1,389	1,139	824	59	781	43	0	717	63 %
Nakagawa ward[a]	5,696	5,088	2,555	45	1,999	554	1	2,338	46 %
Tenpaku ward[a]	4,507	3,735	1,710	38	1,353	356	1	1,475	39 %
Minato ward	7,098	–	3,178	45	2,122	1,055	1	–	–
Moriyama ward	4,575	–	1,864	41	1,138	709	17	–	–
Showa ward	1,438	–	994	69	733	261	0	–	–
Mizuho ward	1,035	–	483	47	415	68	0	–	–
Nishi ward	3,623	–	1,592	44	1,285	304	2	–	–
Chikusa ward	1,638	–	795	49	665	129	0	–	–
Nakamura ward	1,757	–	834	47	703	131	0	–	–
Higashi ward	1,120	–	576	51	537	39	0	–	–
Minami ward	3,593	–	1,350	38	1,084	266	0	–	–
Atsuta ward	2,373	–	1,407	59	1,156	251	0	–	–
Kita ward	1,992	–	893	45	697	191	5	–	–
Meito ward	4,444	–	2,066	46	1,765	300	2	–	–
Midori ward	9,163	–	3,869	42	2,205	1,642	22	–	–
Total	55,441	9,962	24,989	45	18,639	6,299	51	4,530	45 %

[a]Study area

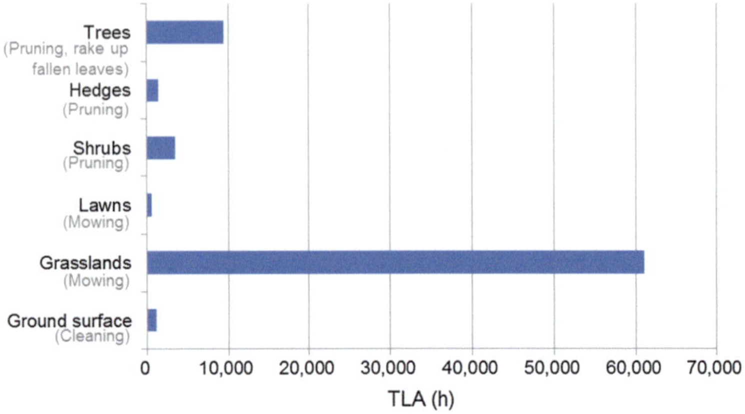

Fig. 10.11 Landscape labor accounts of each item of landscape elements in parks

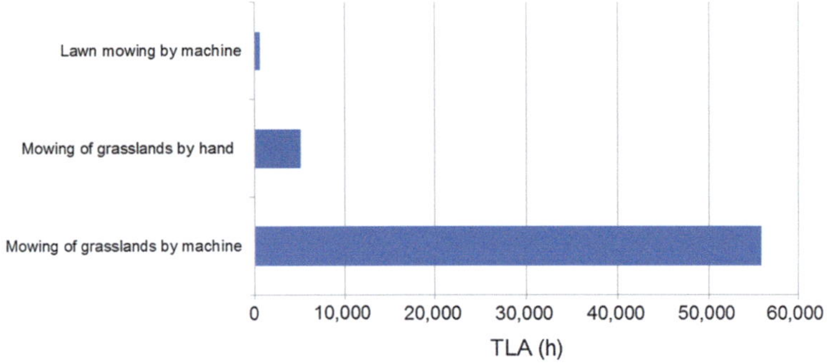

Fig. 10.12 Landscape labor accounts of grasslands and lawns

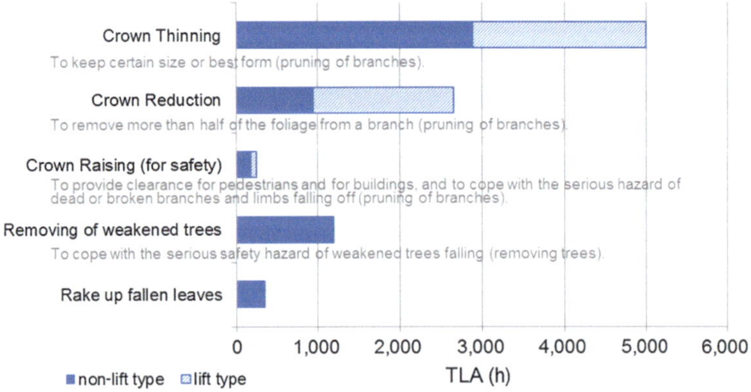

Fig. 10.13 Landscape labor accounts of labor items of trees

Table 10.5 Calculation of the total landscape labor accounts and densities in Nagoya city (parks for neighborhood scale)

	S_l (a)	A_l (a)	LER (%)	A_l (Each type of elements)			TLA_l (h)	NLD_l (h/a)	GLD_l (h/a)	SLI (population: 2,274,511)
				Trees, shrubs, bamboo forests	Grasslands and lawns	Paddy fields, other crop fields, orchards				
Study area	9,962	4,530	45				74,448			
Total area (city scale)	55,441	24,989	45	18,639	6,299	51	410,696	16.43	7.41	0.2

The landscape labor accounts are calculated based on the data and the guidelines for cost estimating for public works (Nagoya City 2015).

Figure 10.14 shows the spatial distribution of roadside trees, shrubs, grasslands and lawns in Nagoya city. Urban streets have many roadside trees in comparison with rural roads. The roadside tree has a heat island mitigation effect by providing a

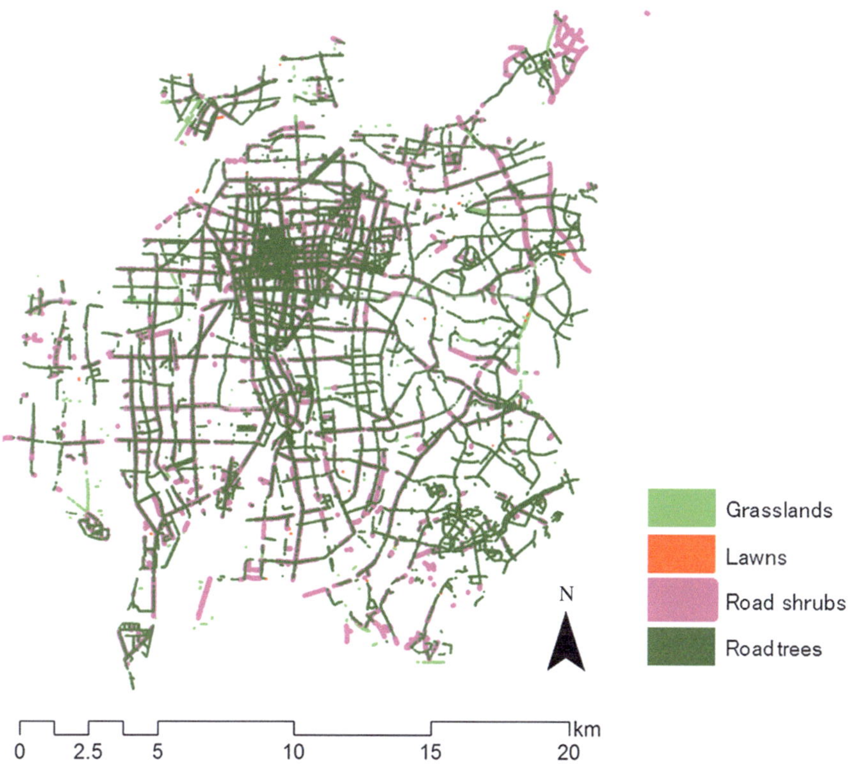

Fig. 10.14 Spatial distribution of roadside trees, shrubs, grasslands, lawns

shady nook to pedestrians. Further, in built-up areas with few green areas, the roadside tree has the important role of complementing the amount of green. Table 10.6 shows the total area size of landscape elements of roads, and the total area size of the landscape unit of roads in each ward. The amount of the green coverage ratio of the city is 3 %, and the highest green coverage ratio is in Naka ward (8 %) which is located in the downtown area.

Figure 10.15 shows the amount of the landscape labor accounts of roadsides of each landscape element, and Fig. 10.16 shows the amount of the landscape labor accounts only for trees. According to the record book, usually pruning of roadside trees is done once a year, although the tree species with the fastest growth are pruned twice a year. Similar to the case of the park, pruning of the tall trees needs a tall mobile vehicle, which requires additional workers to operate the machines. In addition, in the busy streets, traffic guides are necessary.

Table 10.7 shows the results of the amount of the landscape labor accounts and densities for roadside trees. The amount of TLA of the city scale is 169,767 h, NLD

Table 10.6 Total area size of units and elements of roadside trees in Nagoya city

Ward	S (a)	A (a)	LER (%)	A_l (each type of elements) (a)			
				Road trees	Road shrubs	Lawns	Grasslands
Nishi ward[a]	40,001	1,197	3	738	194	17	249
Chikusa ward[a]	34,450	1,039	3	826	120	39	53
Naka ward[a]	29,512	2,413	8	1,929	318	8	157
Nakagawa ward[a]	64,797	1,255	2	780	400	2	73
Nakamura ward[a]	36,075	882	2	626	170	8	77
Tenpaku ward[a]	40,548	958	2	646	236	15	61
Higashi ward[a]	18,013	883	5	631	232	18	3
Kita ward[a]	36,821	918	2	643	137	3	135
Minami ward	37,085	910	2	553	267	5	85
Minato ward	71,432	1,919	3	1,156	545	64	155
Moriyama ward	50,527	1,007	2	427	337	21	222
Showa ward	23,141	590	3	399	94	3	94
Mizuho ward	23,959	562	2	376	60	0	126
Atsuta ward	18,616	688	4	495	168	2	22
Meito ward	39,776	1,308	3	733	331	116	129
Midori ward	66,785	1,254	2	914	191	38	111
Total (study area)[a]	300,216	9,544	3	6,818	1,808	110	808
Total (city scale)	631,537	17,782	3	11,871	3,801	357	1,752

[a]Study area

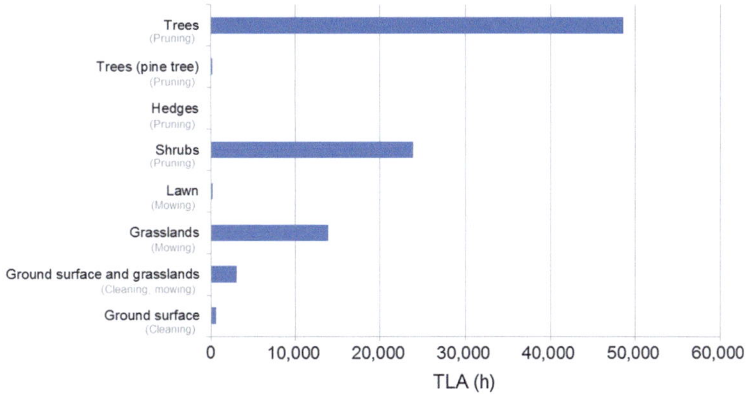

Fig. 10.15 Landscape labor accounts of each landscape element in road

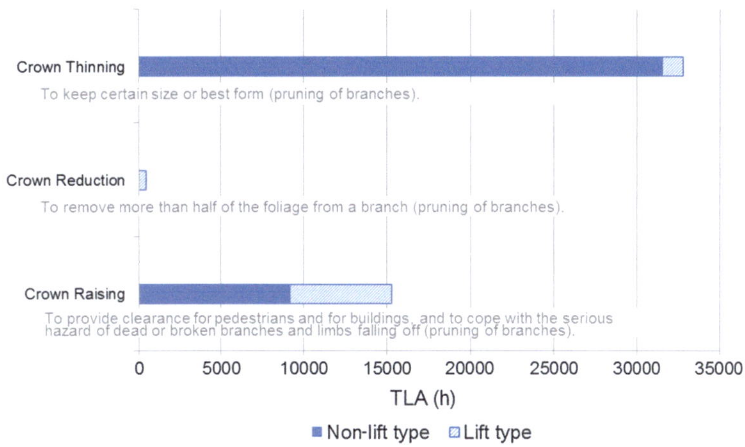

Fig. 10.16 Landscape labor accounts of each labor item of trees in the road

is 9.55 h/a and GLD is 0.30 h/a. Many of the roadside trees were planted several decades ago, and nowadays, they have grown up larger. Such roadside trees cause an increasing need for labor management. Nowadays, the number of Japanese municipalities that reduce the amount of pruning labor by replanting trees with a small size or cutting brunches short (heavy “Crown Reduction”) is increasing. The ideal size of roadside trees should be decided based on the concept of making the function of trees maximum to provide a heat island mitigation effect and shade. Therefore, keeping all trees small in size lacks functionality. In order to manage street trees in a way to keep excellent functionality, it is necessary to create broad spaces for trees along the road.

Table 10.7 Calculation of the total landscape labor accounts and densities in Nagoya city (road)

	S (a)	A (a)	LER (%)	A_l (Each type of element)				TLA (h)	NLD (h/a)	GLD (h/a)	SLI (population: 2,274,511)
				Road trees	Road shrubs	Lawns	Grasslands				
Study area	300,216	9,544	3	6,818	1,808	110	808	91,122			–
Total area (city scale)	631,537	17,782	3	11,871	3,801	357	1,752	169,767	9.55	0.30	0.075

10.2.4 Grassland of Vacant Land and River Bed

Table 10.8 shows the total annual labor density (TLD_{jl}) of grassland management by public entities. In Nagoya city, all river beds are managed by public agencies, and many vacant lands are also managed by them. The landscape labor accounts are calculated based on the guidelines for cost estimating for public works (Nagoya City 2015). As shown in the table, the amount of value of the total annual labor density (TLD_{jl}) is 1.96 h/a. The grasslands of vacant lands and river beds are managed about twice a year, and 95 % of total areas are managed by mowing machines, and 5 % of the total area is managed by hand.

10.2.5 Japanese Garden

10.2.5.1 Overview

In this subsection, as an example of expensive management, a case study of a publically managed Japanese garden is shown. Japanese gardens are loved not only by citizens but also travelers including foreigners, and are very important cultural resources of Japan.

The case study garden is Shiratori Garden in Nagoya. In Sect. 10.2.5, the landscape labor account was calculated based on the data from Yamashita et al. (2015). Most Japanese gardens are traditionally managed by private entities such as

Table 10.8 Total annual labor density of grassland management by public entities in Nagoya city

	The value of per unit area (10a)				(1 a)
	Annual working days (days)	Number of workers	Working hours per day (h)	TLD_{jl} (h/10a)	TLD_{jl} (h/a)
Mowing machine (95 %)	2	1.02	8	15.30	
Mowing of hands (5 %)	2	6.71	8	100.62	
Total				19.57	1.96

temples, shrines and feudal lords, but recently Japanese gardens managed by public entities are gradually increasing because of public values and economic difficulties of private management. Shiratori Garden is a very new garden, and was established in 1991 at the site of the World Design Exposition 1989 alongside of Hori River.

Japanese gardens need elaborative management work such as artistic cut of trees and lawn mowing and so on. They also need professional gardening. Their landscape management labor accounts and labor densities, therefore, are estimated to be high. The management of the garden is made by a private gardening company according to the designated manager system regulated by the Local Government Act. In this subsection, the kinds of management being employed according to the vegetation differences will be examined through interviews with the management company.

10.2.5.2 Landscape Units and Organization

Table 10.9 shows the area size of each type of landscape unit and element of the garden, and Figs. 10.17 and 10.18 show their spatial distribution. Among the total area of about 362 are, about a 79 are area is occupied by water, and walking paths are arranged around the water. Shiratori Park is designed according to the traditional method of "*Chisen Kaiyu-Shiki Teien*", a visitor strolls past a pond and fountain and around the garden, in which people can walk around the garden enjoying the scenery that changes in different seasons. As shown in Fig. 10.18, trees around the water are low, and there are many Japanese pine trees, *Pinus thunberigii,* and Japanese maple trees, *Acer palmatum*. In the outer periphery of the garden, many tall evergreen broadleaf trees are arranged and act as a visual barrier from the outside.

The managing gardening company has its own employees but contracts some of the labor to outsourcing craftsmen, part-time workers and senior workers. The outsourcing craftsmen handle the pruning of trees, and in particular, high skilled outsourcing craftsmen deal with the elaborative treatment of pine trees, which needs very special skills. The employees are mainly in charge of pruning of shrubs, cutting of ground covers and fertilizing of lawns. Senior workers are in charge of the weeding of lawns and grasses, cleaning of water surfaces and walking paths. Part-time workers, who are women farmers, are hired additionally for the weeding in high seasons.

10.2.5.3 Landscape Labor Accounts and Densities

Table 10.10 shows the annual labor accounts, labor contents and types of workers for the maintenance of each type of ground surface. The water surface and its surrounding area are the most important landscape units in this garden, and the amount of its annual landscape labor accounts (TLA_l) is high. Also the walking path has a high amount of TLA_l.

Table 10.9 Area size of each type of landscape unit and element of Shirotori garden

Landscape unit	Landscape elements (ground surface, shrubs, hedges, facilities)		A_l (a)
Water surface	Water surface		78.4
Surrounding area of water surface	Ground cover plants	Bamboo grass (*Sasa nipponica*)	4.6
		Bamboo grass (*Shibataea kumasasa*)	3.5
		Bamboo grass (*Veitchs bamboo*)	3.9
		Lawns	6.4
		Moss plants	1.0
	Bare land or grassland		15.5
	Under growth or grasslands		9.3
	Hedges		0.3
	Shrubs		8.0
Surrounding area of main building	Ground cover plants	Moss plants	2.6
	Bare land or grassland		2.4
	Undergrowth or grasslands		0.7
	Hedges		0.3
	Buildings		0.0
	Shrubs		0.7
Surrounding area of main walking path	Ground cover plants	*Ophiopogon japonicus Ker.f. nanus fort.*	1.9
	Bare land or grassland		0.0
	Undergrowth or grasslands		4.5
Walking path (paved) and other paved surfaces	Walking path (paved) and other paved surface		18.8
	Walking path (paved) and other paved surfaces		0.0
	Bridges		0.4
Walking path (unpaved)	Walking path (unpaved)		46.3
	Shrubs		0.2
Lawn space	Ground cover plants (lawns)		14.2
	Shrubs		0.4
Trees of outer edge of the garden	Bamboo forest undergrowth		19.9
	Undergrowth or grasslands		73.1
	Hedges		11.5
	Shrubs		13.9
Buildings	Buildings		8.6

(continued)

Table 10.9 (continued)

Landscape unit	Landscape elements (ground surface, shrubs, hedges, facilities)		A_l (a)
Bridges	Bridges		2.2
Parking lots	Parking lots		8.0
Total			361.6

Landscape elements (trees)	Horizontal projection area (a)	Number of trees
Japanese maple (*Acer palmatum*)	14.9	181
Pine tree (*Pinus thunberigii*)	7.4	152
Bamboo forest	23.3	–
Other trees	58.3	1,312

Fig. 10.17 Spatial distribution of landscape units

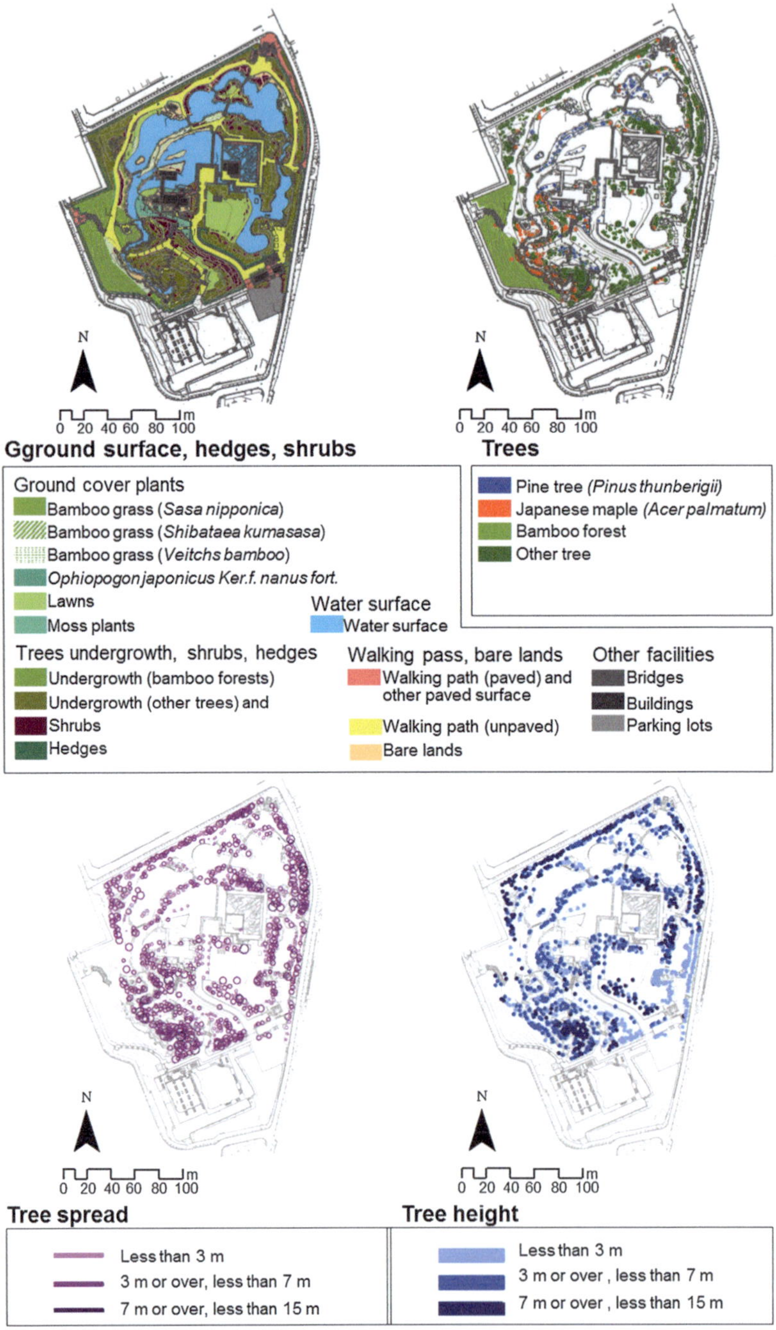

Fig. 10.18 Spatial distribution of landscape elements, tree spread, tree height

Table 10.11 shows the annual labor accounts, labor contents and types of workers for the maintenance of hedges, shrubs and trees. The pine trees need special maintenance skills, and the amount of TLA_l is high (Fig. 10.19).

Table 10.12 and Fig. 10.20 show the calculated results of the annual landscape labor accounts and densities of each landscape element, respectively. Figure 10.21 shows the spatial distribution of the amount of GLD_l of Shiratori garden. The total amount of the landscape labor accounts of all areas (TLA) are 6,407 h and total gross labor densities (GLD) are 17.7 h/a. Maintenance of water surface, the surrounding area of the water surface, the walking path and lawn space have high value of annual labor accounts. Cleaning of the water surface and the walking path are carried out almost every day. Mowing of grasslands and lawns are frequent during the summer season. Such labor contents of cleaning and mowing grasslands do not need special expertise and are maintained by senior workers. On the other hand, the labor content of trimming of trees needs high expertise. In Fig. 10.22, the annual labor accounts and net labor densities of each type of tree are shown. The total amount of the landscape labor accounts (TLA_l) for maintenance of trees is 1,473 h, and these works are done by the employees and outsourcing craftsmen. Pine trees have a very important meaning in Japanese gardens, and need careful treatment, i.e. special pruning, to keep the shape in ideal form. Therefore, their pruning requires extra time and manpower. The net labor density (NLD_l) of pine trees is 187.77 h/a and the annual labor accounts are 1,390 h. On the contrary, the pruning of the other trees requires only cutting twigs and dead branches, and has a low net labor density of 1.16 h/a and low annual labor accounts of 67 h, even though they occupy a wide area (Fig. 10.17).

10.2.5.4 Personal Landscape Labor Accounts and Densities

In Fig. 10.23, personal labor accounts and densities in each landscape element are considered. The senior worker has the highest value of annual labor accounts of 1,129 h, and the second highest value is for garden employees, 480 h. The senior worker also has the highest personal labor density, 0.82 h/a. The senior workers and garden employee work every day. The outsourcing of craftsmen for pine tree work occurs for a short duration between October and November. Use of part time workers for mowing grass in the lawn space takes twenty five days per year.

10.2.5.5 Conclusion

In this subsection, as an example of expensive management, a case study of a publically managed Japanese garden was shown. The characteristics of the labor accounts can be summarized as follows.

In Japanese traditional ponds and fountains, during a stroll around the garden, the views from walking paths and bridges are regarded as very important. Therefore, grasses around the water are cut cleanly, and algae in the water are

Table 10.10 Annual labor accounts of each labor content (ground cover plants and the others) of Shirotori garden

Landscape elements	Labor contents	Operator	Annual working day	Number of workers	Working hours per day	TLA_l (h)	
Water surface	Cleaning 1	S	308	1	0.5	154	904
	Cleaning 2	S	12	1	8	96	
	Cleaning 3	S	12	1	5.5	66	
	Cleaning 4	S	12	1	5.5	66	
	Cleaning 5	S	308	1	0.5	154	
	Cleaning 6	S	308	1	0.25	77	
	Cleaning 7	S	12	1	5.5	66	
	Plants maintenance 1	G and S	2	10	7.5	150	
	Plants maintenance 2	G and S	1	10	7.5	75	
Surrounding area of water surface (ground cover plants (bamboo grass))	Mowing Fertilizing	S	141	2	5.5	1,551	1,923
		G	1	1	4	4	
	Mowing (*Shibataea kumasaca*)	S	12	2	5.5	132	
	Mowing (*Shibataea kumasaca*)	S	2	1	5.5	11	
	Trimming (*Shibataea kumasaca, Sasa veitchii, Sasa nipponica Makino et Shibata.*)	G	6	5	7.5	225	
Surrounding area of main building	Cleaning 1	G	103	1	2	206	212
	Cleaning 2	G	2	1	3	6	

(continued)

Table 10.10 (continued)

Landscape elements		Labor contents	Operator	Annual working day	Number of workers	Working hours per day	TLA_l (h)	
Surrounding area of main walking path	Ground cover plants (*Ophiopogon japonicus*)	Mowing	S	3	2	5.5	33	150
		Affusion	S	60	1	0.33	20	
	Special shrubs	Trimming of shrubs	O	4	2	7.5	60	
		Fertilizing of shrubs	G	1	1	4	4	
		Mowing around shrubs	S	2	1	5.5	11	
	Special hedges	Trimming	O	1	3	7.5	23	
Walking path	Main walking path	Cleaning	S	308	1	1	308	1,254
	Walking path	Mowing (1)	S	2	1	5.5	11	
		Mowing (2)	S	1	2	5.5	11	
		Cleaning	S	308	2	1.5	924	
Lawn space	Lawn space 1	Mowing of grasslands	S and P	47.5	2	8	760	798
		Lawn mowing	S	3	2	3	18	
		Fertilizing	G	1	1	1	1	
	Lawn space 2	Mowing of grasslands	S	2	1	5.5	11	
		Fertilizing 1	G	1	1	4	4	
	Lawn space 3	Fertilizing 2	G	1	1	4	4	

S senior worker, *G* garden employee, *O* outsourcing craftsmen, *P* part time job worker

Table 10.11 Annual labor accounts of each labor content (trees, hedges and shrubs) of Shirotori garden

Landscape elements		Labor contents	Operator	Annual working day	Number of workers	Working hours per day	TLA_l (h)	
Hedges and shrubs	Hedges	Trimming of hedges (inside of garden)	G	3	3	7.5	68	233
		Trimming of hedges (fringe of garden)	G	4	5	7.5	150	
	Shrubs	Trimming of shrubs 2	G	2	1	7.5	15	
Tree	Trees	Trimming of other trees	G	3	3	7.5	68	68
	Pine trees	Trimming(1) of pine trees	O and S	30.4	5	8	920	1,391
		Trimming(2) of pine trees	O	1	1	8	8	
		Snow-protective decoration 1 of pine trees	G and S	7	4	7.5	210	
		Snow-protective decoration 2 of pine trees	G and S	10	5	4	200	
		Snow-protective decoration 3 of pine trees	G	1	7	7.5	53	
	Acer palmatum	*Acer palmatum*-protective	G	1	2	7.5	15	15

S senior worker, *G* garden employee, *O* outsourcing craftsmen, *P* part time job worker

Fig. 10.19 Photos of trimming of pine trees

removed in any time. Trees and shrubs beside the water are maintained in low shapes. As a consequence, labor densities around the water and the open space in the center are high, as shown in the spatial distribution of labor densities in Fig. 10.21. The pine trees have much higher net labor densities as shown in Table 10.12. On the contrary, the labor densities of tall evergreen broadleaf trees of the outer edge of the garden are almost nothing in recent years. The reason is that the ground cover plants under the evergreen broadleaf trees do not grow because of the lack of sunshine and the trees also do not need to be cared for frequently (Fig. 10.24).

Based on the landscape management labor accounts, we consider the future landscape planning for the conservation of the cultural landscape of Japanese gardens as follows. 1) In such style of Japanese gardens as *Chisen Kaiyu-Shiki Teien*, the important landscape units (i.e. pine trees and ground cover plants of the

Table 10.12 Annual labor accounts and gross labor densities of each landscape unit of Shirotori garden

Landscape unit	Landscape elements (plants of ground surface, shrubs, hedges, facilities)		A_l (managed area) (a)		TLA_l (h)	S_l (a)	GLD_l (h/a)
Water surface	Water surface		78.4	904	904	78	11.5
Surrounding area of water surface	Ground cover plants	Bamboo grass (*Sasa nipponica*)	4.6	1,923	2,866	52	54.6
		Bamboo grass (*Shibataea kumasasa*)	3.5				
		Bamboo grass (*Veitchs bamboo*)	3.9				
		Lawns	6.4				
		Moss plants	1.0				
	Trees	Pine tree	4.9	933			
		Japanese maple (*Acer palmatum*)	6.6				
		Bamboo forest	3.1				
		Other trees	10.0				
	Hedges		0.3				
	Shrubs		8.0				
Surrounding area of main building	Ground cover plants	Moss plants	2.6	212	339	7	50.8
	Bare land or grasslands		2.4				
	Undergrowth or grasslands		0.7				
	Trees	Pine tree	0.6	120			
		Japanese maple (*Acer palmatum*)	1.1				
		Other tree	1.2				
	Hedges		0.3	6			
	Shrubs		0.7	0			

(continued)

Table 10.12 (continued)

Landscape unit	Landscape elements (plants of ground surface, shrubs, hedges, facilities)		A_l (managed area) (a)		TLA_l (h)	S_l (a)	GLD_l (h/a)
Surrounding area of main walking path	Ground cover plants	*Ophiopogon japonicus Ker.f. nanus fort.*	1.9	150	233	13	17.8
	Bare land or grasslands		0.0				
	Undergrowth or grasslands		4.5				
	Hedges		1.9				
	Shrubs		4.7				
	Trees	Pine tree	0.4	82			
		Japanese maple (*Acer palmatum*)	0.3				
		Other trees	1.4				
Walking path	Main walking path		7.8	1,254	1,254	8	19.2
	Walking path		57.5			58	
Lawn space	Ground cover plants (lawns)		14.2	798	800	15	54.7
	Shrubs		0.4				
	Trees	Japanese maple (*Acer palmatum*)	0.0	2			
		Other trees	1.4				
Trees of outer edge of the garden	Bamboo forest undergrowth		19.9	0	11	112	0.1
	Undergrowth or grasslands		73.1				
	Hedges		9.6	5			
	Shrubs		9.2	6			
	Trees	Pine tree	1.5	0			
		Japanese maple (*Acer palmatum*)	6.9				
		Bamboo forest	20.2				
		Other trees	44.3				
Other	Buildings		0.0	0	0	9	0.0
	Bridges		0.0	0	0	3	
	Parking lots		0.0	0	0	8	
Total					6,407	362	17.7

water surface and surrounding area of the water surface) need a high level of landscape management labor accounts. 2) Pruning of trees and trimming of ground cover plants should contain a nice shape for keeping the aesthetic landscape. Checking constantly the appearance of the garden from various locations is an important point of view for maintenance of the Japanese garden. It is very important to develop human resources who can create a high-level technical planning for Japanese garden. 3) While the labor contents of the cleaning of the ground surface and the labor contents of the mowing grasslands do not need a high-level skill of engineering, it needs a high amount of landscape management labor accounts. At present, these labor contents are handled by elderly persons. For preparation for the

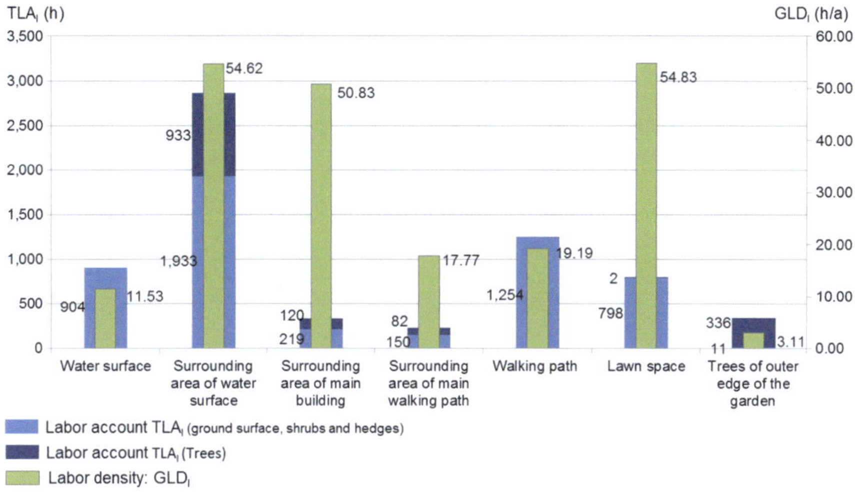

Fig. 10.20 Annual labor accounts and densities of each landscape unit

coming of the future aging society, the Japanese garden management by them might be an effective opportunity to create jobs.

10.3 Towards the Sustainable Land Management—Future Prospects in Planning for Urban Green Managed by Public Entities

In the above subsections, the landscape management labor accounts of urban green managed by public entities are described for the trees of parks, gardens, roadsides, riverbeds and vacant land. In recent years, the maintenance cost for public green has been reduced. In contrast, urban green is growing larger, which leads to the need of a greater amount of landscape management labor. For example, in such parks with a large scale secondary forest as Higashiyama Park, the landscape management labor for prevention of expanding bamboo forests and removal of the risk of fallen trees occur additionally. The neighborhood parks and roadside trees also need landscape management labor for trees which negatively affect buildings and traffic. In such landscape management labor for grown trees, professionals who can handle the appropriate machines are necessary. Such work is difficult for normal citizens. In the future, the concept of landscape management labor accounts could be effectively utilized as a measure of magnitude of order to know the amount of labor for these risks.

Fig. 10.21 Spatial distribution of gross labor densities of units

On the other hand, the management of shrubs and grasslands with relatively easy work might be promoted by the volunteer group and administration together. The vivid activity of citizens is very important to increase the social awareness of urban green.

Suzuki et al. (2014) show the result of the attitude survey on the urban green management (Fig. 10.25). Among them, one question is "Have you ever participated in the volunteer activities of urban green management? (Q.1)", and the answers to the question show that "Currently participating" and "Participated in the past" occupy about 15 % of the total respondents.

To another question that "In the future, would you like to participate in volunteer activities? (Q.2)", about 31 % of the total respondents answer that "I would join the activity". The percentage of those persons suggests the potentiality of the population who may be involved in urban green management in the future.

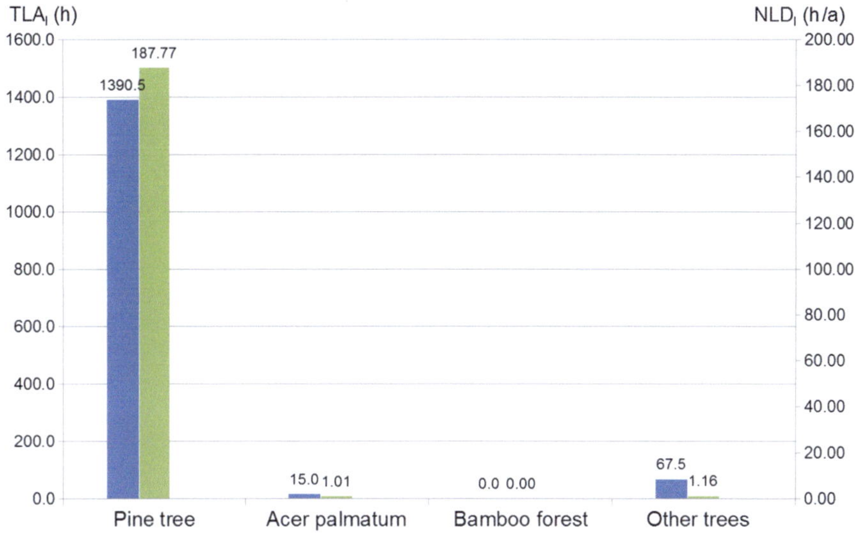

Fig. 10.22 Annual labor accounts and net labor densities of each type of tree

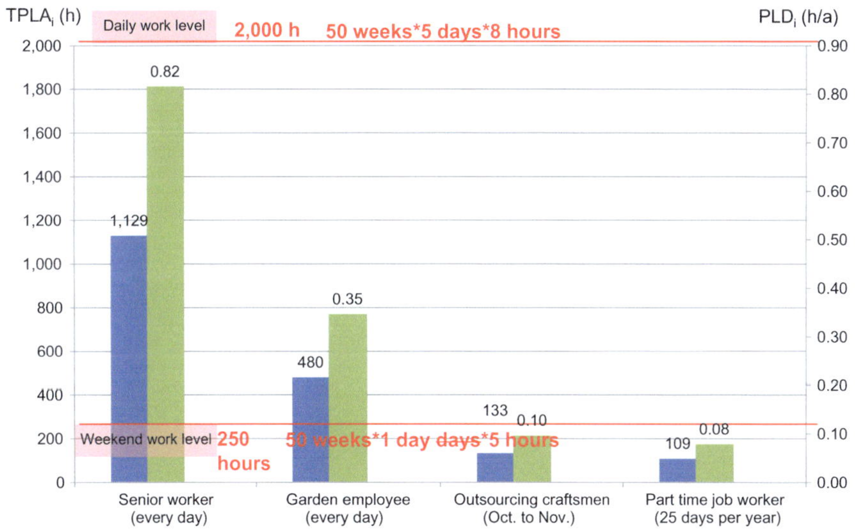

Fig. 10.23 Personal landscape labor accounts and densities

The total landscape labor accounts (TLA) of Nagoya city are estimated by using that percentage and the total population of Nagoya city (about 1.2 million people of the ages over 15 years old), as follows.

In the first estimation, the number of people participating in current activities is estimated at 45,000. When the value of personal annual labor accounts is set at 3 h,

Fig. 10.24 Photos of typical landscape of Shirotori garden. **a** High labor accounts of main walking path. **b** High labor accounts of surrounding area of water surface and main building. **c** Low labor accounts of edge of the garden

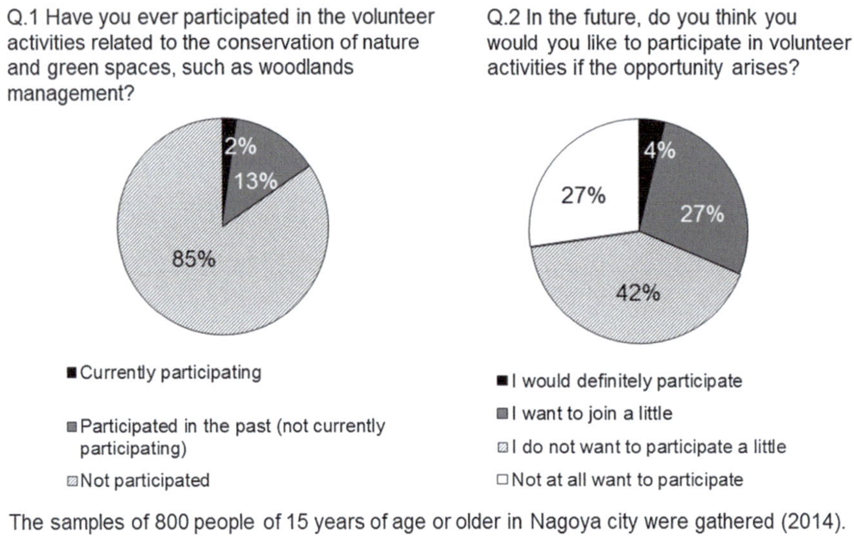

Fig. 10.25 Survey on attitude toward the urban green management

the estimated value of the total annual landscape labor account is 130,000 h in all of Nagoya city. This value of TLA is comparable to about one third of the total annual landscape labor account of the neighborhood park managed by public entities. (See Table 10.7, the amount of the total landscape labor accounts of the neighborhood park for the entire area of Nagoya city is about 41 million hours.)

In the second estimation, the number of the people who want to participate in future activity is estimated as 620,000. As well as in the first estimation, when the value of the personal annual labor account is set as 3 h, the ensured total annual landscape labor account is estimated as 1,860,000 h all of Nagoya city.

As mentioned above, the concept of the landscape management labor account can be a criterion for evaluation of the time when citizens participate in urban green management.

10.4 Private Green Spaces as a Mosaicked Structure of Small Green Sites—the Arako District

10.4.1 Overview

As shown in the introduction of this chapter, private green spaces play a very important role as a basic green element of the urban area. In this section, the Arako district is chosen for the case study of private green spaces as a mosaicked structure of small green sites. The Arako district is located in the western part of Nagoya (Fig. 10.26), and has a population of 7,257 persons in 2015. This district has developed from old agricultural villages to an urban area according to the urban expansion, and historical village landscapes remain in the old town near the Arako Kannon temple. In this area, there are many green covered spaces like kitchen gardens and gardens in residential sites. Recently, green landscapes like hedges by Maki trees, *Podocarpus macrophyllus*, are going to decrease because of the change to artificial walls with low maintenance. However, residents' awareness of good landscapes is gradually growing, and the Association of Village Arako was established as a resident initiative to manage Arako Park.

On the contrary, in the area where farmlands remained until recently, urbanization by using the land readjustment project is progressing rapidly, and the decrease of farmlands has become remarkable. Besides that, unused vacant spaces

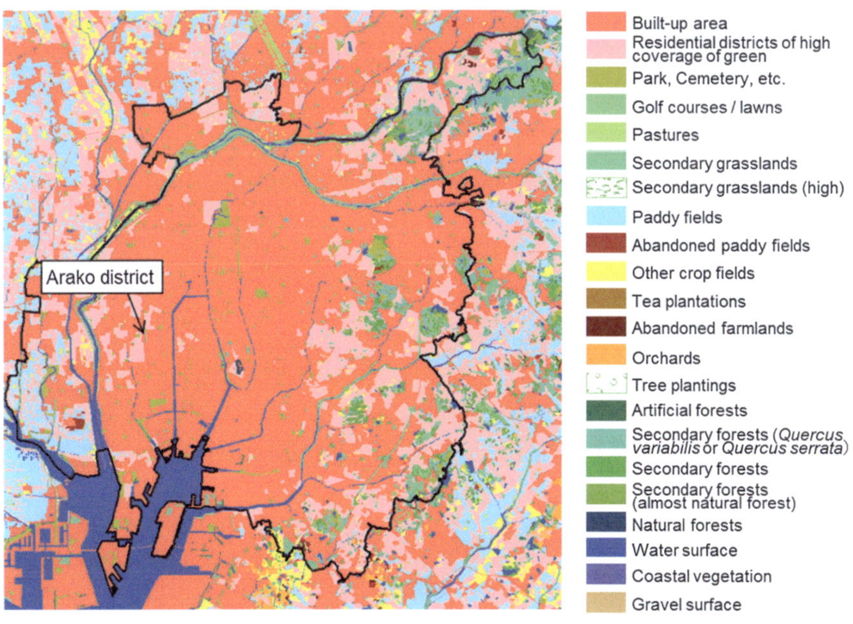

Fig. 10.26 Study areas

are also observed. Some parts of farmlands remain as "productive green", but most farmlands consist of narrow sites and many are used as kitchen gardens. These small scattered farmlands require much management on the part of aged farmers. In the old towns aging of society is progressing, but in the area near to the Arako railway station, young people are increasing according to the increase of high density collective houses.

As shown above, the Arako district is a typical urban landscape complex units of the mixture of old village and new developed residential area with relative rich green spaces. The issues of this area are the conservation of private green sites like residential gardens and kitchen gardens and maintenance of scattered small productive green sites.

10.4.2 Landscape Units and Organizations

The landscape units and organization of the Arako district are shown in Table 10.13 and the spatial distribution of the landscape units and green elements are shown in Fig. 10.27. Figure 10.28 shows the landscape features of the town.

As shown in Table 10.13, the landscape element ratio (LER: nearly equal to the green coverage ratio) of the entire town is 18.3 %. As shown in the distribution of the elements in Fig. 10.27, green elements are scattered in the town as a mosaicked structure in small scales. Among the green elements, the amount of elements of detached housing is the largest (500 are). The elements of vacant open spaces continue as the second largest green elements (230 are). The green elements of farmlands are also large. The total amount of the green elements of private sites reaches 1,238 are and 77 % of the total area of the green elements. Centrally, the amount of green elements of public parks is only 239 are, and it occupies 15 % of total green elements. It will be easily recognized that the maintenance of privately owned green elements is a very important issue of this town.

10.4.3 Labor Accounts and Labor Densities

10.4.3.1 Labor Accounts and Labor Densities in Residential Areas

Table 10.14 shows the annual labor accounts and densities of green elements of gardens and hedges of residential areas in the Arako district of two cases. As easily understood, the labor accounts and densities of green elements in private gardens and hedges are different according to the characteristics and lifestyles of residents. In two cases, both working area sizes are almost the same, but the annual labor accounts and densities differ from 140 to 26 h, and from 50.76 to 9.90 h/a respectively. The person of Case 1 works relatively intensively, once or twice a month, to care for the pruning of trees. Centrally, the person of Case 2 works once a

Table 10.13 Area sizes of landscape units and elements, and management organization of Arako district

Landscape unit	Landscape element of each type of unit		Organization	A_l (a)	S_l (a)		LER (%)
Parks	Tree and grassland		Public	239	534	5,265	44.7
Building lots	Public facilities	Tree and grassland	Public	7	124		5.3
	Shrines and Temples	Tree and grassland	Private owner	64	216		29.6
	Commercial lots	Tree and grassland	Private owner	9	345		2.7
	Industrial lots	Tree and grassland	Private owner	21	440		4.8
	Multiple dwelling house	Tree and grassland	Private owner	32	530		6.1
	Detached houses	Tree and grassland	Private owner	500	2,958		16.9
	Parking lots	Tree and grassland	Private owner	5	118		4.1
Other crop field and orchard	Other crop field and orchard		Private owner	230	276	276	83.3
Rail road	Grassland		Public	15	657	657	2.3
Road	Tree and grassland		Public	100	1,686	1,686	6.0
Vacant land	Tree and grassland		Public and private owner	377	904	904	41.7
Total area size (A, S)				1,600	8,788		18.2

week mowing the grass, but does not use much time at once. The difference of these working styles appears as the differences of the amount of annual labor accounts and labor densities. In the case of the Arako district, total net labor density (TLDij) is 30.82 h/a.

10.4.3.2 Labor Accounts in Farmlands

Table 10.15 shows 4 cases of the annual labor accounts and densities of farmlands in the Arako district. Farmlands in the Arako district consist of mainly dry croplands but partially of orchards. There are several management types, i.e. a type to be managed by a family, and a type of rental farming as a style of allotment gardens. The cases of the survey are all the types of family management without intention to sell the products.

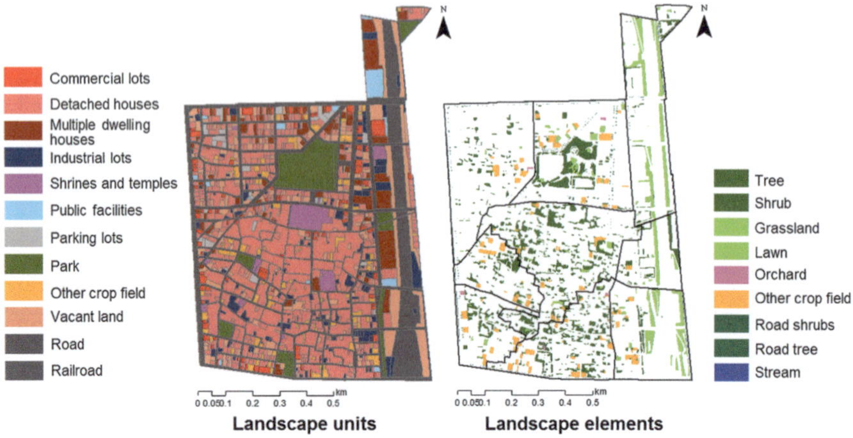

Fig. 10.27 Spatial distribution of landscape units and elements

Fig. 10.28 Pictures of the Arako district

8**Table 10.14** Annual labor accounts and densities of green elements of gardens and hedges of residential areas of Arako district

Case	Operator	Labor content	Annual working day	S_l (a)	A_l (a)	TLA_l (h)		GLD_l (h/a)	TLD_{jl} (h/a)	
1	Landowner	Tree pruning 1	4	7.5	2.8	6	140	18.66	50.76	30.82
		Mowing	25			34				
	Gardener	Tree pruning 2	10			100				
2	Landowner	Tree pruning	16	9.9	2.6	16	26	2.63	9.90	
		Mowing	5			10				

Table 10.15 Annual labor accounts and densities of farmlands of Arako district

Case	Labor content	Operator	Annual working day	S_l (a)	A_l (a)	TLA_l (h)	TLD_{jl} (h/a)	
1	Orchard, other crop field	Landowner	125	6.5	6.5	125	19.23	44.84
2	Other crop field	Allotment garden	44	6.0	6.0	116	19.33	
3	Other crop field	Allotment garden	180	8.0	8.0	720	90.00	
4	Other crop field	Landowner	48	2.0	2.0	48	24.00	

According to the *Agriculture business management statistics investigation farm products life costs statistics* (Ministry of Agriculture, Forestry and Fisheries of Japan 2012), the average of the annual labor density of other crop field in Tokai area is 29.77 (h/a). In Arako district, the total net labor density of the cases is 44.84 (h/a), and relatively higher compared to the average labor density, 90.00 (h/a), and 29.77 (h/a), of vegetables and fruits respectively for the purpose of sales. These management types with high labor densities are observed in the case of the aged farmer after the retirement age and the full-time housewife in the younger generation. These workers can obtain their cultivation times easily in the daytime in a week and weekend.

The purpose of the management of the farmland is primarily to maintain their health. All surveyed farmlands are located in the Arako district, and can be accessed easily within five minutes by bicycle. However, many land owners own several small fields scattered in the district, and the management of all scattered fields are a burden to the owners.

10.4.3.3 Summery of Labor Accounts and Labor Densities

Figure 10.29 shows the average net labor densities of residential areas and farmlands. Average labor densities of farmland and residential gardens count for 38.14 h/a, and 30.33 h/a respectively. It can be inferred to have a high value in the management of farmland, because the farmlands in the Arako district are managed not for sales of produce but for the maintenance of health.

10.4.3.4 Personal Labor Accounts and Working Style Types in the Arako District

Figure 10.30 shows personal labor accounts and labor densities in the Arako district. According to the *Agriculture business management statistics investigation farm products life costs statistics* (Ministry of Agriculture, Forestry and Fisheries of Japan 2012), the average of the annual labor density of paddy field in Tokai area is

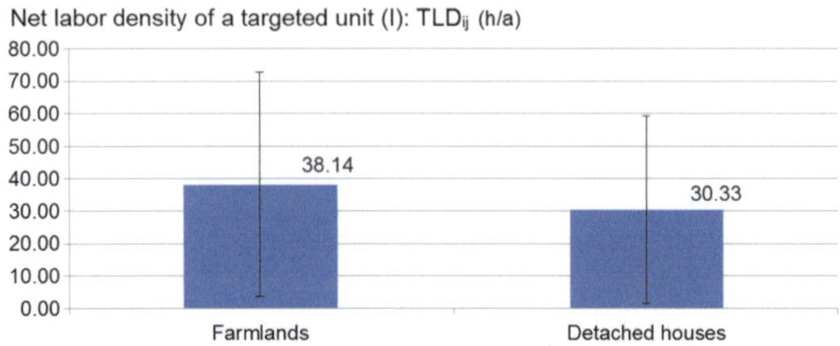

Fig. 10.29 Average of Net labor densities of an element in a landscape unit of residential areas and farmlands

2.63 (h/a). All labor densities of the farmlands in the Arako district are much higher than that of the average farmer in the Tokai area (2.63 h/a). However, the amount of the personal annual labor accounts is not so high. They visit their farmland once a week or a very short time every day. That might be a characteristic of urban farming that is not for selling produce but for personal reasons such as health maintenance. The maintenance of the gardens and hedges has a characteristic of very short working time in a year but relatively high densities.

10.4.3.5 Conclusion

Table 10.16 shows the results of the estimation of the landscape labor accounts and densities in the Arako district. Estimation of TLA value is carried out by multiplying each value of TLD_{ij} and A_l. According to the results, the value of the total annual labor account (TLA) of the district, net labor density of the district

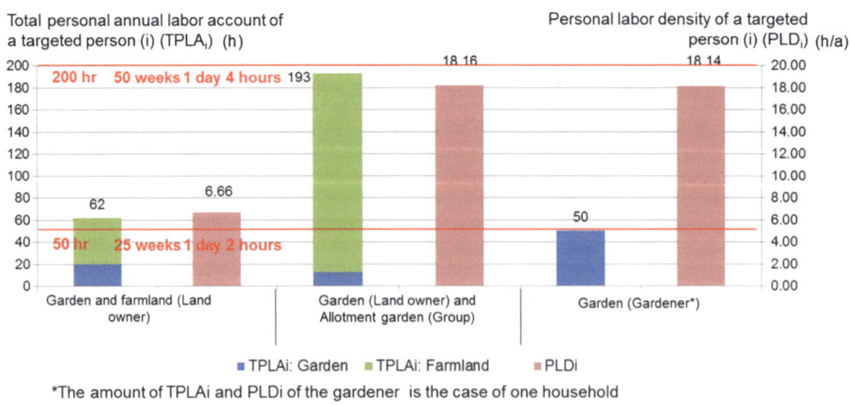

Fig. 10.30 Personal labor accounts and labor densities

(NLD) and gross labor density of the district (GLD) are 24,296 h, 15.19 h/a and 2.76 h/a, respectively. The sum of TLA_l of detached houses and that of farmland accounts for about 89 % of TLA of the Arako district. This district has a population of 7,257 people in 2015, and recently the development of the multiple dwelling houses has increased year after year. The size of the area of green coverage in the multiple dwelling houses of this district is small. Furthermore, in general, the management of green of the multiple dwelling houses is done by outsourcing craftsmen. Therefore, the amount of the total annual landscape labor accounts is low. However, many people live in the multiple dwelling houses with low amount of total annual landscape labor accounts. Naturally, the residents of the multiple dwelling houses do not contribute to the amount of the annual landscape labor accounts of the farmlands and the detached houses. The value of the social labor intensity (SLI) is 3.3, which is extremely low.

10.5 Landscape Labor Accounts in Different Private Land Uses—Detached Houses, Collective Houses and Commercial Land Uses

10.5.1 Overview

In this subsection, landscape labor accounts in different private land uses such as detached houses, collective houses and commercial land uses are described through the interviews with the residents and a professional gardening company.

Basically the management of landscape elements is done by the land owners by themselves or by gardeners hired by the owners. As to the interview to a professional gardening company, the attitude of land owners to the management of their lands has changed greatly, and the opportunity of the participation of professionals has decreased. The reasons for the decrease in professional participation are explained as follows. The first reason is that the garden is reduced by fragmentation into smaller scale multiple sites. The second reason is that the gardens with rich green elements are changed to paved parking places by the alternation of generation of owners. The third reason is that there is a tendency to change garden styles to those of low amount of maintenance, in which elaborative works by professional gardeners are not needed. As the consequence, contracts with professional gardeners with annual based seasonal maintenance of gardens have changed to one-shot contracts. In addition, gardens which require maintenance to be done once every few years or not at all are increasing.

On the other hand, the green management of collective houses and commercial land use sites are done by contracts with professional gardening companies, and the fees are divided among the users.

Table 10.16 Overview of landscape labor accounts and densities of Arako district

Landscape unit		Landscape element of each type of unit	Organization	A_l (a)	S_l (a)		TLD_{jl} (h/a)	TLA_l (h)	NLD_l (h/a)	GLD_l (h/a)	SLI
Parks		Tree and grassland	Public	239	534	534	16.43	3,927	16.43	7.35	3.3
Building lots	Public facilities	Tree and grassland	Public	7	124	4,731	8.45	56	8.45	0.45	
	Shrines and Temples	Tree and grassland	Private owner	64	216		4.88	311	4.88	1.44	
	Commercial lots	Tree and grassland	Private owner	9	345		8.45	78	8.45	0.22	
	Industrial lots	Tree and grassland	Private owner	21	440		8.45	179	8.45	0.41	
	Multiple dwelling house	Tree and grassland	Private owner	32	530		5.77	187	5.77	0.35	
	Detached houses	Tree and grassland	Private owner	500	2,958		21.94	10,975	21.94	3.71	
	Parking lots	Tree and grassland	Private owner	5	118		0.00	0	0.00	0.00	
Other crop field and orchard		Other crop field and orchard	Private owner	230	276	276	29.77	6,857	29.77	24.80	
Railroad		Grassland	Public	15	657	657	1.96	30	1.96	0.05	
Road		Tree and grassland	Public	100	1,686	1,686	9.55	959	9.55	0.57	
Vacant land		Tree and grassland	Public and private owner	377	904	904	1.96	738	1.96	0.82	
Total				1,600	8,788			24,296	15.19	2.76	3.3

10.5.2 Labor Accounts and Densities in Different Private Land Uses

Table 10.17 shows the labor accounts and densities in different private land uses. Cases 1–6 are the results of the interviews with the residents. The site sizes differ from a usual single house size to a big site size with a rich garden. Cases 7–25 are the results of the interviews and questionnaire survey to professional gardening companies. Cases 7–12 are results for detached houses managed by professional gardeners, and cases 13–25 are the results for multiple dwelling houses and commercial land use sites.

Figure 10.31 shows the average and deviation values of each net labor density of detached houses managed by land owners, detached houses managed by professional gardeners, multiple dwelling houses and commercial land use sites managed by professional companies. The average labor densities of detached houses managed by landowners are bigger than that of multiple dwelling houses and commercial land use sites managed by professional hands. It depends on low skills of the nonprofessional workers. Also the owner might manage their green elements for their health promotion, which might push the densities in higher levels.

On the contrary, the average labor densities of the professional workers are almost the same low values. The management by professional workers does not differ so much, because of the professional skills and constant productivity of their work.

10.6 Towards the Sustainable Land Management—Future Prospects in Planning for Private Green

In the Sects. 10.4 and 10.5, the landscape management labor account of various urban green managed by private entities was described.

Based on the above results of the landscape management labor account, the management of private green sites in a sustainable way can be suggested as follows.

(1) In the case of the Arako district, the amount of the green of the detached houses has contributed greatly to the total amount of green in the district. For conservation of the green of the detached houses, the labor force of the residents who can carefully manage their own green is necessary. The aging of landowners causes a large burden for the labor force in treating their own green spaces. It is important that financial aid is prepared to employ the outsourcing of professional gardeners. One case of reference is the subsidy of 7,000 yen per meter paid to the hedges of the traditional style in the downtown area of Matsusaka city, which will be referred in Chap. 11.

(2) In multiple dwelling houses, the amount of the landscape labor accounts can be shared by many residents. Therefore, it may be possible to decide on the

Table 10.17 Labor accounts and densities in different private land uses

Case	Landscape unit type	Operator		Labor content	Annual working day	S_l (a)	A_l (a)	TLA_l (h)		GLD_l (h/a)	NLD_l (h/a)		
											Case	Mean	SD
1	Detached houses	Landowner and gardener	Landowner	T	4	7.5	2.8	6.0	140.0	18.66	50.76	44.43	30.58
				G	25			34.0					
			Gardener	T	10			100.0					
2		Landowner		T	16	9.9	2.6	16.0	26.0	2.63	9.90		
				G	5		10.0	10.0					
3		Landowner		T	38	7.0	1.4	45.2	81.8	11.65	58.23		
				G	40		36.6	36.6					
4		Landowner		T	7	2.6	0.5	6.0	7.4	2.78	13.92		
				G	9		1.4	1.4					
5		Landowner		T	26	2.0	0.4	34.0	37.0	18.50	92.50		
				G	3		3.0	3.0					
6		Landowner and gardener	Landowner	T	13	5.8	1.2	7.9	47.9	8.26	41.29		
				G	5			33.0					
			Gardener	T	5			7.0					
7	Detached houses	Gardener (not including houses managed by landowner)		T	1	2.0	0.4	4.0	7.5	3.75	18.75	36.83	12.73
				L				2.5					
				G				1.0					
8				T	1	4.0	0.8	22.5	37.5	9.38	46.88		
				L				11.0					
				G				4.0					
9				T	1	4.0	0.8	22.5	30.0	7.50	37.50		
				L				5.5					
				G				2.0					
10				T	1	3.5	0.7	22.5	37.5	10.71	53.57		
				L				11.0					
				G				4.0					
11				T	1	7.0	1.4	22.5	37.5	5.36	26.79		

(continued)

Table 10.17 (continued)

Case	Landscape unit type	Operator	Labor content	Annual working day	S_l (a)	A_l (a)	TLA_l (h)		GLD_l (h/a)	NLD_l (h/a)		
										Case	Mean	SD
			L				11.0					
			G				4.0					
12			T	1	3.0	0.6	15.0	22.5	7.50	37.50		
			L				5.5					
			G				2.0					
13	Multiple dwelling houses (small scale)	Gardener	T	2	2.3	0.5	8.0	15.0	6.67	33.33	36.03	5.15
			L				5.0					
			G				2.0					
14			T	4	4.8	1.0	20.0	30.0	6.26	31.32		
			L				8.0					
			G				2.0					
15			T	6	5.1	1.0	33.0	45.0	8.82	44.12		
			L				10.0					
			G				2.0					
16			T	4	3.7	0.7	20.0	30.0	8.15	40.76		
			L				8.0					
			G				2.0					
17			T	1	6.9	1.4	33.0	45.0	6.53	32.66		
			L				10.0					
			G				2.0					
18			T	1	9.9	2.0	52.5	67.5	6.80	33.99		
			L				12.0					
			G				3.0					

(continued)

Table 10.17 (continued)

Case	Landscape unit type	Operator	Labor content	Annual working day	S_l (a)	A_l (a)	TLA_l (h)		GLD_l (h/a)	NLD_l (h/a)		
										Case	Mean	SD
19	Multiple dwelling houses (large scale)	Gardener	T	14	333.8	66.8	270.0	510.0	1.53	7.64	7.43	5.06
			L				210.0					
			G				30.0					
20			T	0	522.5	104.5	180.0	352.5	0.67	3.37		
			L				150.0					
			G				22.5					
21			T	0	358.6	71.7	150.0	292.5	0.82	4.08		
			L				121.5					
			G				21.0					
22			T	0	150.4	30.1	90.0	183.5	1.22	6.10		
			L				75.5					
			G				18.0					
23			T	0	21.3	4.3	30.0	68.0	3.19	15.96		
			L				30.0					
			G				8.0					
24	Commercial land use	Gardener	T	2	62.2	12.4	30.0	105.0	1.69	8.45	30.95	31.83
			L				59.0					
			G				16.0					
25			T	6	141.2	28.2	450.0	1510.0	10.69	53.46		
			L				910.0					
			G				150.0					

The results of cases 7–25 are investigated with the cooperation of Mr. Hayashi Yoshinobu (professional landscape gardener). Interview survey was conducted in July 2015

T trimming of tree

G mowing of grassland

L lawn mowing

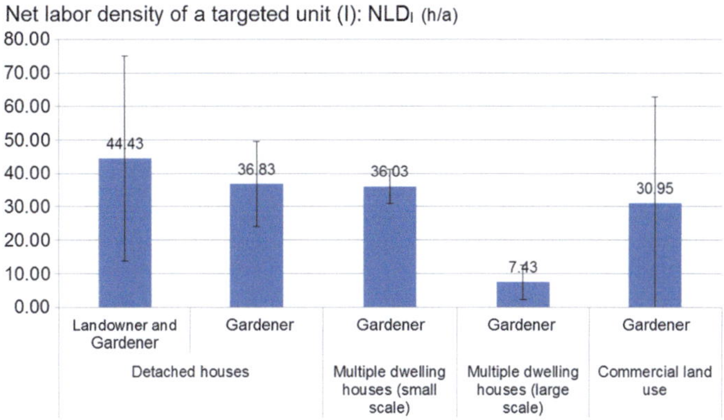

Fig. 10.31 Average of net labor densities of units of detached houses, multiple dwelling houses, and commercial land use

green amount in such a way as corresponding to the number of residents. However, there is a limit to increasing the green only within the site of the multiple dwelling houses.

(3) For preservation and creation of private green spaces, it is important that the neighbors should have the recognition that private green is a common property of the district. When only landowners manage their own green sites themselves, the amount of green of the district will be decreasing. In the case of the Arako district, the farmlands that are lacking in the amount of management labor are able to obtain more hours for management by land sharing. For example, selling the set of multiple dwelling houses and farmlands located in the district may be one of the solutions.

In private properties, the number of persons concerned in green is usually small, even though the district has a large population. As mentioned above, the concept of the landscape management labor account can be the criterion which let us know the tasks in the green management for district residents.

In this chapter, the landscape units were divided into those of the public and private hands, and the landscape management labor account of a variety of the urban green was described.

In this book, the sites or districts with comparatively rich green were selected for the case studies. In other land uses (e.g., the case of the high density built-up area or the small green district) other methods may be required. The remaining issue is the research in various target areas for green management.

References

Kawaguchi N, Murayama A, Shimizu H, Takatori C (2014) The state of scale of open space/green covered space in city blocks and their temporal change, and the characteristic feature of their distribution on Nagoya blocks. J City Plann Inst Jpn 49(3):207–212 (In Japanese with English abstract)

Kawaguchi N, Takatori C, Murayama A, Shimizu H (2016) Proposal and estimation of landscape management labor accounts in urban areas. J City Plann Inst Jpn (in Japanese with English abstract) (under review)

Ministry of Agriculture, Forestry and Fisheries of Japan (2012) Agriculture business management statistics investigation farm products life costs statistics. http://www.maff.go.jp/j/tokei/kouhyou/noukei/. Accessed 1 July 2015

Nagoya City (2010) Green status survey. Licensed by Nagoya city

Nagoya City (2011) Nagoya green master plan 2020

Nagoya City (2012) Basic survey of city planning. Licensed by Nagoya city

Nagoya City (2015) 'Doboku koji sekisan kijunsho' (guidelines for cost estimating for public works)

Suzuki Y, Shimizu H, Takatori C, Kawaguchi N (2014) A statistical analysis of residents perception of ecosystem disservice in Nagoya City (53):473–476. Architectural Institute of Japan (in Japanese)

Yamashita N, Shimizu H, Miyawaki M (2015) Landscape characteristic and management of Japanese gardens with a path around a pond. In: Proceedings of the Tokai chapter architectural research meeting. Architectural Institute of Japan, Tokai Branch (in Japanese)

Chapter 11
Urban Landscape Case Study in a Middle City: Matsusaka Chuo District

Nobuko Kawaguchi

Abstract This chapter considers landscape management labor accounts of a variety of the urban greenspaces in a middle city within an urban landscape: Matsusaka Chuo District, Matsusaka City, Mie Prefecture. Matsusaka Chuo District flourished as a castle town, and in the district, there are many tourist spots such as Matsusaka Castle, small historical towns with traditional wooden houses and wide residential sites with rich green spaces in the western part. The eastern part of the district consists of central shopping streets with a railway station and residential areas. In this chapter, the management of private gardens within historical heritage housing sites, green areas of temples, and the management of small rivers that flow throughout the district are considered.

Keywords Local middle city · Urban green · Japanese garden

11.1 Introduction

This chapter considers landscape management labor accounts in a middle city within an urban landscape. The target city is Matsusaka, located in the center of Mie Prefecture with a population of 168,017 in 2010 that has shown a decreasing tendency. Matsusaka a typical middle city that stretches from the flatland area of the coastal zone to the steeply sloped mountainous area along the Kushida River. It contains a variety of Japanese basic landscape types shown in the Chap. 2: urban (including urban paddy field mixed types), paddy fields, other field types, paddy field Satoyama, other field Satoyama, and nature types. This diversity is the reason that many case studies in this book were carried out in this city. In this chapter, the landscape management labor accounts in an urban landscape type of Matsusaka Chuo District in Matsusaka City will be considered (Fig. 11.1).

N. Kawaguchi (✉)
Graduate School of Environmental Studies, Nagoya University,
Furo-cho, Chikusa-ku, Nagoya, Aichi 464-8601, Japan
e-mail: nobqou@gmail.com

H. Shimizu et al. (eds.), *Labor Forces and Landscape Management*,
DOI 10.1007/978-981-10-2278-4_11

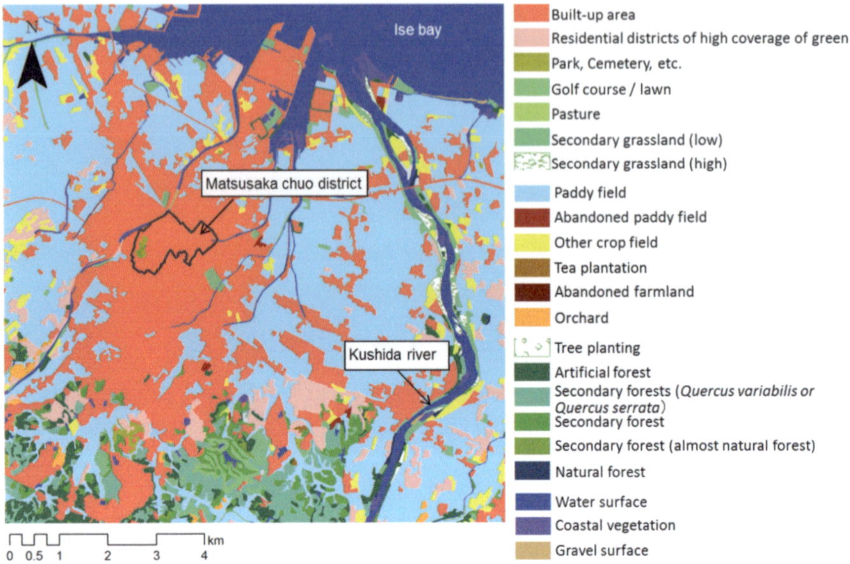

Fig. 11.1 Study area

Matsusaka Chuo District is located in the center of Matsusaka City with a population of 3629 in 2010 and has an established residents' council, which consists of the residents of 10 towns in the district. Matsusaka City once flourished as a castle town, with many tourist spots remaining, such as Matsusaka Castle, and in the western portions, small historical towns with traditional wooden houses and wide residential sites with rich green spaces. The eastern portion of the district consists of central shopping streets with a railway station and residential areas. In this chapter, the management of private gardens within historical heritage housing sites, green areas around temples and castles, and the management of small rivers, which flow in the district, are examined.

11.2 Landscape Units and Managing Organizations

Table 11.1 shows the sizes of landscape units, green coverage ratios, and managing organizations. In addition, Fig. 11.2 shows the spatial distribution of the landscape elements and units and Fig. 11.3 shows typical landscapes throughout the district. In the framed area of Fig. 11.2, the Tonomachi District planning area, there are many residential sites with rich green elements and temple forests. In this area, special district plans have been set in order to conserve these landscapes, including a special subsidy for the pruning of traditional green hedges alongside streets. The public conscious to maintain these landscapes is very high among residents.

Table 11.1 Area size of the landscape units and managing organizations

Landscape unit		Landscape element of each type of units	Organization	A_l (a)	S_l (a)		LER (%)
Parks		Tree and grassland	Public (municipality)	407	743	743	55
Built-up areas	Public facilities	Tree and grassland	Public (municipality)	110	1,183	8,445	9
	[a]Cultural heritage building	Tree and Grassland	Public (municipality)	83	151		55
	[a]Shrine	Secondary forests	Private (landowner)	239	309		77
	[a]Temple	Tree and grassland	Private (landowner)	89	369		24
	[b]Private building lots	Tree and grassland	Private (landowner)	504	5,025		10
	Parking lots	Tree and grassland	Private (landowner)	37	1,407		3
Vacant land		Grassland and tree	Private (landowner)	28	79	79	35
Farmland	Other crop field	Other crop field	Private (landowner)	9	12	23	77
Water surface	Rivers (water surface)	Water surface	Public (municipality)	0	82	300	73
	Dry riverbed	Tree and grassland	Public (municipality)	67	67		
	[a]Water channel	Water surface and grassland	Public (neighborhood associations)	151	151		
Rail road	Rail road	Non-element	Public (municipality)	40	413	413	10
Road	Road	Tree	Public (municipality)	64	1,813	1,813	4
Total				1,828	11,815		15

[a]The survey of the landscape labor account
[b]The survey of the management of the traditional hedges

A major issue for the residents in residential areas near shopping streets is an increase of vacant lots and houses due to demographic changes. Residents believe that the increase of vacant lands might lead to deterioration of the neighborhood character and the landscape as a tourist and shopping destination. Some resident want to transform such vacant lands into community parks under their own local management.

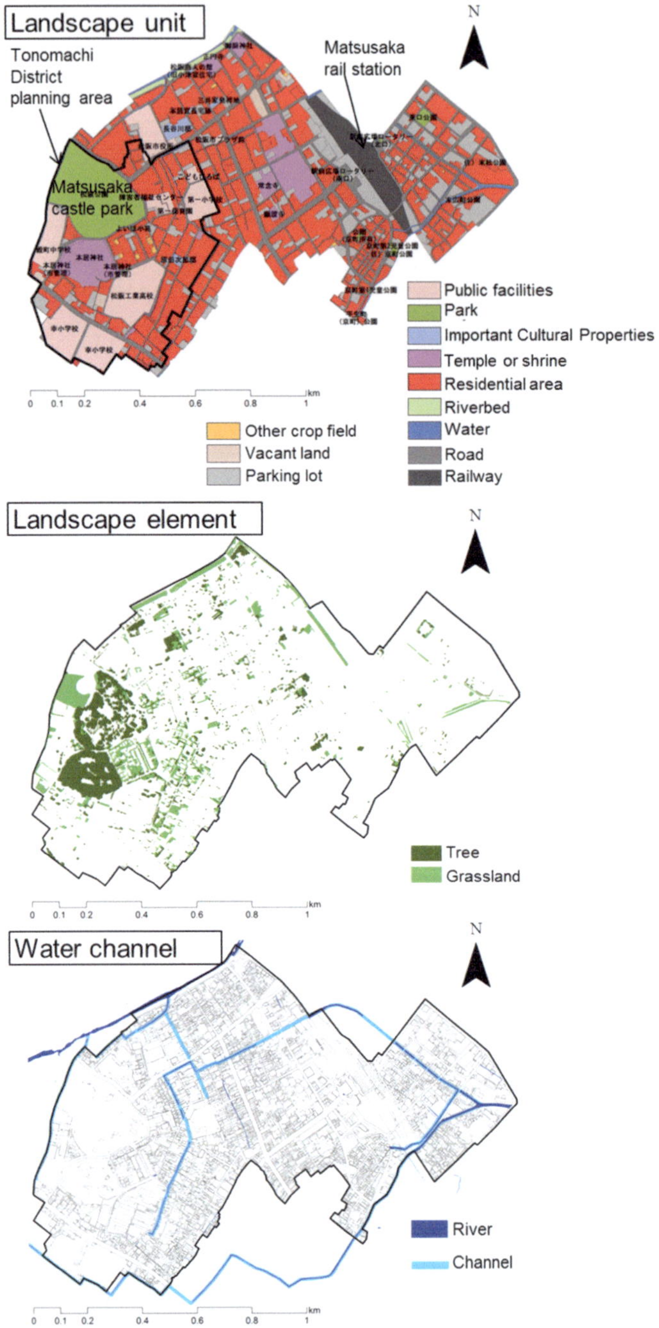

Fig. 11.2 Spatial distribution of the landscape units, elements, and water cannels

Fig. 11.3 Typical landscapes

Another characteristic of this district is the presence of small rivers throughout residential areas, which were originally moats of Matsusaka Castle used for drainage. Now, however, their use as drainage tributaries is unnecessary due to the completion of modern sewage systems and water purification. The management of these rivers is done through the participation of one person per household is obliged, an activity that has become a tradition.

As shown in Table 11.1, the total green coverage ratio is 15 %, which is slightly smaller than that of Arako District shown in Chap. 10. The most unique characteristic of Matsusaka Chuo District (compared to Arako District) is that there are many green elements of cultural heritage such as temples, shrines and a cultural heritage building that is now managed by the municipality but originally was managed as a private house. The city government purchased the land to conserve the high historic value of the buildings and gardens; as such, the green elements in Matsusaka Chuo are highly dependent on private green sites.

The green environments of this area are currently in crisis because of demographic changes, but will continue to be maintained mainly by the motivations and activities of the residents themselves, along with some support of the municipality.

11.3 Residential Area

This section details landscape management labor accounts and densities of the residential area, particularly cultural heritage residential sites, the spatial distribution of which is shown in Fig. 11.4. These residences were established by wealthy

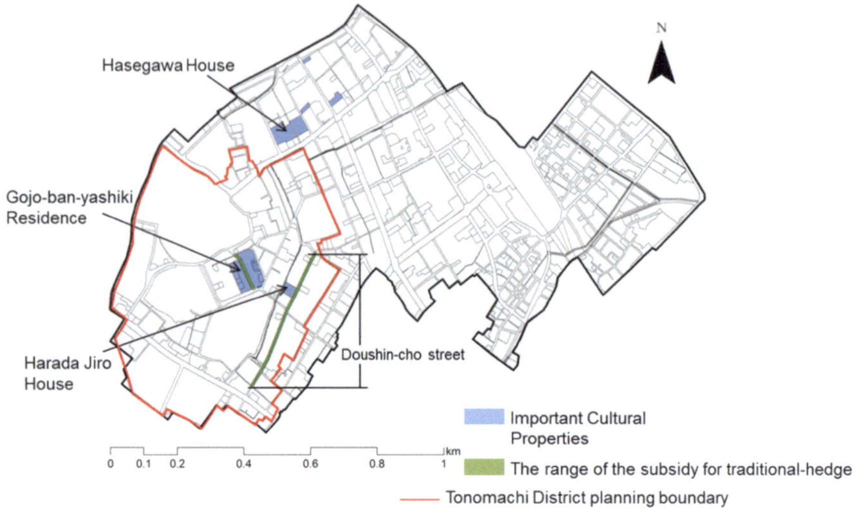

Fig. 11.4 Spatial distribution of the cultural heritage residential sites

merchant families such as Mitsui, Hasegawa, and Ozu. Until recently, the management of these sites was undertaken by the families themselves; after 2000, however, these sites were designated by Matsusaka City to be managed by the municipality.

Table 11.2 shows the labor accounts and densities of Hasegawa House and Harada House. Management is currently done by professional gardeners, with the main labor task of pruning of trees. The labor density of smaller green elements of Harada House is higher than that of Hasegawa House, though the overall amount of the landscape labor accounts is almost same for both houses. It might be considered that the larger the managed area the more the labor density will be diluted.

In addition to these historical heritages, there are many private residential sites in this area with rich green elements. Even though these sites are not designated as historical heritages, they retain the traditional residential landscapes with green hedges, the maintenance of which is supported by the municipality in the form of

Table 11.2 Landscape labor accounts and density of the cultural heritage residential sites

Landscape unit	Labor content	S_l (a)	Non-built-up area (a)	Non-built-up area ratio (%)	LER (%)	TLA_l (h)	GLD_l (h/a)
Hasegawa house	Pruning of trees, shrubs and hedges	49	13	26.5	67.5	138.8	2.8
Harada Jiro House	Pruning of trees, shrubs and hedges	12	2	16.7	56.6	102.5	8.5

1000 yen per one meter. In the Tonomachi District planning area, the style of hedges and outdoor advertisements are restricted under the district plan. Even though the green hedges are regarded as an important element of the landscapes of the district, recently there has been a progression toward iron fences for the relief of management labor.

The number of vacant houses and stores located in the downtown area surrounding the railway station have increased according to *The Second Business Report* (Matsusaka Ekimae dori Shopping District Promotion Association 2006). The number of vacant stores was 46 stores in 2003, and was 56 stores in 2004.

In the spatial distribution of Fig. 11.2, it can be seen that the land use of houses is changing to show increased vacancies and parking lots. It is interesting to note that the characteristic of the shape of the site spatial distributions is long and narrow, showing that the increased vacant house rate is introducing more parking lots and vacant land. Therefore, due in part to the aging society, historical heritage of private residential sites is also at risk of becoming vacant space.

11.4 Temples and Shrines

Figure 11.5 shows the distribution of temples and shrines in the district, and Table 11.3 shows the labor accounts and densities of their landscape elements.

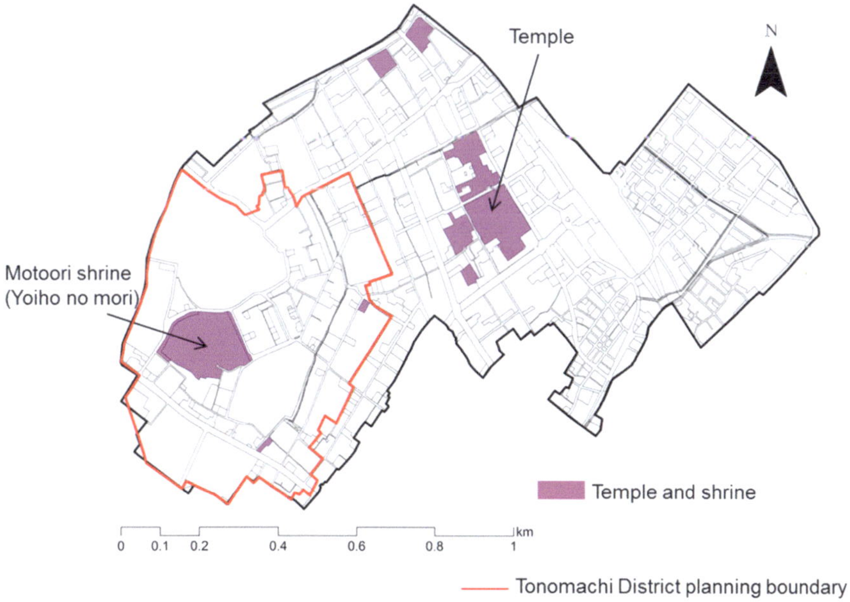

Fig. 11.5 Spatial distribution of temples and shrines

Table 11.3 Landscape labor accounts and density of temples and shrines

Landscape unit		Organizations	Labor content	A_l (a)	S_l (a)	Non-built-up area ratio (%)	LER (%)	TLA_l (h)	GLD_l (h/a)	NLD_l (h/a)
Motoori shrine (Yoiho no mori)	Surrounding area of the secondary forest	Public (municipality)	Mowing of grasslands, pruning of trees	245.8	276	90.5	89.2	56.0	0.20	0.23
	Secondary forest	Private (landowner)	Unmanaged							
Temple		Private (landowner)	Mowing of grasslands, pruning of trees	0.4	14	51.3	2.9	62.0	4.57	160.14

The green elements of Motoori Shrine in the Tonomachi District planning area are called *yoihonomori* and are regarded as a very important landscape element. In the district plan's direction for land use, a sentence, to develop the land use to utilize the natural environment such as the green elements of *yoihonomori* and peripheries. The total area of the Motoori Shrine is 276 (a), and the green coverage ratio is 89.2 %, with most trees more than 20 m high. These green elements are managed by two bodies: the municipality, which manages the peripheries such as mowing grasses and removing dead branches, and the temple, which does not manage any aspects of the green elements. As the result, the labor density of the area is very low, at 0.2 h/a.

Generally, the gardens of temples are managed just like residential gardens, with the managing works primarily consisting of pruning trees in a balanced form. In the studied cases, the management is done by professional gardeners and the labor density of the unit is 4.6 h/a. The difference between the labor density of temples and shrines is due to the differences in attitude to the nature of the two types of structures. Shrines honor gods who are believed to live in the forest, so they tend to retain the forest as it is, but the gardens of temples are representative of utopia and so their management is regarded as a part of training.

11.5 Water Channel

The management of the historical water channel is done by the resident organization and directed by a parent association named Aishinkai. Table 11.4 shows the labor accounts and densities of the management of the water channels done on September 1, 2014. The participants were comprised of the 2000 individuals who represent one person from each household.

The labor contents involve mud recap and mowing grasses alongside the rivers. The collaborative works are done on the main water channels and streams in the back sides of residential sites. The working areas are divided according to the territories of each resident association. Most of the water channels are protected with concrete now but some stretches retain their stone masonry structures. These stone masonry structures are regarded as having high value but it is more effort to keep them clean. The labor density of this work is 39.7 h/a and very high. Even though the time required to do the work is short (approximately 3 h), the character of the mandated collaborative work pushes up the labor density.

Table 11.4 Landscape labor accounts and density of water channels

Landscape unit	Organizations	Labor content	S_l (a)	Length (m)	Number of work people	Working hours per day (h)	Annual Working day (day)	TLA_l (h)	GLD_l (h/a)	GLD_l (Length) (h/m)
Water channel	10 neighborhood associations (1 or more people per household)	Dredging operations, mowing of grassland	151	3568	2000	3	1	6000	39.70	1.68

11.6 Summary of Labor Accounts and Densities

Figure 11.6 shows the gross labor densities of the different landscape units (i.e., historical heritages, temples and shrines, and rivers and water ways in the district), while Fig. 11.7 shows the personal annual labor accounts and densities. According to Fig. 11.6, the landscape labor densities of historical heritages and temples are

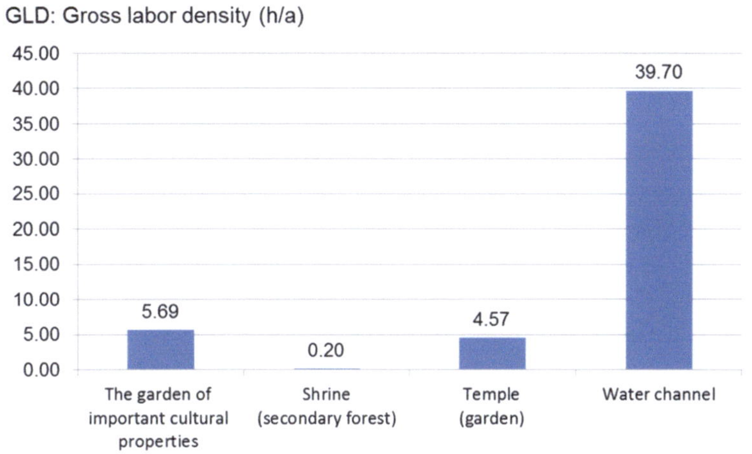

Fig. 11.6 Summary of the amount of gross labor densities

Fig. 11.7 Personal annual labor accounts and density

Table 11.5 Overview of landscape labor accounts and densities in Matsusaka chuo district

Landscape unit		Landscape element	Organization	A_l (a)	S_l (a)		TLD_{jl} (h/a)		TLA_{jl} (h)	NLD (h/a)	GLD (h/a)	SLI
Parks	Tree and grassland	Public (municipality)		407	743	743	0.81		330	0.81	0.44	7.9
Built-up areas	Public facilities	Tree and grassland	Public (municipality)	110	1183	8445	0.32		35	0.32	0.03	
	Cultural heritage building	Tree and grassland	Public (municipality)	83	151		9.65		804	9.65	5.34	
	Shrine	Secondary forests	Private (landowner)	239	309		0.23	4.88	1599	4.88	2.35	
	Temple	Tree and grassland	Private (landowner)	89	369		160.14					
	Private building lots	Tree and grassland	Private (landowner)	504	5025		38.30		19,321	38.30	3.85	
	Parking lots	Tree and grassland	Private (landowner)	37	1407		0.00		0	0.00	0.00	
Vacant land		Grassland and tree	Private (landowner)	28	79	79	1.97		54	1.97	0.69	
Farmland	Other crop field	Other crop field	Private (landowner)	9	12	23	29.77		279	29.77	22.80	
Water surface	Rivers (water surface)	Water surface	Public (municipality)	0	82	300	0.00		0	0.00	0.00	
	Dry riverbed	Tree and grassland	Public (municipality)	67	67		1.97		132	1.97	1.97	
	Water channel	Water surface and grassland	Public (neighborhood associations)	151	151		39.70		6000	39.70	39.70	
Rail road	Rail road	Non-element	Public (municipality)	40	413	413	1.97		79	1.97	0.19	
Road	Road	Tree	Public (municipality)	64	1813	1813	0.00		0	0.00	0.00	
Total				1828		11,815			28,633	15.66	2.42	7.9

almost same because the labor contents carried out by professional gardeners are similar. Conversely, the labor density of shrines is very low due to the attitude toward the nature of shrines, as described in Sect. 11.4 above. The labor density of the water channels is very high because of the obligated collaborative works of residents, which involves approximately 2000 participants. Therefore, the personal labor density of the water channels is very low, shown in Fig. 11.7. Regarding personal labor accounts, most of the amounts are very low, which means that the management of these landscape units is done by residents, with professional gardeners comprising only a small part of the total annual activities.

Table 11.5 shows the results of the estimation of the landscape labor accounts for all of Matsusaka Chuo District, which totals 28,633 h.

11.7 Consideration of Future Landscape Management

The calculated labor accounts were compared to the projected future population in the total area and management problems with methodology in the future were also considered. Figure 11.8 shows the forecasted future population, which was calculated by using individuals born during the same period with population datasets from 2005 and 2010 in Matsusaka Chuo District from the national census. For example, in 2005, the number of people was 3781, and by 2045, the population is expected to decrease by half (about 1657 people). Figure 11.9 shows the calculated social intensities of private sites and water channels in 2005 and 2045 based on the calculated landscape labor accounts in 2014 (Table 11.5).

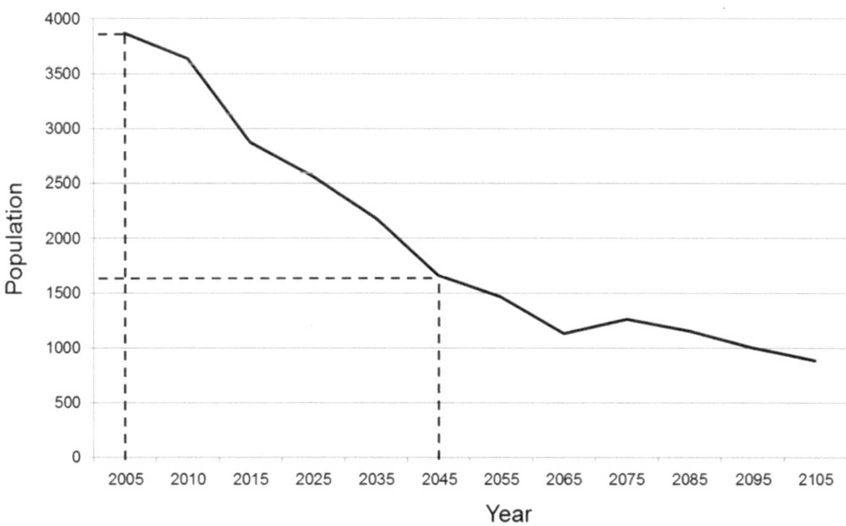

Fig. 11.8 Future population in Matsusaka Chuo District

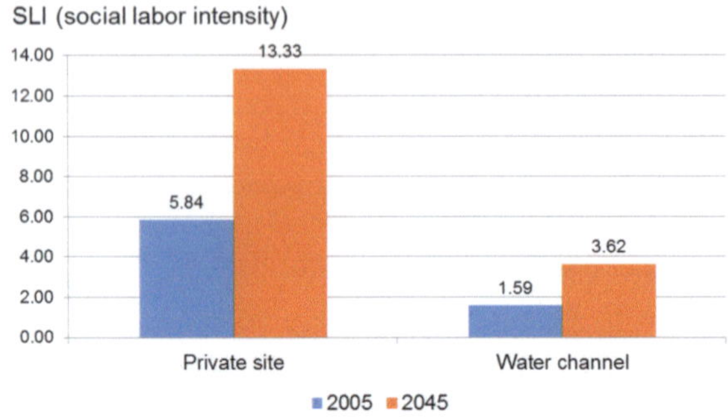

Fig. 11.9 Future social labor intensity of private sites and water channels

The result of the future social intensities reveals a future lack of labor accounts, which are necessary for the conservation of cultural services at private gardens and water channels. Therefore, in the future, gardens that are currently private site should be changed to public conservation sites and water channels will require more participants for maintenance. Some of the private houses will become vacant due to shifts in the labor force and will change to parking lots or vacant land.

For sustainable land management, the following methodologies should be considered:

(1) Land that has already been changed to vacant lots has lost its historic landscape; however, these areas can be managed as public parks by multiple people at the whole city level or neighborhood association level.
(2) The gardens of historical heritage sites should, as far as possible, be conserved. The aging of landowners has created a large burden within the labor force in treating their own green spaces. It is important that financial aid is prepared to employ the use of professional gardeners; in addition, the subsidy for traditional hedges should be increased to 1000 yen per meter (see Sect. 11.3).

In the future, the local downtown area, in which population is decreasing, It is hoped that the above findings regarding landscape management labor accounts will contribute to a better understanding of improvement of future management in the local downtown area, where populations are decreasing.

Reference

Matsusaka Ekimae dori Shopping District Promotion Association (2006) The second business report. Matsusaka Ekimae dori Shopping District Promotion Association, p 22

Chapter 12
Case Studies in Urban Forest Neighborhood Landscape Complex Units: Fujimaki Town

Nozomi Fujiwara and Chika Takatori

Abstract This chapter examines the management of a park and green sites in Fujimaki Town, Nagoya City, Aichi Prefecture, as one example of urban neighborhood landscape complex units with rich green sites. Fujimaki Town is located in the eastern hill of Nagoya, which has rich secondary forests primarily comprised of *Quercus variabilis* and *Quercus serrata*. These are the most common vegetation types in the secondary forests of Satoyama landscapes located in the central part of Japan. This town is designated as a park area in urban planning, but is not well-established due to financial difficulties of Nagoya City. In recent years, more than 400 people and 200 households live in Fujimaki Town. In this chapter, the landscape management of secondary forests by municipalities, volunteers, and residents is clarified and a future sustainable management scenario will be considered from the viewpoint of labor accounts of secondary forests.

Keywords Urban forest neighborhood landscape complex units · Secondary forest · Environmental · Preservation activity · Sustainable management for urban green · Land management

12.1 Introduction

This case study describes the management of a park and green sites in Fujimaki town, Nagoya city, Aichi Prefecture, as one example of the urban neighborhood landscape complex units with rich green sites. Fujimaki town is located in the eastern hill of Nagoya (Fig. 12.1). The eastern hill of Nagoya has rich secondary forests where *Quercus variabilis* and *Quercus serrata* are mainly found. They are the most common vegetation in the secondary forests of Satoyama landscapes located in the central part of Japan. Fujimaki town also has rich secondary forests

N. Fujiwara · C. Takatori (✉)
Graduate School of Environmental Studies, Nagoya University,
Furo-cho, Chikusa-ku, Nagoya, Aichi 464-8601, Japan
e-mail: takatori@corot.nuac.nagoya-u.ac.jp

H. Shimizu et al. (eds.), *Labor Forces and Landscape Management*,
DOI 10.1007/978-981-10-2278-4_12

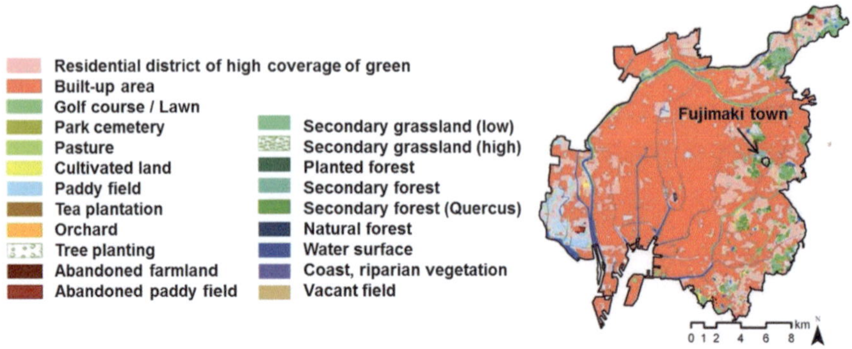

Fig. 12.1 Study area

with mostly *Quercus variabilis*, and the forests have formed the main scenery of this town. In recent years, more than 400 people normally inhabit this town in over 200 households. This town is designated as a park area in urban planning, but has not been established for a long time because of the financial difficulties of Nagoya city. This kind of area is called an "unestablished park". The ratio of private land in Fujimaki town is comparatively higher than that of other unestablished parks in Nagoya. In the recent urban master plan, the Nagoya government specified the dense forests covering from Fujimaki town to the neighboring park, called "Higashiyama park," as needing protection, but actually the management planning and system have not been designed yet. The forests covering Fujimaki town are becoming more and more devastated, and it causes various problems such as a decline in public safety, outbreaks of harmful insects, and worsening of living environments. Given such circumstances, local residents have intensely discussed in town planning meetings of the Fujimaki neighborhood association with the aim to design a plan for green space management with the goal of proposing it to Nagoya city.

12.2 The Landscape Units and Organizations

In this study, we created the land use data of Fujimaki town based upon several sets of data (the Basic Survey of City Planning (Nagoya city 2012a, b), the Fundamental Geospatial Data (Road and building edge data (Geospatial Information Authority of Japan 2009a, b)) and the Digital elevation data (Geospatial Information Authority of Japan 2008)), and clarified the town into types as landscape units, and calculated the area of each landscape unit in the ArcGIS software. As in Figs. 12.2 and 12.3, the landscape units in the town are classified into 5 types; private building lot of high coverage of green, park, secondary forest, bamboo forest, and road. The area and management organization of each landscape unit are given in Table 12.1.

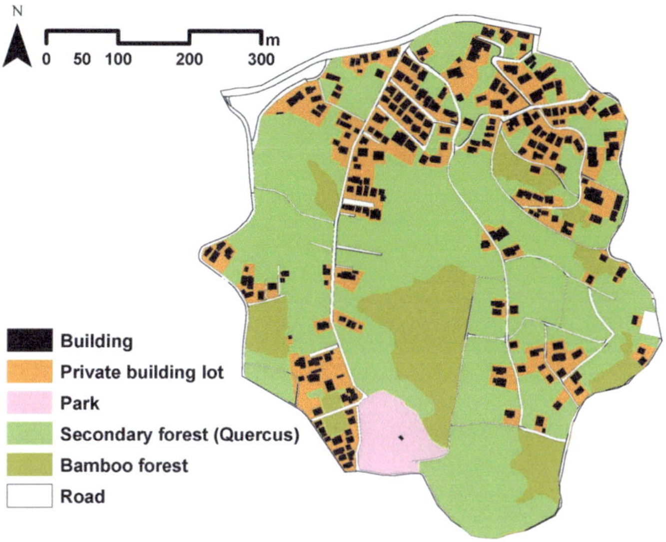

Fig. 12.2 Map of landscape units

The town almost completely consists of private building lots, secondary forests, and bamboo forests. Green spaces in the town are managed by three main organizations: the municipality of Nagoya, volunteers who belong to an NPO called "Team Fujimaki of Higashiyama no Morizukuri no Kai", and residents in the town. As in Table 12.1, gardens in private building lots are managed by residents, and secondary forests are managed by the municipality and volunteers. Bamboo forests are managed by volunteers. Parks are managed by the municipality. Moreover, the activities of volunteers can be mainly classified into 2 types. One is the activity as a team, and the other is the activity as individuals. In this study, we investigated vegetation at Fujimaki town in 2014, and clarified the annual labor account and annual labor density of each landscape unit by interviews with these organizations about how they managed green spaces in 2013.

12.3 Vegetation

Fujimaki town has rich green spaces in forests and private building lots. The green spaces covering the town are getting more and more devastated, and we can see the various problems: a decline of public safety, outbreaks of harmful insects, worsening of living environments and so on. According to the vegetation survey at the town in 2014, it was clarified that these problems were broadly classified into 2 issues. The first issue is that the management of the forest's health is needed. In this town, *Quercus variabilis* are weakened and infirm with age, evergreens are

Fig. 12.3 Photos of landscape units

overgrown which causes dark forests, and rapidly-growing bamboos are spreading throughout the whole town (Fig. 12.4). The second issue is that the management of green spaces around houses and roads is needed to assure safe and comfortable living environments because trees and branches often fall down near houses and roads, and grasses are also growing wildly there. They cause uncomfortable living environments (Fig. 12.5).

12.4 Calculation of Landscape Labor Accounts

12.4.1 Private Building Lot

More than 400 people in over 200 households live in Fujimaki town, and this town has over 250 buildings. Many of them are detached houses with a garden. As in Table 12.1, the total area of private building lots is 728 are as landscape units, and it can be divided into building areas and garden areas as landscape elements, the area of each landscape element is 246 are as building areas, and 482 are as garden areas,

Table 12.1 Landscape units/organizations

Landscape unit		Landscape element of each type of units	Organization	A_l	S_l
Built-up area	Private building lot	Garden	Residents	482	728
	Park	Tree and grassland	Municipality	77	92
Forest	Secondary forest (*Quercus*)	Tree and grassland	Municipality	1,522	1,522
			Volunteer (Team Fujimaki)		
			Volunteer (Individuals)		
	Bamboo forest	Tree and grassland	Volunteer (Team Fujimaki)	391	391
			Volunteer (Individuals)		
Road		—	—	0	293
Total area					3,025

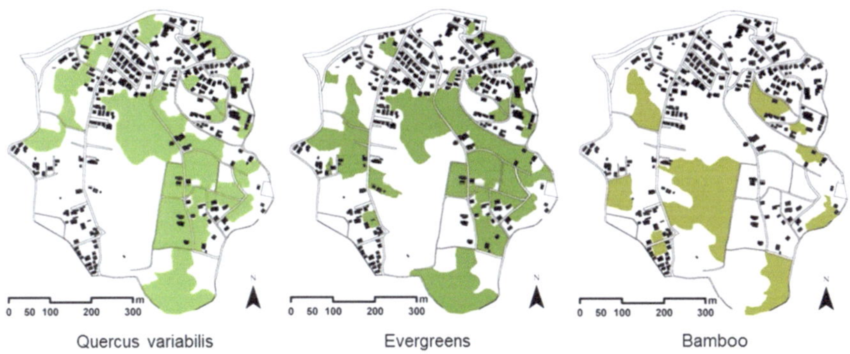

Fig. 12.4 Maps of vegetation

respectively. In this study, we found 6 samples of garden management in the town through interviews. As in Table 12.2, labor contents of garden management are mainly garden tree trimming, mowing, and kitchen garden. The labor account and labor density differ among the 6 samples such as the values of densities are from 6.9 to 85.1 h/a, because the values of labor account and labor density strongly depend on the resident's will for green management. The total area and labor account of 6 samples are 16.7 are and 515 h, and the labor density of a targeted element of 6 samples is 30.7 h/a. As in Table 12.3, this labor density gives a rough estimate of the total annual labor account and gross labor density of whole private building lots, and they were actually estimated at 14,824 h and 20.4 h/a.

Fig. 12.5 Photos of vegetation around houses and roads

Table 12.2 Individual data of 6 samples

Operator	Labor content	Garden area (a)	Labor account (h)		Labor density of each garden (h/a)
Resident 1	Garden tree trimming	1.8	13	13	**6.9**
	Mowing		6		
Resident 2	Garden tree trimming	1.2	10	105	**85.1**
	Mowing		15		
Resident 3	Garden tree trimming	4.6	14	115	24.9
	Mowing		39		
Resident 4	Garden tree trimming	2.0	0	83	40.4
	Mowing		0		
	Kitchen garden		0		
Resident 5	Garden tree trimming	2.5	0	138	55.2
	Mowing		0		
	Kitchen garden		0		
Resident 6	Garden tree trimming	4.5	0	62	13.7
	Mowing		0		
	Kitchen garden		0		
Total		**16.7**	**515**	—	
Labor density of a targeted element (h/a)					**30.7**

12.4.2 Secondary Forest

The secondary forest is the biggest landscape unit in this town. The total area of secondary forests is about 1,522 are, and it accounts for over 50 % of the town (Table 12.1). Private lands and public lands in the town are complexly mixed, and a part of the public forests is managed by the municipality and volunteers. As in Table 12.4, the municipality now and then manages forests around houses and roads to assure safe and comfortable living environments, and the labor contents are mainly mowing, removing weakened trees, weakened branch trimming, and

Table 12.3 Labor account of private building lot

Landscape unit	Landscape element	Organization	Labor content	A_l (managed) (a)	A_l (a)	TLD_{jl} (h/a)	TLA_l (h)	NLD_l (managed) (a)	GLD_l (h/a)
Private building lot	Garden	Residents	Garden tree trimming	482	728	**30.7**	**14,824**	30.7	**20.4**
			Mowing						
		(sample:6)	Kitchen garden						

garbage collection. On the other hand, volunteers periodically manage the inside forests for the forest's health, and the labor contents are mainly tree felling and bush and vine removal. Regarding the annual labor account of each organization, the municipality has 996 h, volunteers (Team Fujimaki) have 90 h, and volunteers (individuals) have 380 h. The total annual labor account and the net labor density of a targeted unit are 1,466 h and 3.42 h/a respectively. But secondary forests that are managed account for about only 35 % of all secondary forests in this town, so the gross labor density of all secondary forests is not relatively high. It is only 0.96 h/a.

12.4.3 Bamboo Forest

The area of the entire bamboo forest is about 391 are, and it accounts for about 13 % of this town (Table 12.1). But bamboo has a strong reproductive power, and the forests are encroaching wider on neighboring forests. The density of bamboo which is not managed is about 100 bamboos per 1 are, therefore there is not much vegetation in the lower layer of these bamboo forests. As in Table 12.5, a part of the public bamboo forests is managed by volunteers, and the labor content of both organizations is mainly bamboo felling. Regarding the annual labor account of each organization, volunteers (Team Fujimaki) have 50 h, and volunteers (individuals) have 92 h. The total annual labor account and net labor density of a targeted unit are 142 h and 2.24 h/a. But many bamboo forests are in private lands and they had been hardly managed because of the absence of landowners or the lack of their will for green managements so the gross labor density of whole bamboo forests is not relatively high. It is only 0.36 h/a.

12.4.4 Park

As mentioned above, Fujimaki town is an unestablished park, but one basic park for the community is located in the town. The area of the whole basic park is about 92 are (Table 12.1). It is relatively bigger than other typical basic parks, and it has park

Table 12.4 Labor account of secondary forest

Landscape unit	Landscape element	Organization	Labor content	A_l (managed) (a)	S_l (a)	TLA_l (h)			NLD_l (managed) (h/a)	GLD_l (h/a)
Secondary forest	Tree and grassland	Municipality	Mowing	429	1,522	630	**996**	**1,466**	**3.42**	**0.96**
			Weakened tree felling			328				
			Weakened branch trimming			28				
			Garbage collection			11				
		Volunteer (Team Fujimaki)	Tree felling			90	**90**			
		Volunteer (Individuals)	Tree felling			144	**380**			
			Bush and vine felling			236				
		–	–	1,093		–			–	

Table 12.5 Labor account of bamboo forest

Landscape unit	Landscape element	Organization	Labor content	A_l (managed) (a)	S_l (a)	TLA_l (h)		NLD_l (managed) (h/a)	GLD_l (h/a)
Bamboo forest	Tree and grassland	Volunteer (Team Fujimaki)	Bamboo felling	63	391	**50**	**142**	**2.24**	**0.36**
		Volunteer (Individuals)				**92**			
		Unmanaged		327		–	–	–	

athletics and plenty of grounds. As in Table 12.6, the park is regularly managed by the municipality, and labor contents are mainly lawn mowing and hedge trimming. Regarding the labor account of each labor content, lawn mowing is 68 h, and hedge trimming is 6 h. The total annual labor account, net labor density, and gross labor density of a targeted unit are 74 h, 0.96 and 0.81 h/a.

12.4.5 Summary of Landscape Labor Accounts

The labor account and density of every landscape unit is given in Tables 12.7, 12.8 and Fig. 12.6. In 2013, the landscape unit which has the maximal labor account is secondary forests. They were managed by the municipality and volunteers. But forests in private lands had been hardly managed because of the absence of landowners or the lack of the management will of them. Actually, secondary forests that are managed account for about only 35 % of all the secondary forests in this town, so the labor density is not relatively high. On the other hand, the landscape unit which has the maximal labor density is garden in private building lots. In fact, the labor account and density of each garden is different from individual to individual, but the values of the labor density are relatively high. The reason is that labor contents are mainly tree trimming and mowing, and these need much labor per one are or one tree. Furthermore, the area of these gardens is not very large, and much labor is invested in the small and limited areas. The total labor account in 2013 of all landscape units is 16,506 h, and the gross labor density of the whole town as a complex unit is 5.46 h/a. Moreover, there are 415 residents in the town,

Table 12.6 Labor account of a park

Landscape unit	Landscape element	Organization	Labor content	A_l (managed) (a)	S_l (a)	TLA_l (h)		NLD_l (managed) (h)	GLD_l (h)
Park	Tree and grassland	Municipality	Lawn mowing	77	92	68	**74**	**0.96**	**0.81**
			Hedge trimming			6			

Table 12.7 Labor densities of urban forest neighborhood landscape complex unit

Landscape unit		Landscape element	Organization	A_l (managed) (a)	A_l (un-managed) (a)	A_l (total) (a)	S_l (a)	TLD_{jl} (h/a)	TLA_l (h)	NLD_l (managed) (h/a)	NLD_l (total) (h/a)	GLD_l (h/a)	NLD (h/a)	GLD (h/a)	SLI
Built-up area	Private building lot	Garden	Residents	482	0	482	728	30.7	14,824	30.7	30.7	20.37	**6.68**	**5.46**	**37.6**
	Park	Tree and grassland	Municipality	77	0	77	92	0.96	74	0.96	0.96	0.81			
Forest	Secondary forest (*Quercus*)	Tree and grassland	Municipality Volunteer (Team/Individuals)	429	1,093	1,522	1,522	3.42	1,466	3.42	0.96	0.96			
	Bamboo forest	Tree and grassland	Volunteer (Team/Individuals)	63	327	391	391	2.24	142	2.24	0.36	0.36			
Road		Paved surface	Municipality	0	0	0	293	0.00	0	0.00	0.00	0.00			
Total				**1,052**	**1,421**	**2,472**	**3,025**		**16,506**						

SLI population: 415

Table 12.8 Net labor densities (managed) of each organization

Landscape unit	Organization	A_l (managed) (a)	TLA_l (h)	NLD_l (managed) (h/a)
Private building lot	Residents	482.4	14,824	30.7
Secondary forest	Municipality	249	996	4.00
	Volunteer (Team Fujimaki)	101.5	90	0.89
	Volunteer (Individuals)	236.1	380	1.61
Bamboo forest	Volunteer (Team Fujimaki)	63.4	50	0.79
	Volunteer (Individuals)	127	92	0.72
Park	Municipality	77	74	0.96

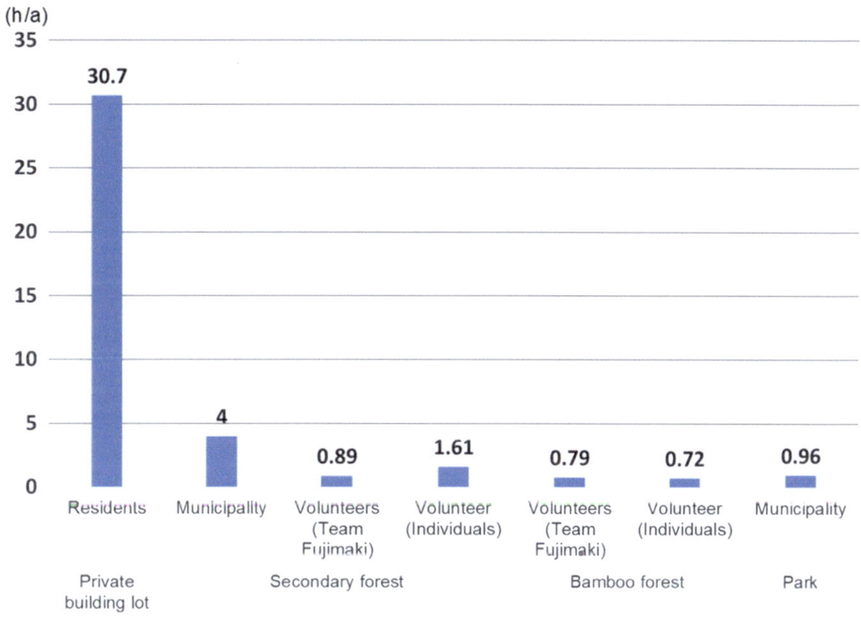

Fig. 12.6 Net labor densities (managed) of each organization

and the social labor intensity is 37.6 h. Compared with the fact that forests around houses and roads have been managed by the municipality, inside forests such as secondary and bamboo forests have been managed by volunteers. In addition, Team Fujimaki is composed of about 20 members, and most of them are residents in Fujimaki town. It was found from these facts that residents in Fujimaki town have strongly contributed to the green management. But compared with the whole area of forests in Fujimaki town, the areas which have been managed are actually only a part of the forests. That means it cannot be said that the present management of forests in Fujimaki town is sufficient.

12.4.6 Personal Labor Accounts and Densities

In this part, the features of a personal labor account and density among 3 types of organizations (municipality, residents, and volunteers) are described. From Fig. 12.7, the three following features are founded. On secondary forest and bamboo forest, it is found that (1) personal labor accounts of municipalities and volunteers (individuals) are comparatively high; they are around or more than 100 h. Meanwhile, personal labor densities of them are low; they are around or less than 1.0 h/a. Furthermore, (2) personal labor accounts and densities of volunteers (Team Fujimaki) are obviously low; they are less than 10 h and around 0.1 h/a. On the other hand, on private building lots, it is found that (3) Personal labor accounts of residents take various values; they are from 13 to 105 h. Meanwhile, personal labor densities of them are high in comparison with secondary forest and bamboo forest; they are e from 6.9 to 85.1 h/a.

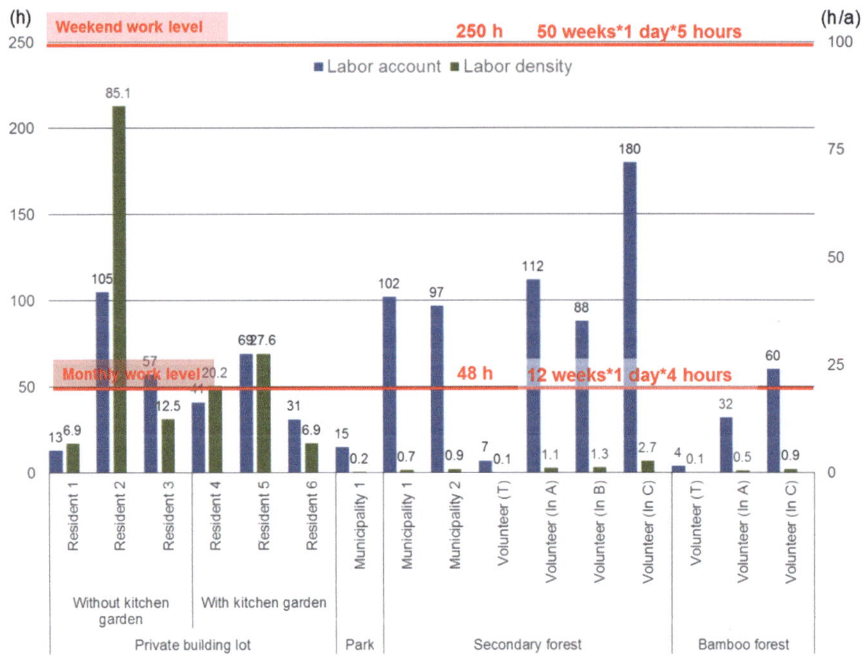

Fig. 12.7 Personal labor accounts and densities in Fujimaki town

12.5 Consideration of Landscape Management in the Future

12.5.1 Method for Creating the Future Scenario

In this study, the labor accounts and labor densities in the future are estimated. The flow of creating the future scenario is given in Fig. 12.8. First of all, Fujimaki town is divided into some areas by using the vegetation data, the Digital elevation data (Geospatial Information Authority of Japan 2008), and the Land-ownership data (Fujimaki neighborhood association 2014), and roles and functions are given to each area by hearing the resident's opinions in the workshop. Secondly, these areas are categorized into some forest types which are specified in the guideline given by the municipality (Nagoya city 2012a, b). This guideline specifies how to manage each forest type. Thirdly, the annual management program is planned along with the guideline. Finally, the labor account and density are calculated along with standard books of green space management (Nagoya city 2014 and the Research society for green space managements 1994).

12.5.2 Making the Area Map of the Future

As the process of making the area map, Fujimaki town is firstly divided into three zones (Fig. 12.9). They are named as "the town where residents exist with nature in harmony", "the bright forest", and "the dark forest". The details of the basic concepts of these zones are shown below; Zone A is the place where forests around houses and roads are managed to assure safe and comfortable living environments, Zone B is the place where large varieties of vegetation and recreation places are

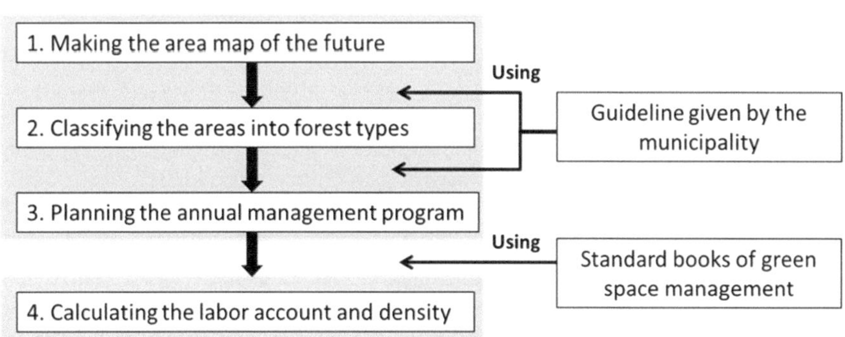

Fig. 12.8 The flow of creating the future scenario

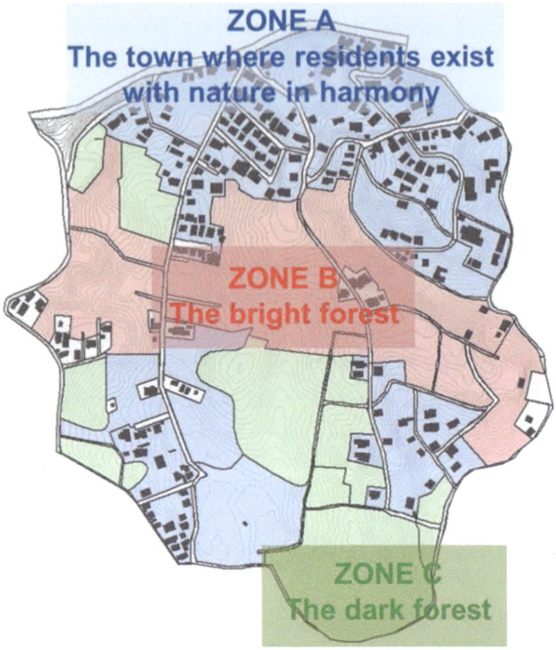

Fig. 12.9 Map of zones in the future scenario

managed for visitors, and Zone C is the place where dense forests are managed to preserve small animals' habitats and characteristic sceneries in the town. And furthermore, these zones are divided into some areas based on these concepts. The details of roles and functions of the areas are described as the name of each area in Fig. 12.10.

12.5.3 Calculating Labor Accounts and Labor Densities

As the process of calculating the labor accounts of the future scenario, firstly the areas which are shown in Sect. 12.5.2 are classified into some forest types by using the guideline (Nagoya city 2014) which specifies how to manage several forest types (Fig. 12.11). The forest types are named as "private building lot", "Satoyama forest", "transition forest", "evergreen forest", "bamboo forest", and "park". Secondly, the annual management programs of forest types are planned along with the guideline, and the labor account and density of each forest type are calculated by using a standard book (Nagoya city 2014; Research society for green space managements 1994) (Table 12.9). And furthermore, the labor contents are classified into 2 types; initial work and routine work. The former type refers to the contents which are needed for the first 3 years at the beginning of management, and the later refers to the contents which are needed at all stages of management. The details of management concepts of forests are described in Table 12.9.

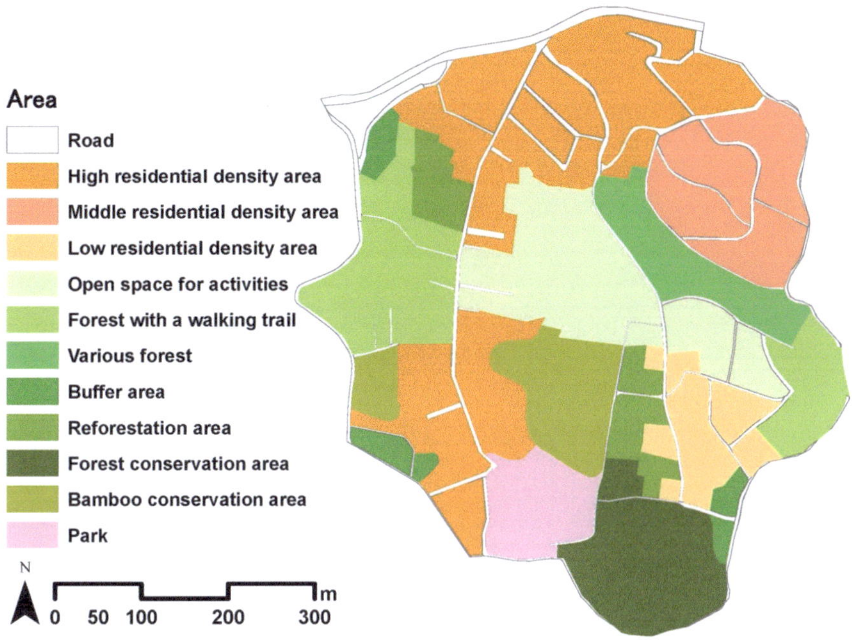

Fig. 12.10 Map of areas in the future scenario

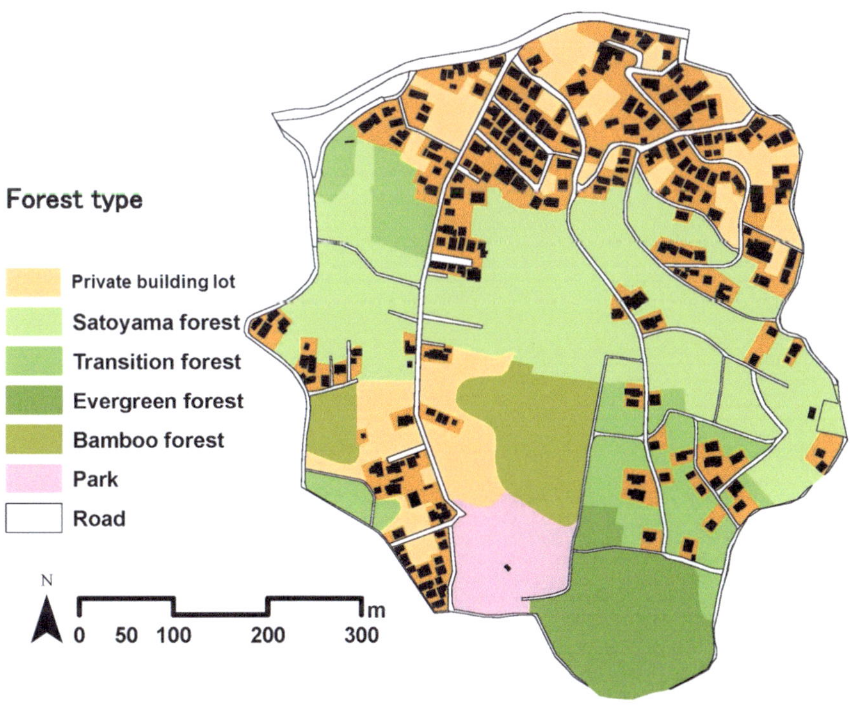

Fig. 12.11 Map of forest types in the future scenario

Table 12.9 The annual management programs of forest types

Landscape unit in present		Forest type in future		A_1 (a)	Management Type	Labor contents	Frequency	A_1 (detail)		TLA (h) (*Ave.*)		TLA (h)	Management concept
Built-up area	Private building lot	Private building lot	Garden	482	R	Branch trimming Mowing Kitchen garden	For 1 year	482	a	30.7	h/a	14,824	Continuing the present management 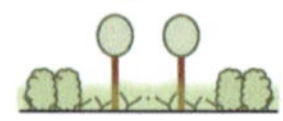
			Buffer	89	R	Mowing Tree felling Branch trimming		89	a	4	h/a	356	
	Park	Park	Forest	115	R	Lawn mowing Scrub trimming	For 1 year	115	a	0.8	h/a	92	Continuing the present management 
			Buffer	10	R	Mowing Tree felling Branch trimming		10	a	4.0	h/a	40	
Forest	Secondary forest	Satoyama forest	Forest	695	I	Thinning	For 3 years	3,475	trees	1.8	h/tree	2,085	Tree density 20 trees/a 15 trees/a 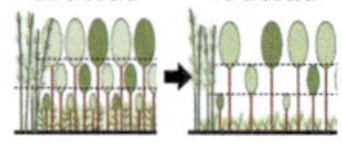 Present Future
					I	Burying barriers		690	m	0.5	h/m	115	
					R	Undergrowth mowing Scrub trimming	For 1/2 year	695	a	0.4 2.5	h/a	4,031	
			Buffer	122	R	Mowing Tree felling Branch trimming	For 1 year	122	a	4	h/a	488	
		Evergreen forest	Forest	222	I	Deciduous tree felling	For 3 years	888	trees	1.8	h/tree	533	Felling deciduous trees (4 trees/a) 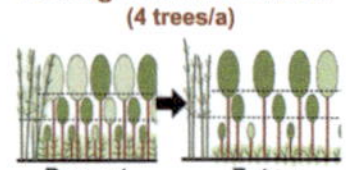 Present Future
					I	Burying barriers		200	m	0.5	h/m	33	
					R	Scrub trimming		222	a	2.5	h/a	555	
			Buffer	15	R	Mowing Tree felling Branch trimming	For 1 year	15	a	4	h/a	60	
		Transition forest	Forest	210	I	Burying barriers	For 3 years	120	m	0.5	h/m	20	Tree density 20 trees/a 20 trees/a 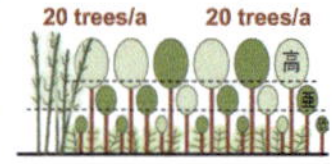
					R	Undergrowth mowing	For 1 year	210	a	0.4	h/a	168	
			Buffer	63	R	Mowing Tree felling Branch trimming		63	a	4	h/a	252	
	Bamboo forest	Bamboo forest	Forest	185	I	Thinning	For 3 years	7,400	trees	0.2	h/tree	493	Suppressing bamboo invasion Tree density 100 trees/a 50 trees/a 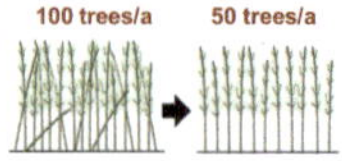 Present Future
					I	Burying barriers		420	m	0.5	h/m	70	
					R	Undergrowth mowing	For 1 year	185	a	0.4	h/a	74	
					R	Bamboo felling		1,850	trees	0.2	h/tree	370	
			Buffer	8	R	Mowing Tree felling Branch trimming		8	a	4	h/a	32	

R Routine work
I Initial work
Ave. average

12.5.4 Comparison of the Annual Labor Accounts and Labor Densities

As in Table 12.10 and Fig. 12.12, the present labor account of the whole town is 16,506 h, and the future labor account of that is 21,342 h. It seems possible that the

Table 12.10 The results of annual labor account and labor density of forest types

Present									Future										
Landscape unit		A_l (man aged) (a)	A_l (unman aged) (a)	A_l (total) (a)	S_l (a)	Present work			Forest type	A_l (man aged) (a)	A_l (unman aged) (a)	A_l (total) (a)	S_l (a)	Initial work		Routine work		GLD (1-3 years) (h/a)	GLD (after 3 years) (h/a)
						TLA_l (h)	TLD_{jl} (man aged) (h/a)	GLD (h/a)						TLA_l (h)	TLD_{jl} (man aged) (h/a)	TLA_l	TLD_{jl} (man aged) (h/a)		
Built-up area	Private building lot	482	0	482	728	14,824	30.7	5.46	Private building lot	571	0	571	1,087	0	0.0	15,180	26.6	8.16	7.06
	Park	92	0	92	92	74	0.8		Park	125	0	125	125	0	0.0	132	1.1		
Forest	Secondary forest	429	1,093	1,522	1,522	1,466	3.4		Satoyama forest	817	0	817	817	2,200	2.7	4,519	5.5		
									Evergreen forest	237	0	237	237	566	2.4	615	2.6		
									Transition forest	273	0	273	273	20	0.1	420	1.5		
	Bamboo forest	63	327	391	391	142	2.2		Bamboo forest	193	0	193	193	563	2.9	476	2.5		
Road		0	0	0	293	0	0.0		Road	0	0	0	293	0	0.0	0	0.0		
Total		1,066	1,421	2,487	3,025	16,506			Total	2,216	0	2,216	3,025	3,349		21,342			

differences of these values are not big because the labor account of private building lots is too much. But if you pay attention to other forest types without private building lots in Fig. 12.13, it is obvious that the labor accounts of the present and future make a big difference. The present labor account of the whole town without private building lots is 1,682 h, and the future labor account of that is 9,511 h. The future is about 5.7 times more than the present. Especially, the labor account of the 3 forest types, “Satoyama forest”, “bamboo forest”, and “evergreen forest”, shows a significant increase, and these forest types are also the top three forest types with the greatest labor account of the future as in Fig. 12.14. The reason is that further management areas and labor contents are added in the management programs. The three forest types comparatively require much labor during the initial work. That means these forest types seem to need a lot of labor for the first 3 years at the beginning of management. The gross labor density of the whole town is 5.46 h/a in the present and 8.16 h/a in the future. But in this study, initial works are set as labor contents carried out within the initial 3 years, therefore the calculated results, 24,692 h and 8.16 h/a, mean the labor account and the labor density which are needed for the first 3 years at the beginning of management, and then they will be 21,342 h and 7.06 h/a after 3 years as routine work.

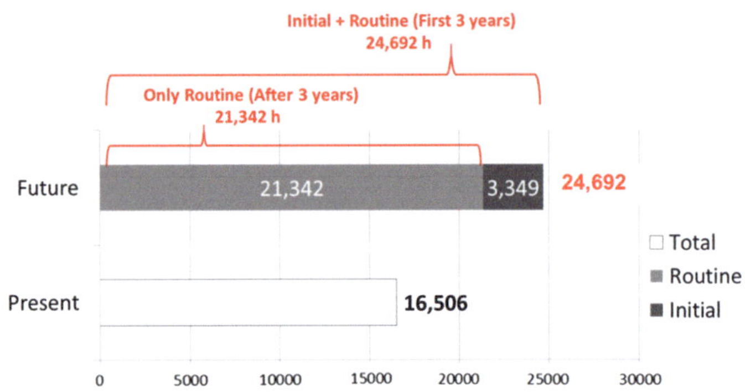

Fig. 12.12 Annual labor account of the whole town

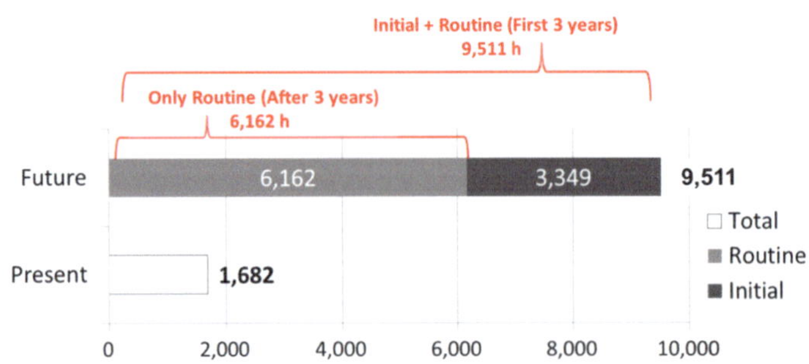

Fig. 12.13 Annual labor account of the whole town without private building lot

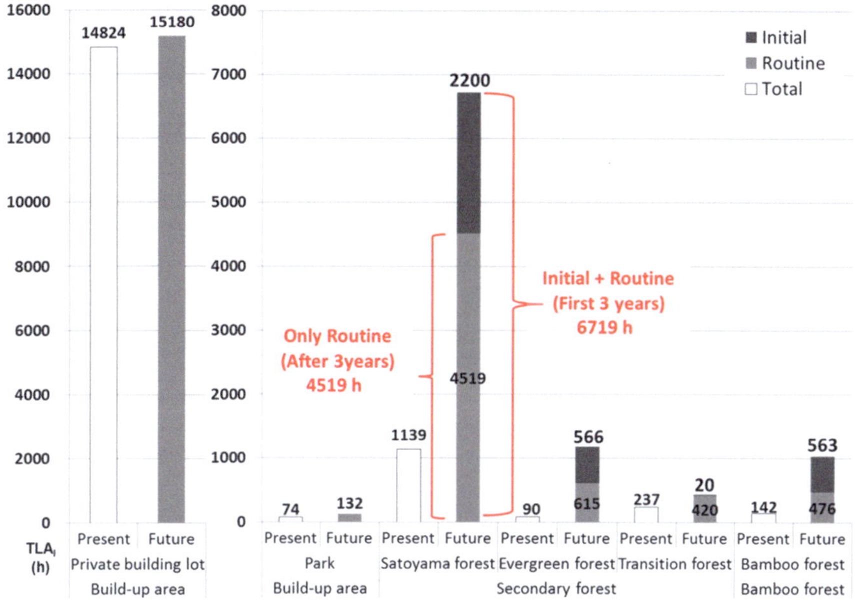

Fig. 12.14 Comparison with the annual labor account of the present and future

12.6 Summary of This Study

In this study, it is revealed that the labor account of the future, especially for the first 3 years, is about 5.7 times as much as that of the present. The social labor intensity of the present is about 40 h, and those of the future might be 59 h as management for the first 3 years, and 51 h as management after the first 3 years as in Table 12.11. Therefore it seems to be impossible that only the municipality and volunteers can take care of all labor accounts of the future. With these facts, more efforts to promote residents' participation in green management and a new management system are very important. Actually, it can be said that residents in Fujimaki contribute to the green management currently because 11 of the 13 members of Team Fujimaki are living in the town as mentioned above, but 11 is only 2.7 % of all the residents. In addition, most of them are the residents who live inside the forest, and they tend to participate in the management, compared with the

Table 12.11 Comparison with SLI of the present and future

Population 415 in 2011	Present	Future	
	2013	First 3 years	After 3 years
TLA (h)	16,506	24,692	21,342
GLD (h/a)	5.46	8.16	7.06
SLI (h)	40	59	51

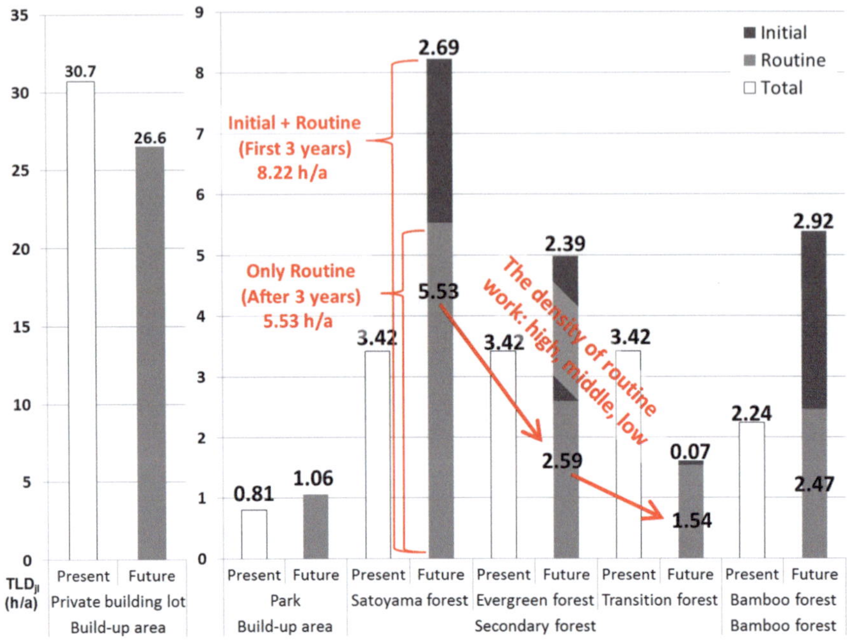

Fig. 12.15 Comparison with the annual labor density of the present and future

residents who live outside the forests. As in Fig. 12.15 of this study, the secondary forests are broken down into 3 forest types based on the concepts of management intensity, namely the places where manpower should be invested are selected and concentrated. This plan is useful and important for reducing manpower as long as sustainable management with civic collaboration is realized.

References

Fujimaki Neighborhood Association (2014) Land-ownership data. Licensed by Fujimaki neighborhood association

Geospatial Information Authority of Japan (2009a) Fundamental geospatial data, road edge data. http://fgd.gsi.go.jp/download/menu.php. Accessed 2014

Geospatial Information Authority of Japan (2009b) Fundamental geospatial data, building edge data. http://fgd.gsi.go.jp/download/menu.php. Accessed 2014

Geospatial Information Authority of Japan (2008) Fundamental geospatial data, digital elevation data. http://fgd.gsi.go.jp/download/menu.php. Accessed 2014

Nagoya city (2012a) Basic survey of city planning, land use data. Licensed by Nagoya city

Nagoya city (2012b) Nagoya no Mori-dukuri guideline (The reforestation guideline)

Nagoya city (2014) Doboku koji sekisan kijunsho (Guidelines for cost estimating for public works)

Research Society for Green Space Managements (1994) Standard book of green space managements, Economic Research Association

Chapter 13
Case Studies in Flatland Farmland Neighborhood Landscape Complex Units: Asami District

Yuki Murase, Jyun-Ichi Kitamura, Nobuko Kawaguchi, Hiroyuki Shimizu and Chika Takatori

Abstract This chapter discusses the case study of Asami District, located in the eastern part of Matsusaka City in Mie Prefecture between the Kushida River in the east and the Kongo River in the west, as a flatland farmland neighborhood landscape complex unit. Asami District was developed during the 8th century, and in some parts still uses Jori-sei, a traditional land allotting system, that encourages and maintains rich biodiversity in the area. However, as traditional paddy field allotting based on Jori-sei imposes a stronger level of labor to aged farmers, the modern agricultural land improvement project is progressing rapidly in this area. The loss of these undeveloped fields with ecological, historical, and cultural value could negatively impact the area; as such, a well-considered improvement method for sustainable land management of the area is required. As a consequence, the relationship between biodiversity and landscape management labor accounts is clarified and future scenarios that are balancing them are considered.

Keywords Flatland farmland neighborhood landscape complex units · Land improvement · Biodiversity · Socio-ecological production landscape · Traditional irrigation system

13.1 Introduction

In this chapter, landscape management labor accounts in a flatland farmland neighborhood landscape complex unit will be discussed. Farmlands in flatlands in Japan, especially those in alluvial plains, mainly consist of paddy fields and create

Y. Murase · N. Kawaguchi · H. Shimizu · C. Takatori (✉)
Graduate School of Environmental Studies, Nagoya University,
Furo-cho, Chikusa-ku, Nagoya, Aichi 464-8601, Japan
e-mail: takatori@corot.nuac.nagoya-u.ac.jp

J.-I. Kitamura
Mie Prefectural Museum, 3060 Isshiden-Kouzubeta, Tsu, Mie 514-0061, Japan
e-mail: kitamura@bio.sci.toho-u.ac.jp

H. Shimizu et al. (eds.), *Labor Forces and Landscape Management*,
DOI 10.1007/978-981-10-2278-4_13

typical Japanese countryside paddy field landscapes. The targeted area of this chapter is the Asami district. This district is located in the east part of Matsusaka City in Mie Prefecture between the Kushida River in the east and the Kongo River in the west (Fig. 13.1) and has a typical feature of a flatland paddy field neighborhood landscape complex unit.

During the 20th Century, most Japanese paddy fields have been, and are being, transformed to productive efficient allotments by the "Agricultural Land Improvement Projects", which were designated in the Land Improvement Act in 1949 and are being led by the Ministry of Agriculture, Forestry and Fisheries. The aims of the projects are improvement of agricultural productivity, an increase in total agricultural production and contribution to the selective expansion of

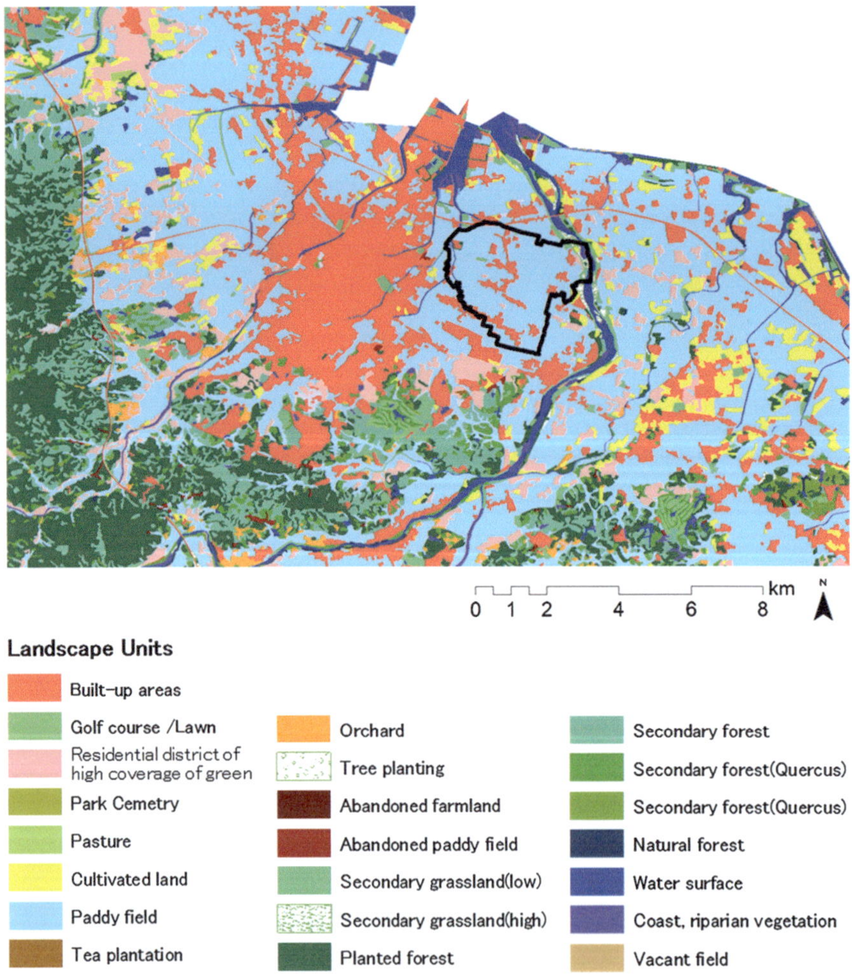

Fig. 13.1 Location of Asami district

agricultural structures. The main physical methods to improve paddy fields are the rectangular shaping of lands, rationalization of waterways by the separation of headraces and drainages, pipelining of headraces, covering of drainage with concrete, and constructing the level difference of paddy fields and drainage in order to improve early drying of fields. These physical improvements have succeeded in helping reduce labor forces and heavy lifting of farmers. Recently the need for intensification of paddy fields by these Land Improvement Projects has increased, due to the decline and aging of farmers and the decline in rice prices. Traditionally, the maintenance of waterways and farm roads was done by the local communities (Okamoto 2003). However, the intensification by this Land Improvement Act may collapse these systems and weaken local communities today.

On the other hand, these Land Improvement Projects have dramatically changed ecosystems. For example, a paddy field is important as a breeding, growth, and evacuation place for some types of fish (Saito et al. 1988). The paddy field of a traditional type has a low gap between the paddy field and waterways, and the fish could easily come into the paddy field. However, in the modern type paddy field, fish cannot move to the allotments due to separation by headraces and drainages and creation of gaps between paddy fields and waterways, and this has made it difficult for some types of wildlife to live (Katano et al. 2001). These defects are becoming well known, and some compensation engineering is employed, such as fish ladders, catchment ditches and so on (Natuhara 2013). However, separate introduction of such compensation engineering is not sufficient for the creation of a comprehensive and integrated preservation system at the landscape level. It is necessary to introduce an ecological landscape management system in which the distinction between the area where productivity efficiency is promoted by introducing some kinds of compensation technologies and the area where traditional paddy fields with rich biodiversity should be maintained.

The total area of Asami district, which is the focal site in this chapter, is 73,651 are, and the population and households in 2015 are 2146 and 845, respectively. The Asami district is a typical aged society and is suffering from a declining birthrate and aging population. The percentage of the population of age 65 or older is 30.8 % and is higher than that of the Japanese average of 25.06 % in 2013. It is said that the aging of the Asami district is progressing ten years earlier than the Japanese average. This district was developed in an ancient era in the 8th century, and still remains Jori-sei, a traditional land allotting system in some parts, and this traditional nature-near water supply system maintains rich biodiversity even today. However, as traditional paddy field allotting based on the Jori-sei system requires harder labor by aged farmers, the modern agricultural land improvement project is progressing rapidly in this area. The loss of these undeveloped fields with ecological, historical, and cultural value has high risk, and a well-considered improvement method for sustainable land management of the area is required. As a consequence, in this chapter, the relation between biodiversity and landscape management labor accounts will be clarified and the future scenarios will be considered.

13.2 The Landscape Units and Organizations

Figure 13.2 shows the landscape units in the Asami District. The management organizations of landscape units are shown in Table 13.1. There are three paddy field types in this district, i.e. a traditional Jori-sei paddy field type in the eastern area (Type I), a modern land improvement paddy field type adjusted in the early 20th century in the northwest area (Type II), and a new land improvement paddy field type which was completed in 2014 or is still undergoing development in the southeast area (Type III), shown in Fig. 13.2. Figure 13.3 shows the typical landscape photos of three types. According to Table 13.1, it can be well recognized that this district's traditional paddy field type, Type I, remains in large amounts. The paddy fields are managed by farmers, either an individual farmer or a corporate farmer, and the waterways and farm roads are managed by local communities in each village. As shown in Fig. 13.4, there are also two types of paddy fields in usage in the district. One is for only wet rice cultivation, and the other is the field for changing crops from rice cultivation to wheat and soybean cultivation. Table 13.2 shows the ratio of each crop. The other landscape units in this district are kitchen gardens and gardens inside and outside of residential land which are managed by residents, and the forests in temples and shrines of each village which are managed by local communities.

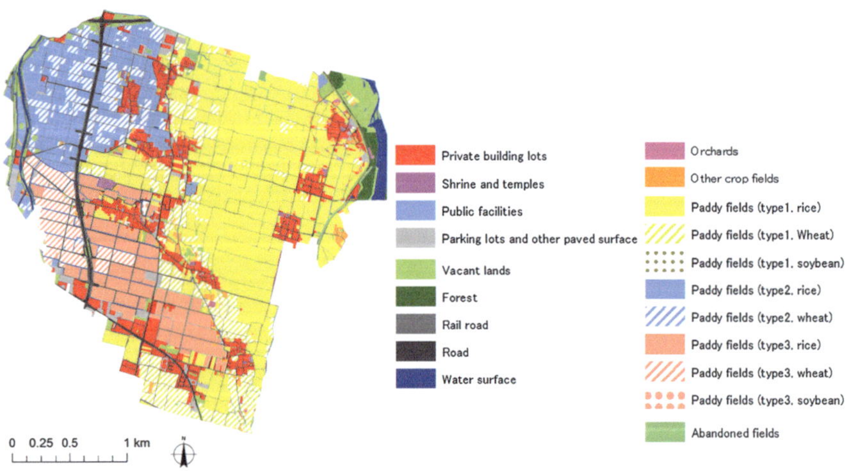

Fig. 13.2 Distribution of landscape units in Asami district

Table 13.1 Areas of each landscape unit in Asami district

Landscape unit		Landscape element	Organization	A_l (a)	S_l (a)	
Paddy fields (change of crops on paddy field of 3 products)	Paddy fields (type1)	Rice	Self-employment	21,645	25,000	37,759
			Farm corporation	3,355		
	Paddy fields (type2)		Self-employment	4,280	6,321	
			Farm corporation	2,041		
	Paddy fields (type3)		Self-employment	4,986	6,438	
			Farm corporation	1,453		
	Paddy fields (type1)	Wheat	Farm corporation	6,645	6,645	12,999
	Paddy fields (type2)		Farm corporation	3,394	3,394	
	Paddy fields (type3)		Farm corporation	2,960	2,960	
	Paddy fields (type1)	Soybean	Self-employment	19	19	70
	Paddy fields (type3)		Farm corporation	51	51	
	Grasslands		Self-employment	4	4	
Other crop fields		Other crop fields	Self-employment	1,078	1,141	
		Tree	Self-employment	63		
Orchards		Orchards	Self-employment	127	127	
Abandoned fields		Grasslands	Land owner	228	228	
Vacant lands		Tree	Land owner	37	3,168	
		Grasslands	Land owner	2,844		
Private building lots		Tree	Land owner	741	7,353	
		Kitchen garden		646		
Shrine and temples		Tree	Agricultural communities, neighborhood association, land owner	127	299	
Public facilities		Tree	Public	150	586	
Parking lots and other paved surface			Land owner	0	1,459	

(continued)

Table 13.1 (continued)

Landscape unit	Landscape element	Organization	A_l (a)	S_l (a)
Road		Public	0	5,545
Rail road	Grasslands	Public	21	21
Riparian forest	Tree	Public	646	646
Water surface	Ponds and water channels	Agricultural communities, neighborhood association, land owner	0	1,108
	River	Public	0	1,138
Total			58,068	73,651

Fig. 13.3 Photos of three types of paddy fields

13.3 The Characteristics of Three Types of Paddy Fields

13.3.1 Spatial Characteristics of Three Types of Paddy Fields

In this section, spatial and ecological characteristics and differences of labor accounts in three types of paddy fields are clarified. The shape and the system for irrigation and drainage of the three types are shown in Table 13.3. Type I, the traditional Jori-sei paddy field, remains as it was in the Heian era (9–12th centuries), where the land of about one hectare is divided into several sections. One cultivating lot of Type II becomes larger than that of Type I. In Type III, the modern land improvement paddy field, one cultivating lot becomes larger than that of Type II in order to use contemporary farming machinery easily. Their average areas are 17 are, 23.3 are and 61.2 are for Type I, II and III, respectively.

The irrigation and drainage systems are also different. In Type I and II, the irrigation waterways and drainage systems are not clearly separated. And the sheathing boards are fitted to draw water into the paddy fields by slightly raising the water surface of the waterway when the water supply is needed. On the other hand, in Type III, water was pumped from the main irrigation channels through the underground pipeline and drained out in a lower position than the paddy fields, and

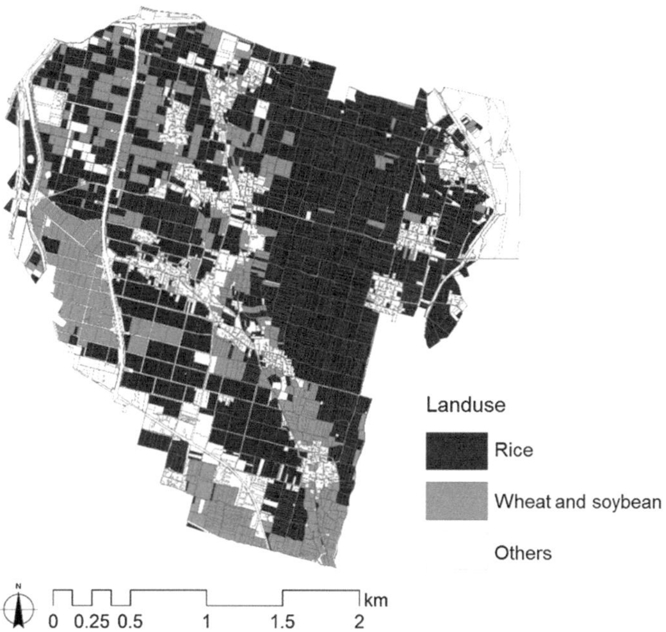

Fig. 13.4 Distribution of cultivated land according to products (paddy field, wheat, and soybeans)

Table 13.2 Crops area size and ratio

	Rice (a) (%)	Wheat and soybean (a) (%)	Total area size (a)
Type I	**250.1**	**66.9**	**316.9**
	79 %	21 %	
Type II	**63.5**	**34.1**	**97.6**
	65 %	35 %	
Type III	**64.4**	**29.8**	**94.2**
	68 %	32 %	
Asami district	**377.9**	**130.8**	**508.7**
	74 %	26 %	

Area (ha)

the open channels have wider width for the highly efficient water supply. Regarding the difference of structural materials, Type I consists of mud and Type II and III are of both concrete. As a consequence, Type I has a good environment for vegetation and wildlife because of the natural structural material and low water speed. However, both Type II and III have severe environments for wildlife because of the high water speed and the loss of connectivity between paddy fields and waterways, even though some compensation engineering is employed, such as fish ladders, catchment ditches and so on in Type III.

Table 13.3 Spatial and Ecological characters of three types of paddy field

			I. Traditional Jorisei paddy field		II. Modern land improvement paddy field		III. New land improvement paddy field	
Spatial characters of paddy field	Shapes of 3 paddy field types		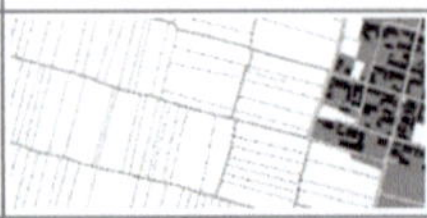		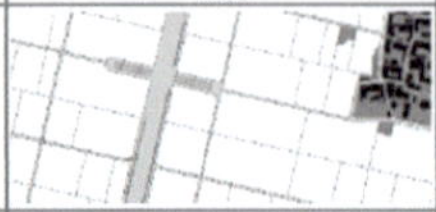		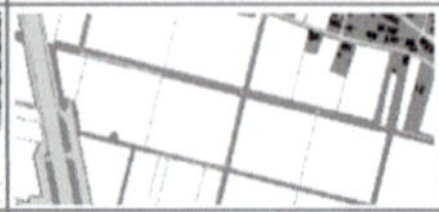	
	Total area (a)		31694.2		10130.5		9419.6	
			Ave.	*SD*	*Ave.*	*SD*	*Ave.*	*SD*
	Perimeter (m)		208.5	61.9	197.8	54.4	318.5	92.6
	Area (a)		17	11	23.3	14.2	61.2	35.4
	Complexity (m/a)		14.8	5.5	10.4	3.9	7.9	12.9
	Graph		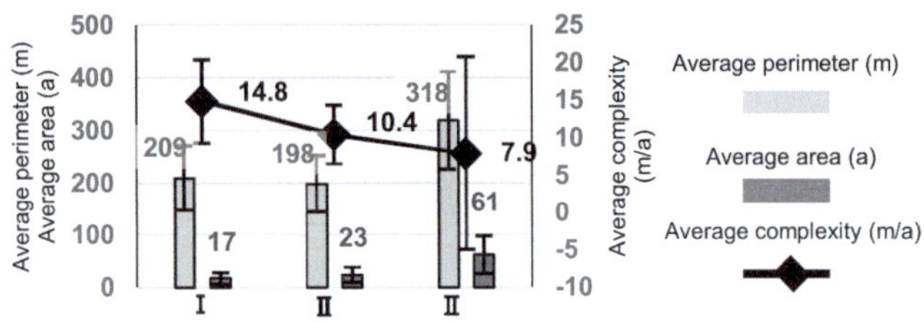					
Waterways and Farm roads	Waterway types		Dual purpose cannels		Dual purpose canals		Separated canals	
	Water intake method		Water level control by weir plate		Water level control by weir plate		Pipeline direct water intake	
	Features of waterways	Pavement	Mainly no pavement or paving side-walls. Various width		Paved with concrete or U-shaped gutter 0.5~1m width		Large waterway and drainage are paved with concrete 1~4m width	
		Water	The water level happens to come to that of the paddy field. Flow rate is slow.		Gaps of water levels and flow rate depend on climate and water intake condition.		Gaps of water levels and flow rate depend on climate and water intake condition. Some have no water.	
		Intake port Drain port	Just digging soil, or attaching U-shaped gutter or diversion valve in temporary. Many are essential to re-pair.		Intake; mainly diversion valve. Drain; diversion valve or concrete frame		Intake; depends on the valve. Drain; concrete frame. Large gaps to water-ways in height	
		Plants	Various plants including trees and shrubs both inside and outside of wa-terways.		Little soil and few plants.		Few plants. Some have environ-ment-friendly catchment mass and side wall.	
	Photos of waterway							
	Features of farm roads		Mainly no pavement. Covered with weeds. They're not wide enough for heavy machineries to go pass.		The half are unpaved and covered with weeds. The other half are paved with asphalt. They're wide enough for heavy machineries to pass.		Mainly paved with stones. Weeds are growing thickly on the slope between waterway and paddy field Widest roads among 3 types	
	Photos of farm roads							
Living things	biological survey	Waterway	Average numbers of fish species observed					
			3.1		1.6		0.8	
		Paddy field	Ratio of number of paddy fields where fishes are observed					
			44.9		0		2	
	Comments		Fishes inhabit both waterways and paddy fields		Fishes inhabit only waterways		Fishes don't inhabit waterways or paddy fields	

Ave. average / *SD* Standard deviation

13.3.2 Ecological Characteristics of Three Types of Paddy Fields

Next, the differences of ecological characteristics of three types of paddy fields were investigated by the field survey taken in June, 2015. Table 13.4 shows the results. Figure 13.5 shows the existence of fish in three types of paddy fields. Figure 13.6 shows the number of species existing in irrigation waterways of three types of paddy fields.

The clear difference was observed in the fish distribution. Fish could be found only in paddy field Type I. Also, many fish existed in the waterways of Type I. In Type II, many fish were also found in the waterways but there were no fish in the paddy field. Fish living in the channels of irrigation and drainage can come in and out of paddy fields of Type I. On the other hand, Type III blocks them from passing into the fields owing to the big gap of elevation and high water speed. Especially, *Cobitis minamorii*, a kind of fish which is designated as an endangered kind of the *Red List* (Ministry of the Environment 2015) (type IB), could be found only in Type I. However, there are very few fish except the *Misgurnus anguillicaudatus* in the waterway of the southern area of Type I, which is far from the Zeze River, as shown in Fig. 13.6. It is assumed that the Zeze River is a waterway where water flows throughout the year and has a good environment for the fish habitats in the winter season.

In conclusion, Type I has the best ecological environment and Type III has the lowest ecological value. It is also important to preferentially preserve the Zeze River and its surrounding environment.

13.3.3 Calculation of Landscape Labor Accounts

Table 13.5 shows the typical examples of labor contents, labor accounts and labor densities of corporate and individual farmers for Type I, II and III paddy fields. Labor densities are 5.56, 3.06 and 2.41 h/a for Type I, Type II and Type III, respectively. According to the statistical survey, the average of landscape labor densities of paddy fields in the Tokai region of Japan is 2.67, which shows that Type II and III are closer to the average value than Type I. The differences of labor densities among them are considered to be caused by the difference of intensities in labor contents of water management, planting, and reaping. The old preparation of farmland like Type I and II needs a large amount of labor density, because paddy fields with small areas are not easy to cultivate with big machines, and a lot of work is imposed on farmers in such fields. Also, the farmers need to move frequently among the small and dispersed paddy fields and a lot of waterways and farm roads in Type I and II. On the other hand, corporate farmers cultivate the land more efficiently and have a smaller value of labor density than individual farmers

Table 13.4 Number of creatures in the paddy field and irrigation waterways in three types

Paddy field type		Total number of surveyed place		Freshwater fishes									
				Japanaese rice fish (*Oryzias latipes*)	Field gudgeon (*Gnathopogon elongates elongates*)	Prussian carp (*Carassius* sp.)	Amur catfish (*Silurus asotus*)	Common carp (*Cyprinus carpio*)	Pond loach (*Misgurnus anguillicaudatus*)	Small striated spined loach (*Cobitis minamorii tokaiensis*)	Slender bitterling (*Tanakia lanceolata*)	Largemouth black bass (*Micropterus salmoides*)	Rosy bitterling (*Rhodeus ocellatus ocellatus*)
I	Paddy field	245	A	27	7	20	2	0	5	0	0	–	–
	Water channel	60	B	79	59	100	6	65	79	23	13	8	3
			C	32	28	36	6	22	29	14	7	7	3
			A	53	47	60	10	37	48	23	12	12	5
II	Paddy field	48	A	0	0	0	0	0	0	0	0	–	–
	Water channel	46	B	98	15	19	0	5	88	0	1	5	0
			C	26	8	7	0	3	25	0	1	3	0
			A	57	17	15	0	7	54	0	2	7	0
III	Paddy field	49	A	0	2	2	0	0	0	0	0	–	–
	Water channel	8	B	7	3	0	0	0	2	0	0	0	0
			C	2	3	0	0	0	1	0	0	0	0
			A	25	38	0	0	0	13	0	0	0	0

(continued)

Table 13.4 (continued)

Paddy field type		Total number of surveyed place						Frog				
				Freshwater minnow (*Opsariichthys platypus*)	Amur goby (*Rhinogobius* sp.)	Floating goby (*Gymnogobius urotaenia*)	Oily shiner (*Sarcocheilichthys variegatus variegatus*)	Tadpole	Japanese tree frog (*Hyla japonica*)	Indian rice frog (*Fejervarya kawamurai*)	Black-spotted pond frog (*Pelophylax nigromaculatus*)	Rana porosa brevipoda (*Pelophylax porosus brevipodus*)
I	Paddy field	245	A	–	–	–	–	36	27	39	6	1
	Water channel	60	B	8	2	1	1	0	1	0	3	0
			C	6	2	1	1	0	1	0	3	0
			A	10	3	2	2	0	2	0	5	0
II	Paddy field	48	A	–	–	–	–	54	56	29	4	0
	Water channel	46	B	0	0	0	0	1	0	0	1	0
			C	0	0	0	0	1	0	0	1	0
			A	0	0	0	0	2	0	0	2	0
III	Paddy field	49	A	–	–	–	–	33	22	67	6	0
	Water channel	8	B	0	0	0	0	0	0	1	2	0
			C	0	0	0	0	0	0	1	2	0
			A	0	0	0	0	0	0	13	25	0

A The ratio of number of places where living things are observed (%)
B Total observed number of living things
C Total number of places where living things are observed

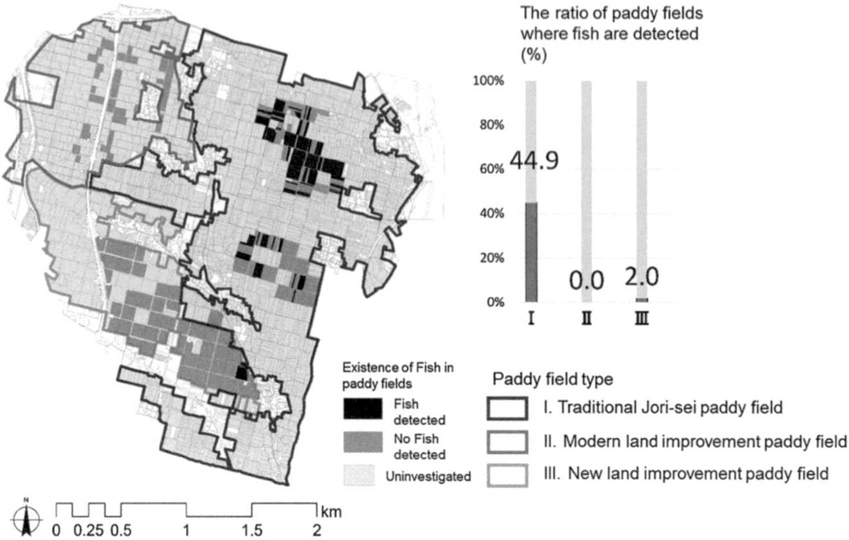

Fig. 13.5 Existence of creatures in three types of paddy field

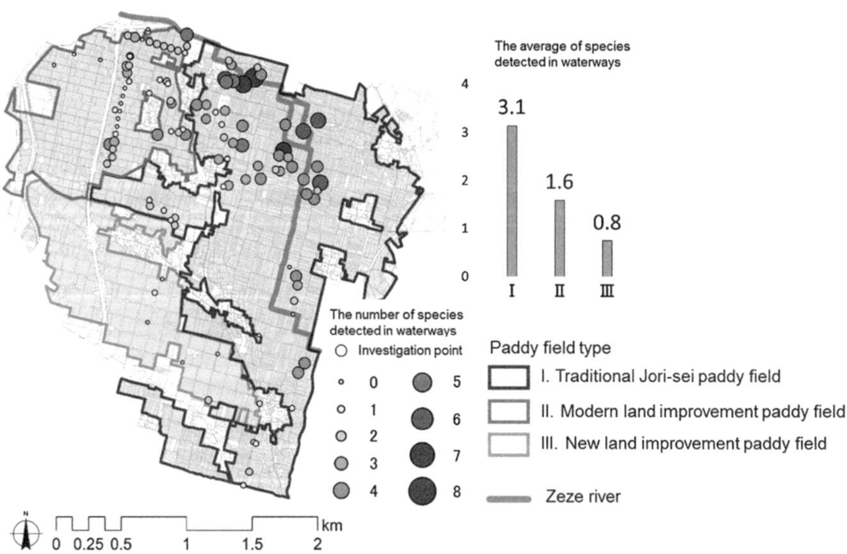

Fig. 13.6 Number of creatures existing in irrigation waterways of three types of paddy fields

Table 13.5 Labor accounts and densities of each labor contents in three types of paddy field

	I			II			III		
Labor contents	$TCLA_{jkl}$ (h)	A_l (a)	TLD_{jl} (h/a)	$TCLA_{jkl}$ (h)	A_l (a)	TLD_{jl} (h/a)	$TCLA_{jkl}$ (h)	A_l (a)	TLD_{jl} (h/a)
Plowing	110	240	0.46	96.0	180	0.53	35.2	117	0.30
Raising seedling	155.5	240	0.65	62.0	180	0.34	20.0	117	0.17
Border coating	15	240	0.06	6.7	180	0.04	0.0	117	0.00
Land preparation	21	240	0.09	35.0	180	0.19	8.5	117	0.07
Rice-planting	94	240	0.39	53.4	180	0.30	24.5	117	0.21
Mowing	240	240	1.00	156.0	180	0.87	35.0	117	0.30
Water management	570	240	2.38	73.8	180	0.41	56.3	117	0.48
Harvest	75	240	0.31	28.0	180	0.16	27.0	117	0.23
Drying and Hulling rice	36	240	0.15	25.3	180	0.14	34.5	117	0.29
Others	18	240	0.08	14.5	180	0.08	40.5	117	0.35
Total	1334.5	240	5.56	550.7	180	3.06	281.5	117	2.41
Area per a paddy field	240a/20 = 12a			180a/11 = 16.4a			117a/1 = 117a		

(Fig. 13.7). However, such corporate farmers also have few employees and they ask the local people to do mowing and water management.

13.3.4 Management of Irrigation Channels and Roads

Table 13.6 shows the organizations and labor contents of the collaborative works in three types of paddy fields. Such collaborative work is called "Deai" (meaning confluence). Figure 13.8 shows the distribution of irrigation channels. Each family supplies one participant for collaborative work. It is a custom for accomplishing

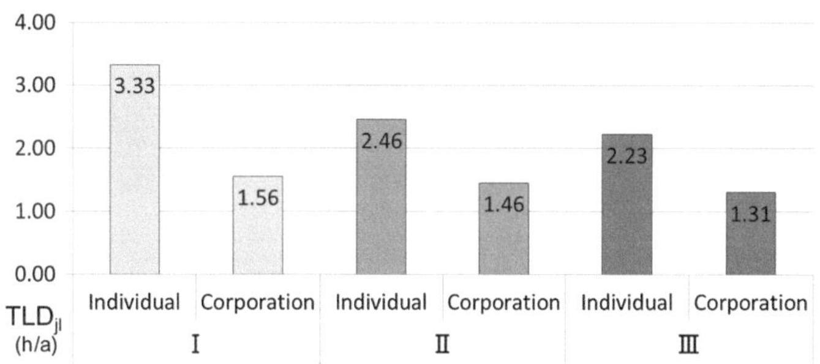

Fig. 13.7 Labor densities of individual and corporate farmers in three types of paddy field

Table 13.6 Organizations and labor contents of collaborative works "Deai" in three types paddy field

Paddy field type	Landscape unit	Landscape element		Operator	Labor content	Number of work people	B_{jk}	C_j	$TCLA_{jkl}$ (h)	TLA_{jl} (h)	A_j (a)	TLD_{jl} (h/a)	TLA_l (h)	A_l (a)	GLD_l (h/a)
Type I	Water channel	Agricultural water channel	Channel	Agricultural communities (Farmers)	Mowing 1	23	4	1	92	276	58	4.7	672	8,008	0.08
					Mowing 2	23	4	1	92						
					Sweeping	23	4	1	92						
			Main channel	District scale neighborhood association (Residents)	Cleaning out mud from a drain	38	4	1	152	152	7	22.1			
		Domestic wastewater channel[a]		Neighborhood association (Residents)	Cleaning out mud from a drain	38	4	1	152	152	4	38.7			
	Agricultural road			Agricultural communities (Farmers)	Cover with gravel	23	4	1	92	92	107	0.9			
Type II	Water channel	Agricultural main water channel		District scale neighborhood association (Residents)	Cleaning out mud from a drain	36	4	1	144	144	4	34.3	348	6,988	0.05
		Domestic wastewater channel[a]		Neighborhood association (Residents)	Cleaning out mud from a drain	36	4	1	144	144	1	101.8			
	Agricultural road			Agricultural communities (Farmers)	Cover with gravel	15	4	1	60	60	180	0.3			

(continued)

Table 13.6 (continued)

Paddy field type	Landscape unit	Landscape element		Operator	Labor content	Number of work people	B_{jk}	C_j	$TCLA_{jkl}$ (h)	TLA_{jl} (h)	A_j (a)	TLD_{jl} (h/a)	TLA_l (h)	A_l (a)	GLD_l (h/a)
Type III	Water channel	Agricultural water channel	Pump	Agricultural communities (Farmers)	Sweeping 1	2	1	139	278	748	15	50.8	1,512	10,175	0.15
					Pump Inspection	2	4	5	40						
					Sweeping 2	10	4	1	40						
			Channel		Cleaning out mud from a drain	30	7	1	210						
			Channel		Cleaning out mud from a drain	45	4	1	180						
			Main channel	District scale neighborhood association (Residents)	Cleaning out mud from a drain	64	2.5	1	160	160	3	47.3			
	Agricultural road			Agricultural communities (Farmers)	Mowing 1	34	4	1	136	604	423	1.4			
					Mowing 2	36	4	1	144						
					Mowing 3	34	4	1	136						
					Mowing 4	31	4	1	124						
					Cover with gravel	16	4	1	64						

[a]Domestic waste water channel area is included in the total element area of agricultural water channel

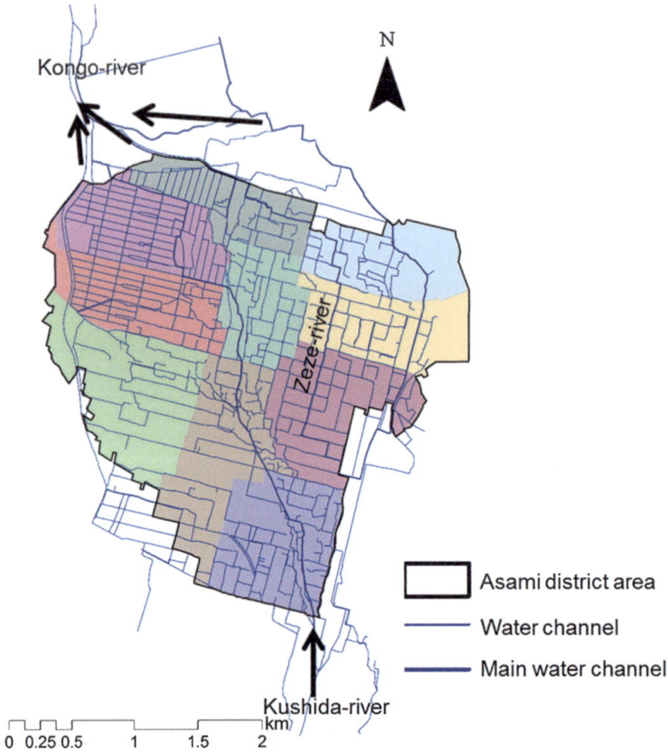

Fig. 13.8 Distribution of irrigation channels

work effectively by a large number of participants which is hard to do by a small number of workers. Because the number of farmland owners who do not cultivate but lend the land has been increasing recently, the Deai for such farmland is undertaken by the borrowing agricultural corporations. Figure 13.9 shows a picture of collaborative work in cleaning a channel.

Irrigation channels and roads are common places where work, such as mud clean up, mowing, and repairing by putting sand into depressions, are done several times per year by participants from the families who join the agricultural cooperative, residents' association and council for community development. The situations of these types of management are totally different among the types of paddy fields.

In Type I, some of the labor for mowing and cleaning up mud from small channels around paddy fields is done by private farmers on their own. Most of the channels are managed by collaborative work. The total number of hours for the collaboration in village A, which manages the Type I paddy fields, is 672 h. The total number of hours for the collaboration in village B, which manages the Type II paddy fields, is 348 h, which is smaller than the amount of Type I. The Type II fields' small channels are covered by concrete and need less management work. Furthermore, in this case, the burden of the residents for the management of all

Fig. 13.9 Collaborative work "Deai" by local community

channels and the roadsides has reduced. In village C of the Type III paddy fields, the total number of hours for the collaboration increases to 1,512 h. It seems curious because the burden of the management of channels and roadsides might have been reduced by the land consolidation project. It might be because the collaboration system has changed to this system in which all members of landowners including those that are not farmers should participate in the work. This change is said to be due to the new additional subsidies from land consolidation. Perhaps labor efficiency might have decreased.

13.4 Labor Accounts of Other Landscape Units

13.4.1 Cultivated Land Other Than Three Paddy Fields Types

The first target is cultivated land for private use. This land is not for growing food to be sold but is food for each family and the neighborhood. Many kinds of vegetables are planted in the farmland in a small area. The labor is mainly done in spring and summer or throughout the year. Table 13.7 shows labor accounts and densities and work areas of four farmers who engage in cultivation for private use. As shown there, labor densities are a large value of 20–100 h/a in the field for private use, while it is only 2–3 h/a in the productive paddy field. Notwithstanding such a situation, cultivation for private use helps keep landowners healthy, gives joie de vivre and improves the quality of their lives.

Table 13.7 Landscape labor accounts and labor densities of cultivated land

Operator	Number of work people	S_l (a)	TLA_l (h)	GLD_l (h/a)
Self-employment1	1	1.5	150	100.00
Self-employment2	1	2.3	150	65.22
Self-employment3	1	25.1	540	21.52
Self-employment4	1	1.7	150	88.24

13.4.2 Garden

There are many residential places that have rich gardens which are managed mainly by local residents. The annual labor accounts and densities of the gardens are shown in Table 13.8. Most labor involves pruning, which is done one or two times a year, and routinely weeding. These labor densities are also high between 20 and 60 h/a and have almost the same values as those of cultivated lands of private use. However, total labor accounts are smaller than those of the cultivated lands.

13.4.3 Temples and Shrines

Table 13.9 shows the landscape labor accounts and labor densities of temples and shrines. Each village has a temple and shrine, and they are managed by each local community. The temples and shrines have a strong relation with the lifestyles of the local people and contribute to maintaining the unity of the local community. The management is done by the collaborative work, Deai, like the management of channels and roadsides. The labor contents are mainly pruning and weeding. The management of temples and shrines needs a lot of labor accounts because they need periodical management by a fixed number of people.

Table 13.8 Landscape labor accounts and labor densities of gardens

Operator	Number of work people	S_l (a)	TLA_l (h)	GLD_l (h/a)
Self-employment1	1	1.3	72	57.14
Self-employment2	1	1.0	60	60.00
Self-employment3	1	2.2	60	27.27
Self-employment4	1	0.3	6	20.00

Table 13.9 Landscape labor accounts and labor densities of temples and shrines

Landscape unit	Operator	Number of work people	$TPLA_i$ (h)	A_l (a)	TLA_l (h)	NLD_l (h/a)
Shrine	Neighborhood association 1	27	8	8.3	216	26.1
Temple		60	4	103.9	240	2.3
Shrine	Neighborhood association 2	16	6	8.2	96	11.7
Temple		16	8	8.2	128	15.6
Shrine	Neighborhood association 3	36	3	12.4	108	8.7
		40	1	12.4	40	3.2
Temple		25	4	8.5	100	11.7
Shrine	Neighborhood association 4	14	12	8.7	168	19.3
Temple		14	10	6.0	140	23.3
Shrine	Neighborhood association 5	18	8	14.4	144	10.0
Temple		20	16	11.1	320	29.0
Shrine	Neighborhood association 6	10	4	12.4	40	3.2
Temple		11	16	12.2	176	14.5
Shrine	Neighborhood association 7	10	32	1.5	320	207.8
Temple		10	36	7.0	360	51.8
Cemetery		10	10	7.0	100	14.2
Shrine	Neighborhood association 8	12	50	16.3	600	36.7
Temple		13	4	16.3	52	3.2
Shrine	Neighborhood association 9	17	29	19.4	493	25.4
Temple		18	29	12.2	522	42.9
Shrine	Neighborhood association 10	10	11	5.7	62.15	11.0
Park		12	18	3.9	70.2	18.0
Temple		8	9	6.8	72	10.6

13.5 Summary of Landscape Labor Accounts and Densities

The results of labor accounts and density in each landscape unit are shown in Table 13.10. Net labor density (NLD), i.e. the total labor accounts of the district per the total area of landscape elements, is 3.98 h/a and gross labor density (GLD), i.e. the total labor accounts of the district per the total area of the district is 3.12 h/a. NLD and GLD have similar values because most parts of the district area consist of ecological landscapes. The social labor intensity (SLI), i.e. the total labor accounts in the district per the population, is 108, which is a little higher than the average SLI of paddy field types of the Chubu metropolitan area shown in Chap. 17.

Figure 13.10 shows the labor densities in the Asami district. The labor densities in paddy fields in different types I, II and III in the table show only for rice cultivation in order to compare the difference between the three types. The paddy field has a large gap among the types and management stakeholders. On the other

Table 13.10 Overview of landscape labor accounts and densities in Asami district

Landscape unit		Landscape element of each type of units	Organization	A_l (a)	S_l (a)		TLD_{jl} (h/a)	TLA_l (h)		GLD_l (h/a)	NLD (h/a)	GLD (h/a)	SLI (population: 2146)
Paddy fields (change of crops on paddy field of 3 products)	Type 1	Rice	Self-employment	21,645	25,000	37,759	3.33	72,076	117,208	3.09	3.98	3.12	108
			Farm corporation	3,355			1.56	5,234					
	Type 2		Self-employment	4,280	6,321		2.46	10,528		2.14			
			Farm corporation	2,041			1.46	2,980					
	Type 3		Self-employment	4,986	6,438		2.23	11,118		2.02			
			Farm corporation	1,453			1.31	1,903					
	Type 1	Wheat	Farm corporation	6,645	6,645	12,999	1.02	6,797		1.02			
	Type 2		Farm corporation	3,394	3,394		1.02	3,472		1.02			
	Type 3		Farm corporation	2,960	2,960		1.02	3,028		1.02			
	Type 1	Soybean	Self-employment	19	19	70	1.02	20		1.02			
	Type 3		Farm corporation	51	51		1.02	52		1.02			
	Grasslands		Self-employment	4	4		0.00	0		0.00			
Other crop fields	Other crop fields		Self-employment	1,078	1,141		29.77	32,078	32,078	28.12			
	Tree			63			0.00	0					
Orchards	Orchards		Self-employment	127		127	28.26	3,588		28.26			
Abandoned fields	Grasslands		Land owner	228		228	0.00	0		0			
Vacant lands		Tree	Land owner	37	3,168		0.00	0		0.00			
		Grasslands		2,844			0.00	0					
Private building lots		Tree	Land owner	741	7,353		41.10	30,452	78,499	10.18			
		Kitchen garden		646			68.74	44,429					

(continued)

Table 13.10 (continued)

Landscape unit	Landscape element of each type of units	Organization	A_l (a)	S_l (a)	TLD_{jl} (h/a)	TLA_l (h)	GLD_l (h/a)	NLD (h/a)	GLD (h/a)	SLI (population: 2146)
Shrine and temples	Tree	Agricultural communities, neighborhood association land owner	127	299	28.10	3,569	11.92			
Public facilities	Tree	Public	150	586	0.32	48	0.08			
Parking lots and other paved surface		Land owner	0	1,402	0.00	0	0.00			
Road		Public	0	5,545	0.00	0	0.00			
Rail road	Grasslands	Public	21	21	0.00	0	0.00			
Riparian forest	Tree	Public	646	646	0.00	0	0.00			
Water surface	Ponds and water channels*	Agricultural communities, neighborhood association, land owner	0	1,108	0.00	0	0.00			
	River	Public	0	1,138	0.00	0	0.00			
Total			58,068	74,160		231,373				

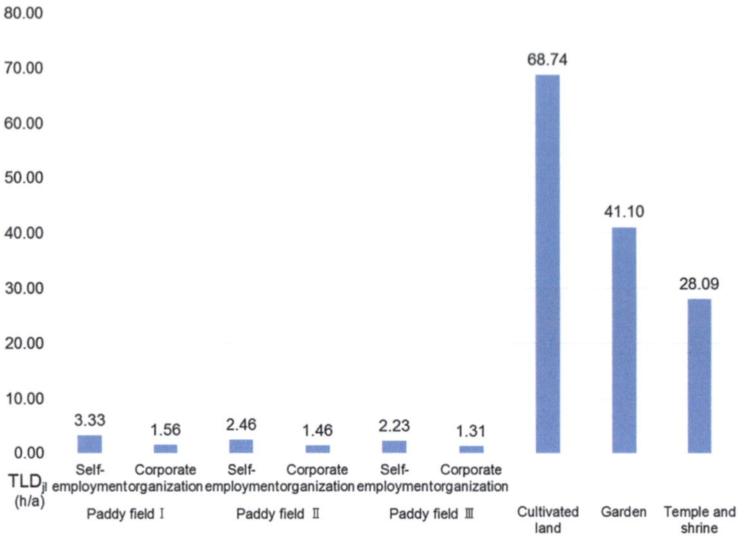

Fig. 13.10 Summary of Landscape labor densities in Asami district

hand, the cultivated land, garden, and temples and shrines have much larger labor densities than the paddy field.

13.6 Personal Labor Accounts and Personal Labor Densities

Figure 13.11 shows the personal labor accounts and labor densities in the Asami district. The annual personal labor accounts and personal labor densities are calculated. The labor accounts of corporate farmers is the highest, because they work every day as professional employees. However, their labor densities are very low because of their efficient working style.

On the paddy field, which is the main land use of this district, the tendency is clarified that the Type I paddy field with small divided cultivating lots of farmland and the grass covered small agricultural channels push up the amount of annual personal labor accounts and density more than those of the other types because of the inefficiency of farming. In other words, they create a greater burden on the farmers. However, as shown above, the Type I paddy fields have very high biodiversity and therefore high ecological values. It is very important to conserve these environments from now on. Some kind of measures for compensating the imbalance between the high labor burdens and hidden economic externality for the conservation of biodiversity will be needed.

Fig. 13.11 Personal labor accounts and personal labor densities in Asami district

Personal labor densities of hedge/garden and kitchen garden are higher than that of productive farmlands of rice and crops. People tend to work more intensively at these works, because these area sizes are small and people work not for pursuing working efficiency but for their annual routines and for their health maintenance or as a hobby.

13.7 Consideration of Landscape Management in the Future

By the consideration between biodiversity surveys and landscape management labor accounts, it is clarified that the richness of the biodiversity and working efficiency have a trade-off relationship. The compromise points between them will be considered, an optimal landscape planning for the future will be proposed, and the landscape management labor accounts under such conditions will be examined.

According to the *Census of Agriculture and Forestry* (Ministry of Agriculture, Forestry and Fisheries 2005, 2010) and the *Population Census* (Statistics Bureau 2005, 2010), while the ratio of the total number of farmer families in 2010 has decreased to 88.5 % of that in 2005, the number of families who lent their farmlands without engaging in cultivation has increased to 119.4 % between the periods, as shown in Fig. 13.12. Strength of the community as an agricultural settlement has been weakened. It should be pointed out that the suggested population of the total district will decline to half in 2065 (Fig. 13.13). Because of the decreasing birthrate and aging population, the number of agricultural workers is also supposed to decrease more from now on.

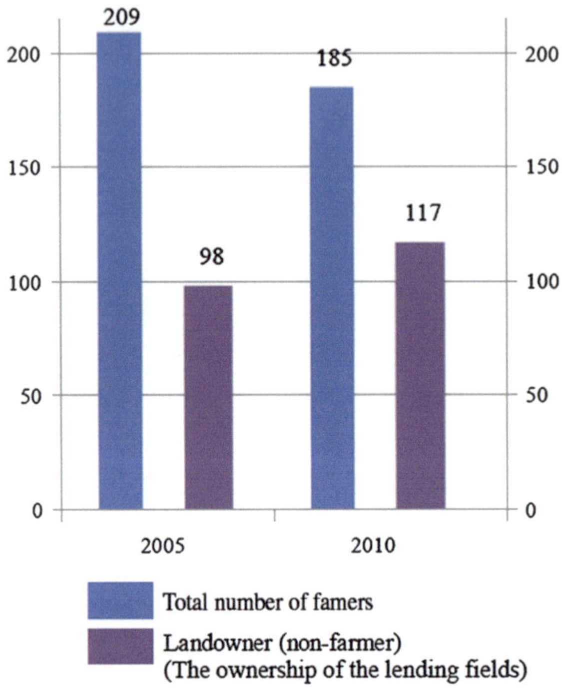

Fig. 13.12 Number of famers (2005–2010)

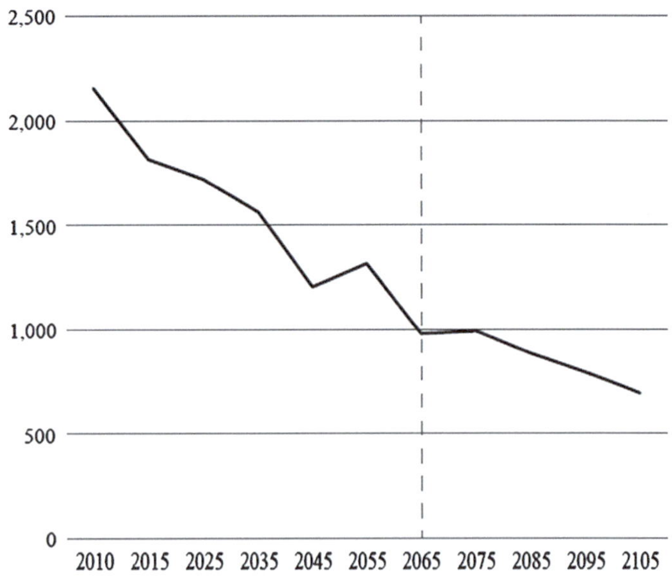

Fig. 13.13 Estimate of the population in future

From the above analysis, the prospect of planning on future preparation of farmland is reduced as follows. The results are shown in Table 13.11.

Scenario 1: Promotion of corporative farming of paddy fields corresponding to population decrease of private farmers in half

This scenario is considered on the assumption that the percentage of the area of three types of paddy fields remains as it is today. The number of private farmers will reduce to half of today, and the lack of the labor of private farmers will be compensated by promoting corporative farming. The sum of the private farmers will decrease from 176 to 88. The percentage of the decrease of private farmers in three types of paddy fields is assumed to be the same. Therefore, more than 11 employees in the corporate farming are needed additionally.

Scenario 2: Preservation of ancient Jori-sei system in a restricted important area and promotion of corporative farming in the rest of the field corresponding to population decrease of private farmers in half

The paddy fields in the Asami district are divided into five categories according to the importance of biodiversity of the area shown in Fig. 13.14. Irrigation and drainage of the Type I paddy fields contain rich biodiversity. Therefore, the areas alongside the Zeze River are the most important places. In scenario 2, the area of priorities 1 and 2, which has a total area size of 12,518 are, is selected to be preserved (Fig. 13.15). The percentage of the managed area by private farmers and corporate farming is assumed to remain as it is today. Therefore, the number of private farmers needed for the maintenance of the preservation area will be calculated as 39 persons. The rest of the private farmers, 49 persons, are assumed to be able to manage the Type III paddy fields of 5766 are. The rest of the area, i.e. 33,143 are, must be managed by corporative farming. The needed employees will be 20 persons and 9 persons are needed additionally than that of today.

The needed employees in scenario 1 are 2 persons higher than that of scenario 2, even though the size of the preserved Jori-sei area of scenario 2 is smaller than scenario 1. The reason is that the percentage of private farmers of Type III paddy fields is higher and has less labor density, i.e. working burden, than in scenario 2. Figure 13.16 shows the comparison of total labor accounts of 2 future scenarios. These scenarios can indicate the sustainable way by promoting corporative farming to compensate for the shrinking situation of private farmers, even conserving the important Jori-sei area to some extent.

The decrease of private farmers might be compensated to some extent by the development of efficient corporative farming. However, never forget the fact that Japanese paddy field landscapes are managed by many kinds of collaborative work, Deai, among village residents. Such collaborative work should be also important and maintained for creating the unity of the village. It can be supported by subsidies shown in the case of the collaborative work. However, it might not be essential. The most important thing is to promote the sense and motivation of collaboration by

Table 13.11 Calculation of labor accounts, densities and necessary persons of future scenarios

Basic data				Present			Scenario 1 (Remaining present field type ratio)			Scenario 2 (Compacting preservation area:)		
Paddy field types	Private or corporative	GLD_l (h/a)	$TPLA_i$ (h)	Managing area size (a)	TLA_l (h)	Necessary persons	Managing area size (a)	TLA_l (h)	Necessary persons	Managing area size (a)	TLA_l (h)	Necessary persons
Paddy field type I	Private	3.3	822.7	24,529.5	81,613.2	99.2	12,264.8	40,806.6	49.6	9,675.7	32,192.4	39.1
	Corporative	1.6	2,340.0	7,205.4	11,259.6	4.8	19,470.1	30,425.2	13.0	2,842.2	4,441.3	1.9
Paddy field type II	Private	2.5	342.4	4,662.2	11,445.7	33.4	2,331.1	5,722.8	16.7	0.0	0.0	0.0
	Corporative	1.5	2,340.0	5,459.4	7,959.2	3.4	7,790.5	11,357.7	4.9	0.0	0.0	0.0
Paddy field type III	Private	2.2	262.1	5,131.1	11,421.8	43.6	2,565.6	5,710.9	21.8	5,766.2	12,835.6	49.0
	Corporative	1.3	2,340.0	4,439.7	5,811.0	2.5	7,005.3	9,168.9	3.9	33,143.2	43,379.7	18.5
–	Sum of private workers	–	–	–	–	176	–	–	88	–	–	88
–	Sum of corporate workers	–	–	–	–	11	–	–	22	–	–	20

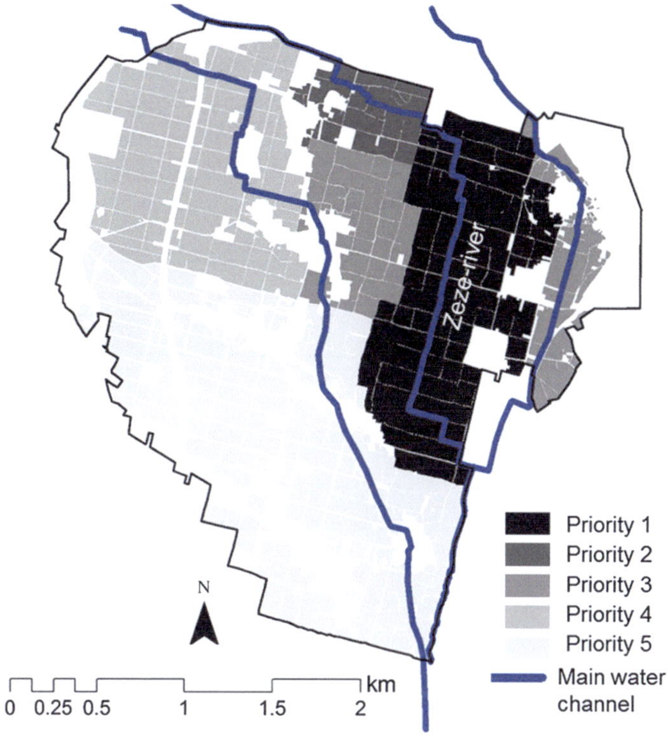

Fig. 13.14 Case study of conservation priority

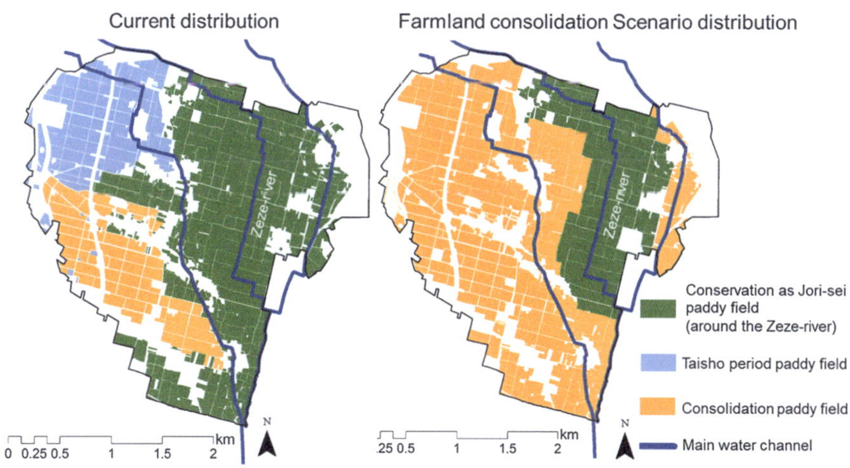

Fig. 13.15 Allocation of paddy field types in future scenario 2 of farmland consolidation plan that considered conservation priority

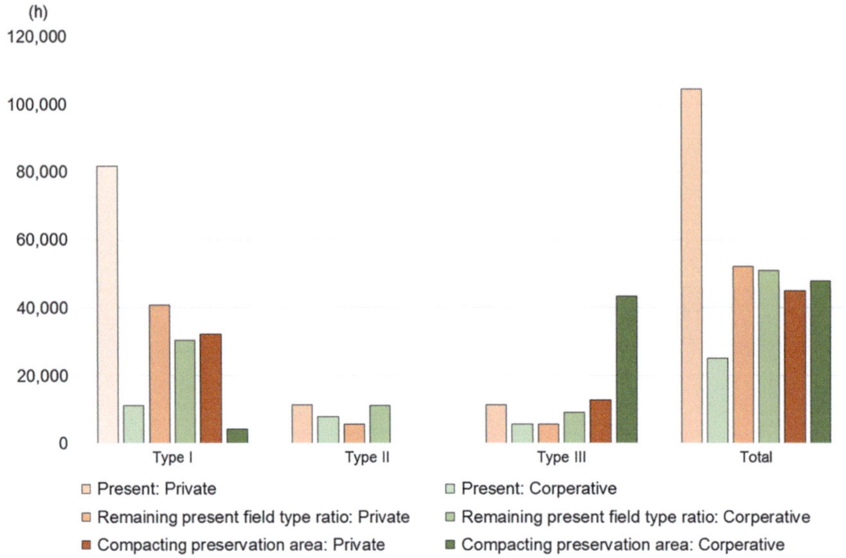

Fig. 13.16 Comparison of total labor accounts of future 2 scenarios

sharing the vision and hope to improve lives and the environment among the village residents. In other words, the pride or sense of attachment to the living environment is important. In the case of the Asami district, the conservation of biodiversity of the Jori-sei paddy fields might act as the core of the action.

The scenarios shown above might serve as a base for the citizen participation processes to incubate such atmosphere among the residents.

References

Katano O, Hosoya K, Iguchi K, Aonuma Y (2001) Comparison of fish fauna among three types of rice fields in the Chikuma River basin. Jpn J Ichthyol 48(1):19–25. (The Ichthyological Society of Japan)

Ministry of Agriculture, Forestry and Fisheries of Japan (2005) Census of Agriculture and Forestry 2005. http://e-stat.go.jp/SG2/eStatGIS/page/download.html. Accessed 1 March 2015

Ministry of Agriculture, Forestry and Fisheries of Japan (2010) Census of Agriculture and Forestry 2010. http://e-stat.go.jp/SG2/eStatGIS/page/download.html. Accessed 1 March 2015

Ministry of the Environment, Japan (2015) Red list. Ministry of the Environment. http://www.biodic.go.jp/english/rdb/rdb_f.html. Accessed 1 March 2016

Natuhara Y (2013) Ecosystem services by paddy fields as substitutes of natural wetlands in Japan. Ecol Eng 56:97–106

Okamoto M (2003) Irrigation organizations and inhabitants in their command area in Japan today. J Rural Plann Assoc 22(3):183–187. (the Association of Rural Planning)

Saito K, Katano O, Koizumi A (1988) Movement and spawning of several freshwater fishes in temporary waters around paddy fields. Jpn J Ecol 38:35–47. (the Ecological Society of Japan)

Statistics Bureau (2005) 2005 Population Census. Director-General for Policy Planning (Statistical Standards) and Statistical Research and Training Institute. http://e-stat.go.jp/SG2/eStatGIS/page/download.html. Accessed 1 March 2015

Statistics Bureau (2010) 2010 Population Census. Director-General for Policy Planning (Statistical Standards) and Statistical Research and Training Institute. http://e-stat.go.jp/SG2/eStatGIS/page/download.html. Accessed 1 March 2015

Chapter 14
Case Studies in Paddy Field Satoyama Neighborhood Landscape Complex Units: Nyu District

Chika Takatori, Chiaki Nakatsuji and Hiroyuki Shimizu

Abstract This chapter discusses the case study of Nyu District, Taki Town, Mie Prefecture, as a Satoyama neighborhood complex unit type. Nyu District is located in Satochi-Satoyama and occupied by 55 forest and 18 % farmland. The current population is about 1,000 people. It is located in the watershed of the Kushida River, a primary river, though there are branch rivers and the Tachibai water channel that introduce water from Kushida River to the area. The villagers have a high interest in the landscape and ecology of the site and a strong network of the local community, whose locally organized volunteer group actively manages the environment. In recent years, however, this village has been dealing with an increase in abandoned farmland and dilapidation of forests. As a consequence, this chapter discusses a method for managing the Satochi-Satoyama environment in context of depopulation and aging from the viewpoint of landscape management labor accounts.

Keywords Paddy field Satoyama neighborhood landscape complex units · Rural planning · Community farming · Irrigation system · Forestry

14.1 Introduction

This chapter shows the case study of the Nyu District, Taki Town, Mie Prefecture as a Satoyama neighborhood complex unit type (Fig. 14.1). The areas of landscape elements in this village is given in Table 14.1. Nyu district is located in *Satochi*-Satoyama and about 55 % of the district is covered by forest and 18 % by farmlands. Currently the population is about 1000 people. It is located in the watershed of the Kushida River, which is a primary river, and there are branch rivers and the Tachibai water channel that introduce water from the Kushida River. The village

C. Takatori (✉) · C. Nakatsuji · H. Shimizu
Graduate School of Environmental Studies, Nagoya University,
Furo-cho, Chikusa-ku, Nagoya, Aichi 464-8601, Japan
e-mail: takatori@corot.nuac.nagoya-u.ac.jp

H. Shimizu et al. (eds.), *Labor Forces and Landscape Management*,
DOI 10.1007/978-981-10-2278-4_14

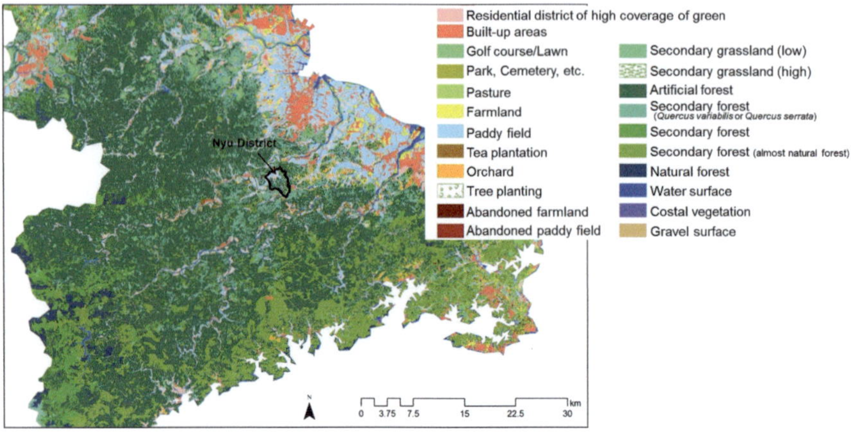

Fig. 14.1 Study area

Table 14.1 Landscape units/organizations

Landscape unit		Organization	Area (a)	
Agricultural field	Cultivated field	Farming union, Agricultural land owners	14,313	14,828
	Abandoned field	Agricultural landowners	515	
Residential area	Garden	Residents	513	3,471
	Kitchen garden	Residents	517	
Forest	Planted forest	Forest union, forest owners	27,018	44,917
	Natural forest	–	17,899	
	Managed forest	–	2,887	
Water system		NPO, residents, volunteers	5,664 (m)	
Total area		–	81,785	

people have great interest in the landscape and ecology of the site and a strong network of the local community (Osawa et al. 2008). Thus, the volunteer group organized by the local people manages the environment actively. On the other hand, in recent years this village has basic problems such as the increase in abandoned farmland and dilapidation of the forest caused by depopulation, a decreasing birthrate and an aging population. Thus, we chose this area as a research site.

14.2 The Landscape Units and Organizations

First of all, we created the land use data in Nyu village, clarified it into farmland, forest, water, and residential area, and then calculated the area of each land use type in the ArcGIS software (Fig. 14.2 and Table 14.1). The type of farmland was made by the map of farmland consolidation made by the land improvement district and a map of the block rotation plan made by the Nyu farming association. The forest type was made by the ArcGIS data of the Mie forest book. The water types such as water

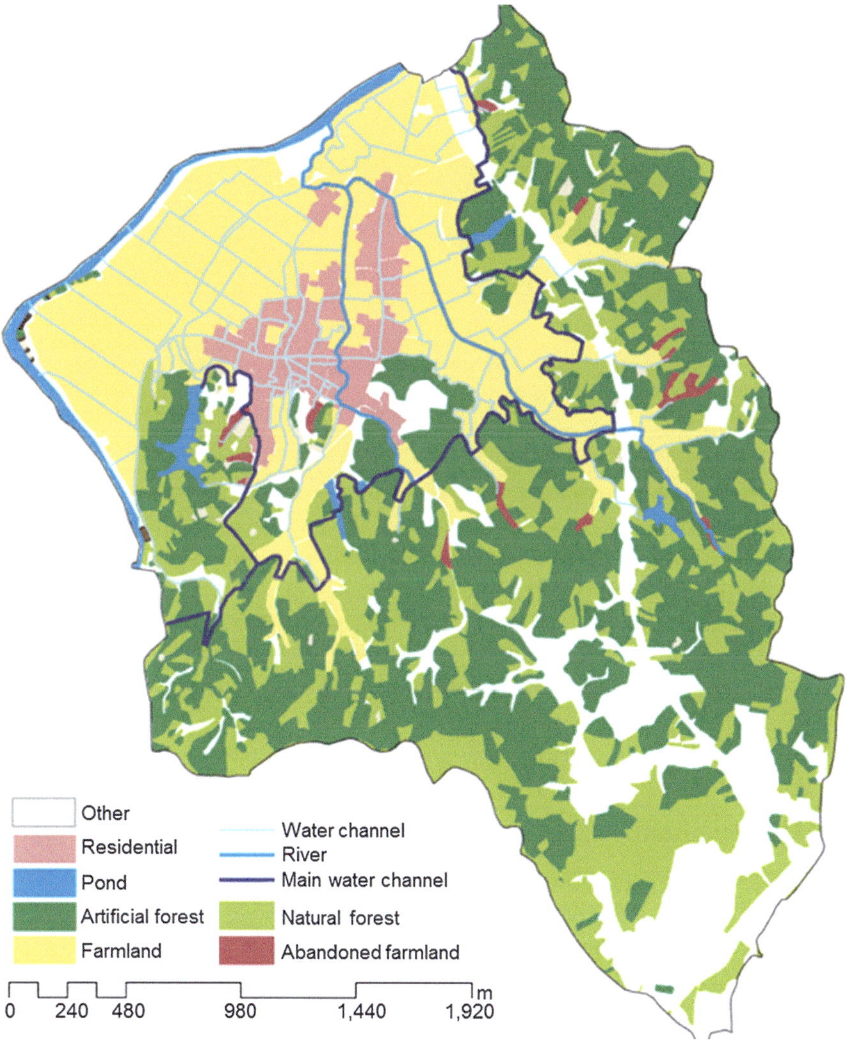

Fig. 14.2 Map of landscape units

Fig. 14.3 Photos of landscape units

channels and ponds were created by the water channel maps and field study, and the residential area was made by speaking to the head of the village. Figure 14.2 shows the management stakeholders of each land use type, and Fig. 14.3 shows the photos of each landscape unit.

Next, a questionnaire to all households in the Nyu villages (293 households and the return rate was 85 %) was carried out. The management stakeholders were chosen by the result of the questionnaire. We asked the detailed contents and time of the production and management work which are conducted routinely by each management stakeholder of each land use. We removed the work such as large-scale repair of facilities and initial cost of construction that can be not assigned to the routine management work by local people. Finally, we searched for additional studies of documentation and literature of labor time.

14.3 Farmland

First of all, the labor accounts of the farmland were calculated. In Nyu village, all paddy-rice cultivation was done by individual farmers, and all dry field farming, which is done by the block rotation of 4 products for 3 years such as wheat, ginger, and soybean was done by the farming association (Figs. 14.4 and 14.5). The abandoned farmlands were found to have increased in the valley places where

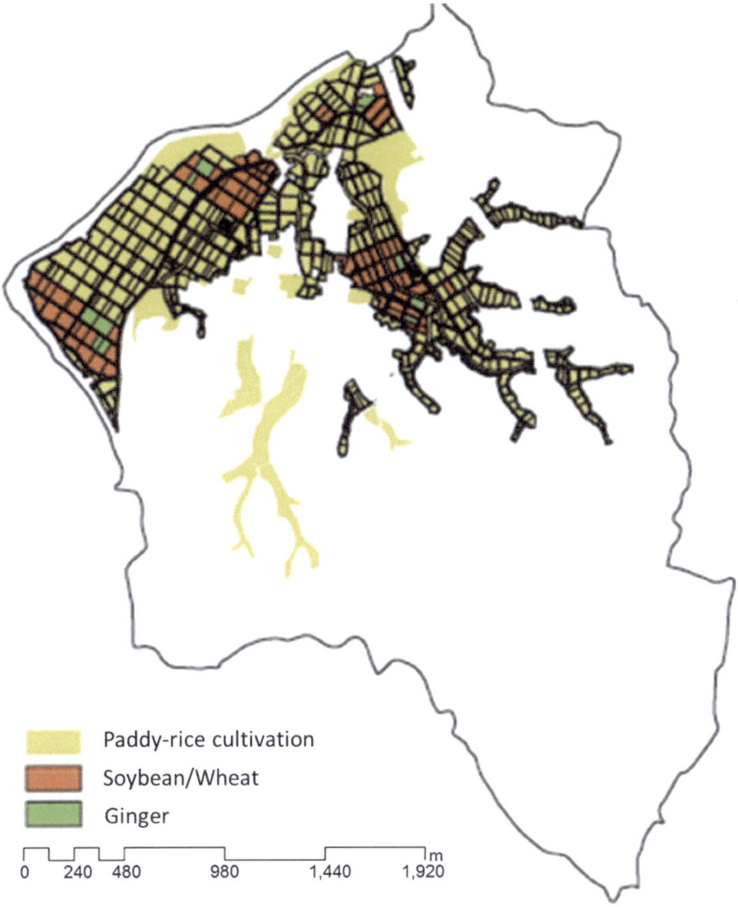

Fig. 14.4 Location of farmland

altitudes are higher than the irrigation water channel (Fig. 14.2), and it caused aggravation of the landscape and wildlife harm indirectly.

Table 14.2 shows the labor account of paddy-rice cultivation. We compared the density of labor accounts according to the management purpose of the paddy field. As a result, 3.43 h/a in the active paddy field, 3.84 h/a in the management of fallow paddy fields, and 1.65 h/a in the management of biotope type which is a fallow paddy field with water. The reason is that a fallow paddy field without water needs periodic plowing and mowing, but a fallow paddy field with water (Biotope type) does not need mowing but needs annual management of water. This shows that there are big gaps between the labor accounts according to management purpose. The largest labor type of paddy fields is water management, and the reason why the

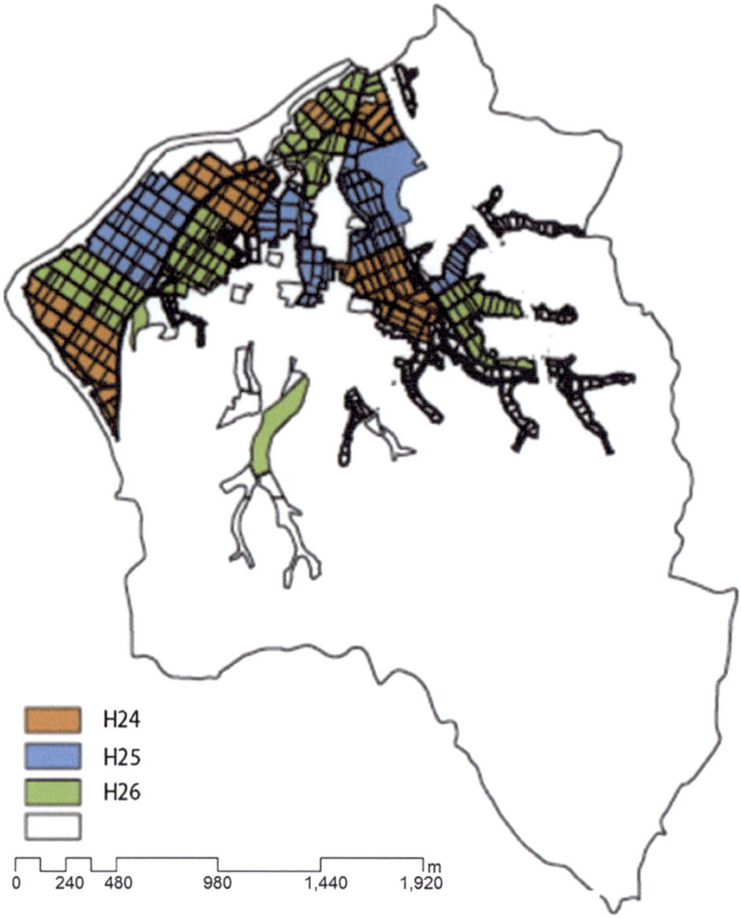

Fig. 14.5 Block rotation

abandoned farmlands are increasing in the valley can be assumed to be the difficulty of water management. Next, we compared the landscape labor densities of the two active paddy fields (3.43, 2.13 h/a) to the average farmer in the Tokai area data from the *Agriculture business management statistics investigation farm products life costs statistics* (Ministry of Agriculture, Forestry and Fisheries of Japan 2012) (2.32 h/a), and there were few gaps between them. One reason for this is the mechanization and generalization of the paddy cultivation are progressing.

In Table 14.3, the landscape labor density of wheat and soybean cultivation were 0.43 and 0.41 h/a, which are more efficient than the labor accounts of paddy cultivation that is 3.43 h/a.

Table 14.2 Labor account of paddy-rice cultivation

Paddy field type and organization	Labor content	Number of work people	Working hours per day (h/day)	Annual working day (day)	S_l (a)	TLA_l (h)		TLD_{jl} (h/a)	GLD_l (h/a)
Active paddy field self-employment 1	Seeding	4	7.00	1	100	28	343	0.28	**3.43**
	Watering seedlings	1	0.50	30		15		0.15	
	Water management	1	0.50	120		60		0.60	
	Puddling	1	7.00	3		21		0.21	
	Rice planning	2	6.00	3		36		0.36	
	Herbicide	1	6.00	1		6		0.06	
	Seeding	2	5.50	3		33		0.33	
	Drying	3	6.00	3		54		0.54	
	Powder shaved	3	7.00	2		42		0.42	
	Mowing	1	4.00	5		20		0.20	
	Plowing	1	7.00	4		28		0.28	

(continued)

Table 14.2 (continued)

Paddy field type and organization	Labor content	Number of work people	Working hours per day (h/day)	Annual working day (day)	S_l (a)	TLA_l (h)		TLD_{jl} (h/a)	GLD_l (h/a)
Active paddy field self-employment 2	Seeding	2	3.00	1	100	6	194	0.2	**2.13**
	Watering seedlings	1	2.00	1		2		0.07	
	Water management	1	0.50	120		60		0.60	
	Puddling	1	7.50	2		15		0.15	
	Rice planning	3	4.50	2		27		0.27	
	Herbicide	1	4.00	1		4		0.04	
	Seeding	2	6.00	2		24		0.24	
	Drying	–	–	–		–		–	
	Powder shaved	–	–	–		–		–	
	Mowing	1	8.00	4		32		0.32	
	Plowing	1	8.00	3		24		0.24	
Fallow paddy field	Mowing	1	4.00	24	50	96	192	1.92	**3.84**
	Plowing	1	4.00	24		96		1.92	
Biotope	Water management	1	0.10	220	20	11	33	0.55	**1.65**
	Mowing	2	3.00	3		18		0.90	
	Plowing	1	1.00	1		4		0.20	
Average in the Tokai region (statics data of active paddy field)		–	–	–	–	–	–	–	**2.32**

Table 14.3 Labor account of other crop cultivation

Crops	Organization	Labor content	Number of work people	Working hours per day (h/day)	Annual working day (day)	S_l (a)	TLA_l (h)		TLD_{jl} (h/a)	GLD_l (h/a)
Wheat	Corporate organization 1	Seeding	3	6.70	7	2,720	141	1,178	0.05	0.43
		Fertilizing	4	7.30	2		58		0.02	
		Herbicides spraying	5	5.50	2		55		0.02	
		Pesticide spraying	3	8.20	3		74		0.03	
		Mowing	2	2.20	1		4		0.00	
		Plowing	3	6.90	19		393		0.14	
		Ridging	2	6.10	6		73		0.03	
		Treading wheat plants	2	6.80	7		95		0.04	
		Reaping	6	7.90	6		284		0.10	
Ginger	Corporate organization 2	Seeding	14	4.50	4	13	252	496	19.38	38.14
		Watering	2	4.30	3		26		1.98	
		Herbicides spraying	2	6.00	3		36		2.77	
		Pesticide spraying	2	4.20	3		25		1.94	
		Mowing	3	5.70	3		51		3.95	
		Ridging	1	1.00	1		1		0.08	
		Covering and management	3	3.90	6		70		5.40	
		Drying prevention	7	4.90	1		34		2.64	

(continued)

Table 14.3 (continued)

Crops	Organization	Labor content	Number of work people	Working hours per day (h/day)	Annual working day (day)	S_l (a)	TLA_l (h)		TLD_{jl} (h/a)	GLD_l (h/a)
Soybean	Corporate organization 3	Seeding	4	6.20	6	2,370	149	808	0.06	0.34
		Fertilizing	2	4.00	2		16		0.01	
		Pesticide spraying	2	5.30	12		127		0.05	
		Plowing	2	5.90	6		71		0.03	
		Inter plowing	2	4.80	15		144		0.06	
		Harvesting	3	5.90	17		301		0.13	

14.4 Artificial Forest

In Nyu village, a large amount of cedars and Japanese cypress were planted after WW2, and 60 % of the forest is over 30 years old now (Figs. 14.2 and 14.6). According to *2007 the forest and forestry paper* (Forestry Agency, Ministry of Agriculture, Forestry and Fisheries of Japan 2008), the managed forest is only 10 %, and the large amount of forest abandoned now is artificial forest (Fig. 14.7). The questionnaire shows that 75 % of forest owners do not manage the forest and the reasons are "I cannot manage them routinely", and "It cannot become an income source", and so on, that are the same factors as found in the national report. Next, we asked four forest owners whose ages of forest were different and who manage the forest differently about the purpose of management and density of labor accounts. Table 14.4 shows that the labor accounts are greatly different according

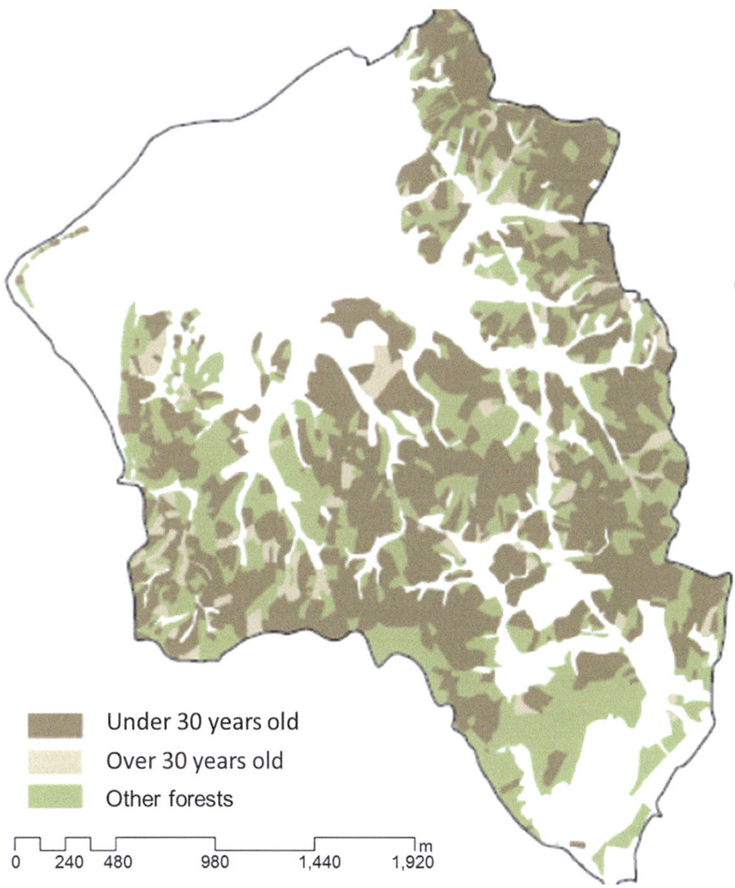

Fig. 14.6 Age of each artificial forest

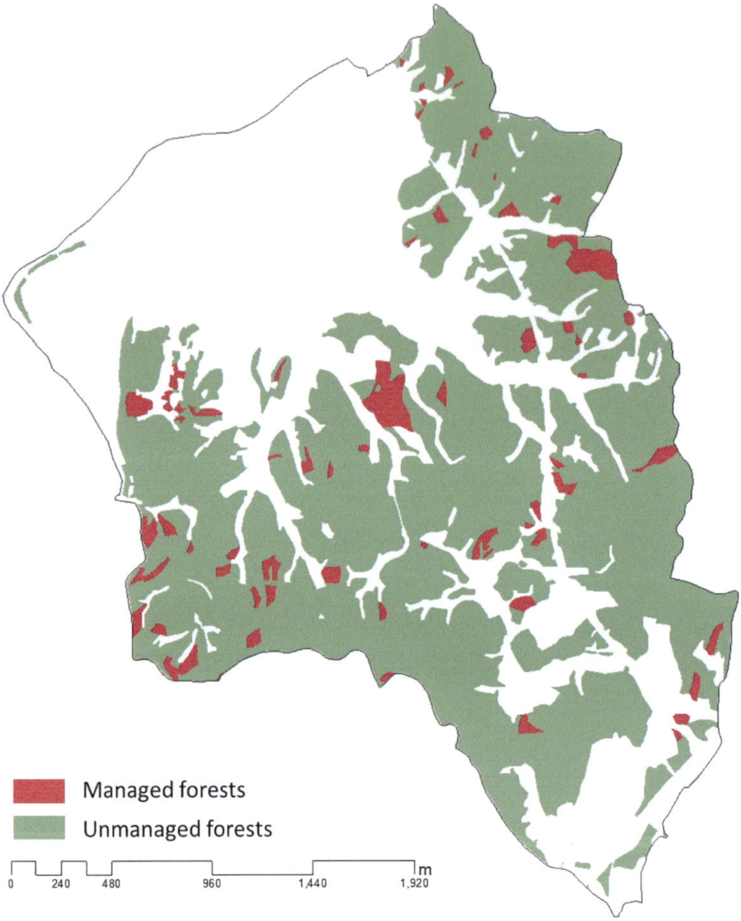

Fig. 14.7 Management record of artificial forest

to "whether the forest is over 30 years old". The labor accounts of forest owner 3 who manages the forest which is over 30 years old is 2.5 h/a, and it is almost the same as the labor accounts of the NPO shown in the existing research by Terada et al. (2010). It is also close to the average labor accounts of paddy fields in the Tokai area which is 2.32 h/a (Table 14.2). On the other hand, the labor accounts of forest owner 1 whose forest is under 30 years old and who produces the wood for himself is very large (10.5 h/a), and 80 % of labor accounts are especially dominated by the thinning. In other words, the labors of forest owners whose forest is under 30 years old are mainly occupied by the non-profitable ones which contribute to the preservation of the ecosystem. This is one of the factors that cause the difficulty of growing and managing the forest. The purpose of owners is only leisure or health.

Table 14.4 Labor accounts of artificial forest

Conifer plantation	Organization	Labor content	Number of work people	Working hours per day (h/day)	Annual working day (day)	S_l (a)	TLA_l (h)		TLD_{jl} (h/a)	GLD_l (h/a)
Under 30 years old	Private forest owner 1	Thinning	1	5.00	70	40	350	579	8.75	14.49
		Undergrowth cutting (conifer tree)	1	5.00	7		35		0.88	
		Undergrowth cutting (broad leaf tree)	1	5.00	1		5		0.13	
		Vine cutting	1	0.50	70		35		0.88	
		Pruning	1	5.00	2		100		2.50	
		Plantation	4	4.50	3		54		1.35	
Under 30 years old	Private forest owner 2	Thinning	5.5	5.00	70	300	1,925	2,035	6.42	6.78
		Undergrowth cutting	5.5	5.00	2		55		0.18	
		Pruning	5.5	5.00	2		55		0.18	
Over 30 years old	Private forest owner 3	Thinning	1	5.00	240	500	1,200	1,250	2.40	2.50
		Undergrowth cutting	1	5.00	10		50		0.10	
Over 30 years old	Private forest owner 4	Thinning and undergrowth cutting	1	4.00	7	100	28	28	0.28	0.28
Over 30 years old	NPO	Thinning and undergrowth cutting	–	–	–	300	990	990	3.30	3.30

14.5 Water System

The irrigation channel network in Nyu village has been preserved and managed by the residents and local community who are the main and branch irrigation system management organizations (Midori-Net Nyu, and Midori-Net Tachibai water channel). Figure 14.8 shows the main and branch channels in Nyu district. In particular, the Tachibai irrigation channel which is the main irrigation system and supplies water from the Kushida River to the villages has multiple functions of not only being the supply of agricultural and life water but also preservation of the ecosystem. This is called "social common capital" (Uzawa 2000). In general, management of social common capital is conducted by the social system and practice about the usage, preservation, and control of resources such as the irrigation

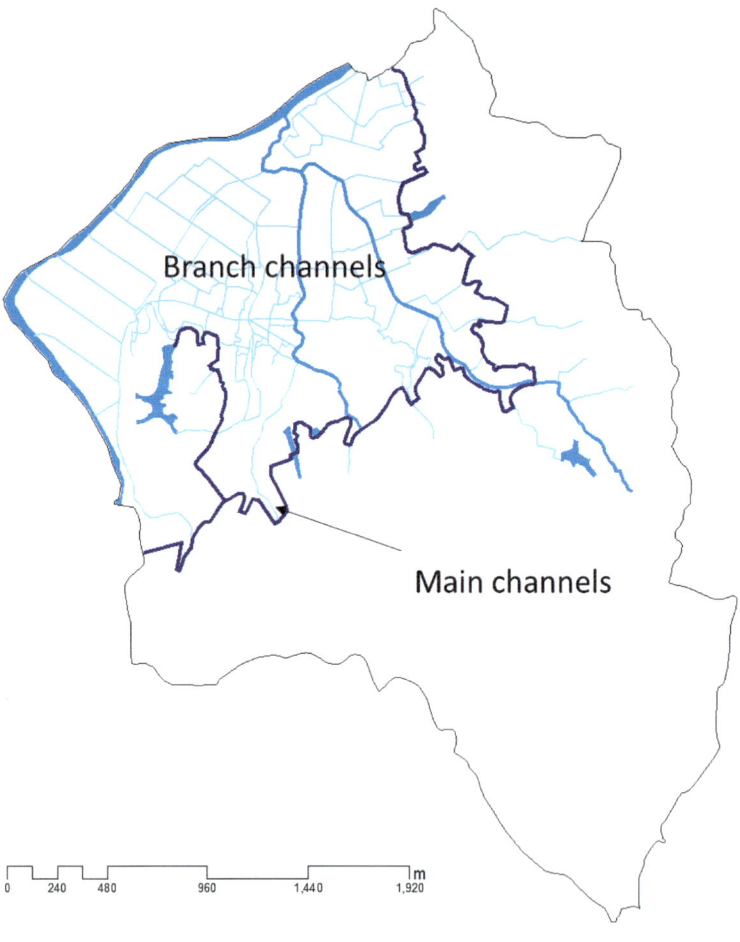

Fig. 14.8 Main and branch channels

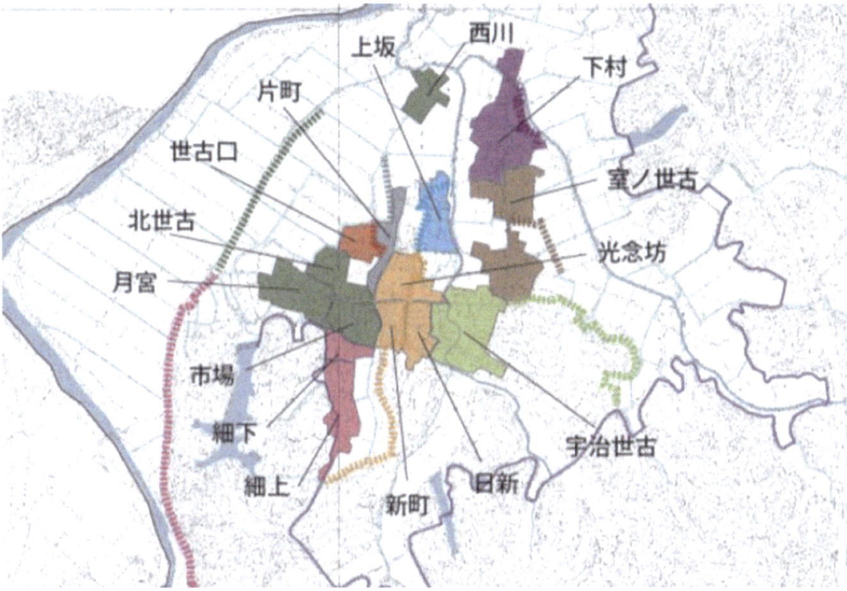

Fig. 14.9 Managed area by each group of residents

channels and farm roads, is different from the market mechanism. In Nyu village, the management organizations routinely check and repair the agricultural water facilities. The maintenance cost is the labor of farmers, and cooperative group work, Deai, is done by the all residents including non-farmers even today. According to the questionnaire, the villages are divided into 15 groups in Deai (Fig. 14.9), and the labor such as cleaning and mowing of the open water channel is almost 0.1 h/a which is equally done by each group. Table 14.5 shows the contents of the labor accounts for the management of water by each stakeholder.

14.6 Residential Area

The elements which create the landscape and ecosystem environment in Nyu village are the gardens such as hedges and private vegetable gardens in the residential area. Figure 14.10 shows the residential area in Nyu district, and Fig. 14.11 shows the location of gardens and kitchen gardens. The data was gathered by the questionnaire and aerial photographs. According to the questionnaire, 69 % of residents manage the gardens and 62 % manage the vegetable gardens. The area of managed gardens was calculated by union of the rate and the extracted garden area from the aerial photographs (Table 14.1). Table 14.6 shows the landscape labor accounts of residential area. There are gaps among the individuals, but all managers prune the

Table 14.5 Labor accounts of water system

Landscape unit	Organization	Labor content	Number of work people	Working hours per day (h/day)	Annual working day (day)	S_l (a)	TLA_l (h)		$TLDj_l$ (h/a)	GLD_l (h/a)
Irrigation water	Residents	Dredge up mud	278	1.5	1	31.6	417	973	13.2	30.79
		Mowing	278	2.0	1		556		17.59	
Hydrangea	Volunteer	Composting	36	2.0	1	22.4	71	426	3.17	19.02
		Mowing	36	2.0	4		284		12.68	
		Pruning	36	2.0	1		71		3.17	
Branch irrigation system	Branch irrigation system management organization	Garbage removal	6	9.0	1	66.9	54	1,663	0.81	36.50
		Water management	2	4.0	123	66.9	984		14.7	
Drainage		Garbage removal	2	5.0	43	66.9	430		6.43	
Pond		Water management	2	0.5	123	442.9	123		0.28	
Hydrangea		Composting	6	2.0	1	5	12		2.38	
		Mowing	6	2.0	4	5	48		9.52	
		Pruning	6	2.0	1	5	12		2.38	

(continued)

Table 14.5 (continued)

Landscape unit	Organization	Labor content	Number of work people	Working hours per day (h/day)	Annual working day (day)	S_l (a)	TLA_l (h)		$TLDj_l$ (h/a)	GLD_l (h/a)
Biotope type paddy field	Main irrigation system management organization	Water management	1	0.1	220	20	11	1,897	0.55	85.58
		Mowing	2	3.0	3	20	18		0.90	
		Puddling	1	4.0	1	20	4		0.20	
		Pruning	3	7.0	4	1.6	84		52.5	
Main irrigation system		Dredge up mud	5	7.0	5	56.6	175		3.09	
		Mowing	5	7.0	5	56.6	175		3.09	
		Water management	1	6.5	220	56.6	1,430		25.25	

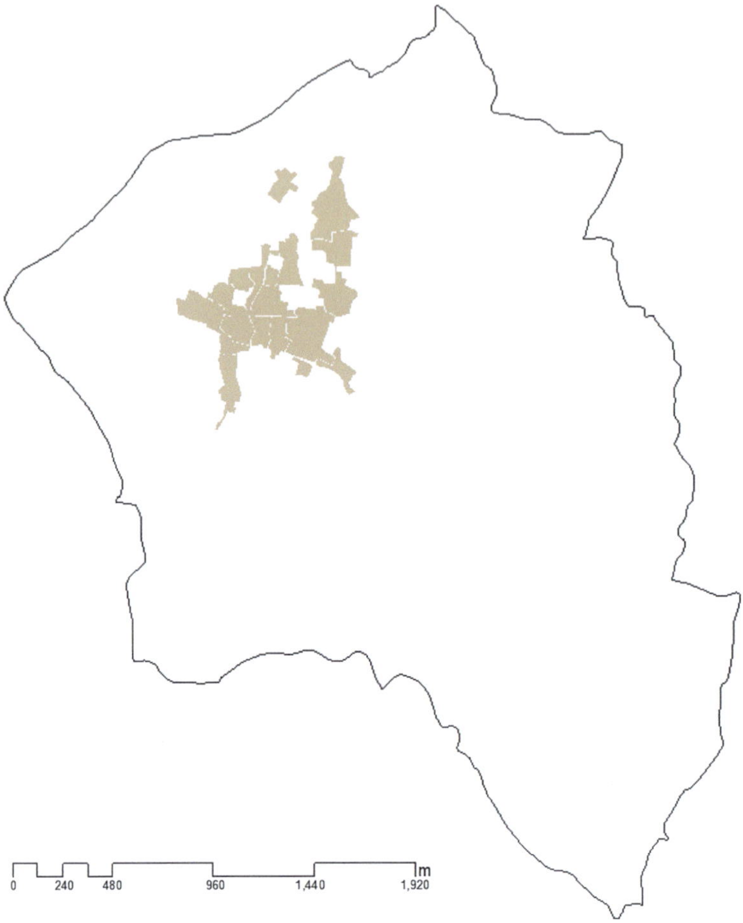

Fig. 14.10 Residential area in Nyu district

garden trees once a year and the labor densities were 3–6 h/a. The kitchen gardens were managed on a small scale and the labor density was high (12 h/a).

14.7 Summary of Landscape Labor Accounts

Finally, from the above the current labor accounts of the total area of Nyu village were calculated. In the calculation, we used the average labor accounts of the managed paddy fields, the average accounts of fields (plowed fields) of each crop, the average accounts of each forest type of over/under 30 years old which has the record of management in the most recent 5 years, and the average accounts of gardens and

Fig. 14.11 Garden and kitchen garden

kitchen gardens in residential areas and crossed them to the area of each land use type. Water land use type was calculated in the total area managed by irrigation system management organizations and residents (Table 14.6 and Fig. 14.12).

Table 14.7 shows the landscape labor densities of paddy field Satoyama neighborhood landscape complex unit. As a result, the total labor accounts of managed land use in Nyu villages were 58,821 h, and the gross density of labor accounts of the total area in Nyu village (81,785 a) was 0.72 h/a. Especially, the labor density was high in the farmlands and residential areas which included the gardens and kitchen gardens (even if a large amount of them were served by a subsidy), and almost all of them were well managed. The abandoned farmlands were increasing in the valley, but the farmers corresponded to it by improving the efficiency of labor by working as an association. On the other hand, we calculated the case that we cultivate abandoned farmlands as paddy-rice cultivation and the total labor accounts were 1,432 h.

Table 14.6 Labor accounts of residential area

Landscape Unit	Organization	Labor content	Number of work people	Working hours per day (h/day)	Annual working day (day)	S_l (a)	TLA_l (h)		TLD_{jl} (h/a)	GLD_l (h/a)
Garden	Pruning	Resident 1	1	3.0	1.5	1.0	4.5	6	4.50	6.0
	Disinfection		1	1.0	1.5		1.5		1.50	
Garden	Pruning	Resident 2	1	3.0	1.5	1.7	4.5	5	2.65	2.9
	Disinfection		1	1.0	0.5		0.5		0.29	
Kitchen garden	Various works		1	0.6	50	5.0	30	60	6.00	12.0
	Mowing		1	0.6	50		30		6.00	

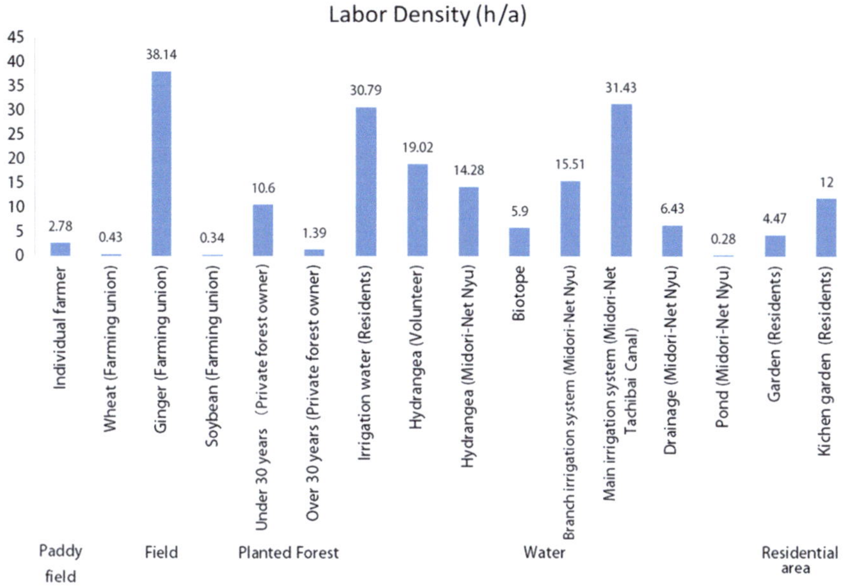

Fig. 14.12 Labor density of landscape unit types

The total labor accounts of managed forests are few, even though the forest occupies 55 % of the total area in Nyu village, and it is managed mainly for the purpose of hobby and health care. We calculated that if we manage all forests which are now abandoned, the total labor accounts of 144,786 h will be additionally needed. This means the labor of 1,957 h per person (the number of current forest employees is 74 people) is additionally needed and it is not realistic. If the artificial forest will be managed in the future, the measures of improving the efficiency of basic labor by committing the work externally and collecting the land owners will be needed.

The management of irrigation systems is maintained well because they have a strong relation with the agriculture, and they are managed in the community work, Deai, as a social common capital in the local community management of Nyu village.

14.8 Labor Accounts and Densities Per Person

In this section, the features of personal labor accounts and densities among all types of organizations are described. From Fig. 14.13, the three following features are found. On secondary forest and bamboo forest, it is found that (1) personal labor accounts of forest owners whose artificial forest is under 30 years old are comparatively high; they're around or more than 500 h. Meanwhile, personal labor

Table 14.7 Labor densities of paddy field Satoyama neighborhood landscape complex unit

Landscape unit		Landscape element	Organization	A_l (a)		TLD_{jl} (h/a)	TLA_l (h)	NLD (h/a)	GLD (h/a)	SLI
Farmland	Paddy field (change of crops on paddy field; the block rotation of 4 products per three years)	Paddy field	Self-employment	9,210	14,828	2.78	25,604	0.89	0.72	56.3
		Wheat	Corporate organization	2,370		0.43	1,019			
		Ginger	Corporate organization	13		38.14	496			
		Soybean	Corporate organization	2,720		0.34	925			
	Abandoned farmland		Farmland owners	515		0.00	0			
Built-up areas	Garden		Residents	513	1,030	4.47	2,293			
	Kitchen garden		Residents	517		12.00	6,024			
Forest	Planted forest	Managed forest	Forest owners	2,887	44,917	6.00	17,322			
		Unmanaged forest	–	24,131		0.00				
	Natural forest		–	17,899		0.00				
Water surface (irrigation channels)			NPO, residents, volunteers	5,664(m)		87.60	4,958			
Total				81,785			58,821			

SLI population 1,045

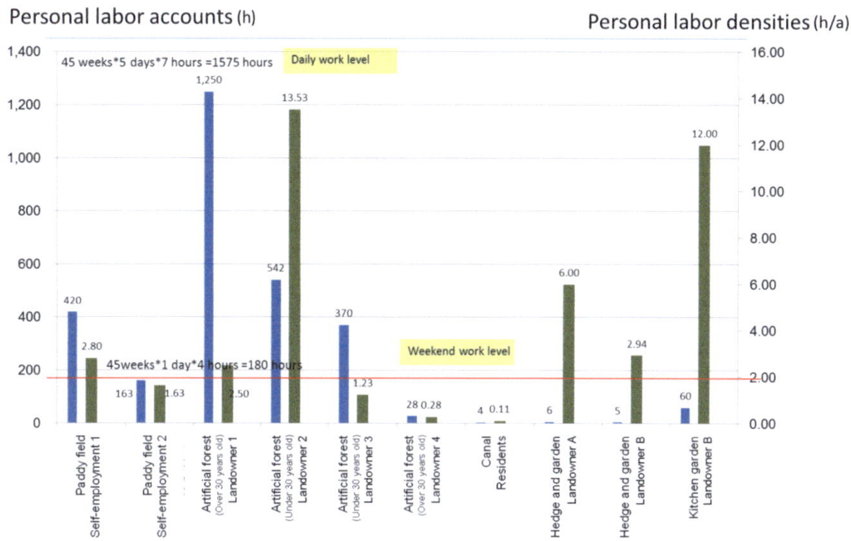

Fig. 14.13 Working style types in Nyu district

densities of them are low; they're around or less than 1.0 h/a. Further, (2) personal labor accounts and densities of self-employees of paddy fields are relatively low; they're less than 500 h and 3.0 h/a. On the other hand, on kitchen garden or garden, it is found that (3) personal labor accounts of residents are low; they are under 60 h. Meanwhile, personal labor densities are high in comparison with farmlands and artificial forests; they're over 3 h/a.

14.9 Consideration of Landscape Management in the Future

Finally, the calculated labor accounts were compared to the future population in the total area and considered the management problem methodology in the future. We forecasted the future population by using the individuals born during the same period method, with population datasets from 2005 to 2010 in Nyu village from the national census (Fig. 14.14). In 2010, the number of people was 989 (the number of men was 465 and women was 524), and the average labor time was 116 h per person which was calculated by breaking the total labor accounts in Nyu villages in the total population. In 2050, the population will decrease to half of the current population (about 500 people) and the total labor time will also decrease to half which in total labor accounts is 58,300 h. This means the future lack of labor accounts which are necessary for the preservation of the local landscape and ecosystem.

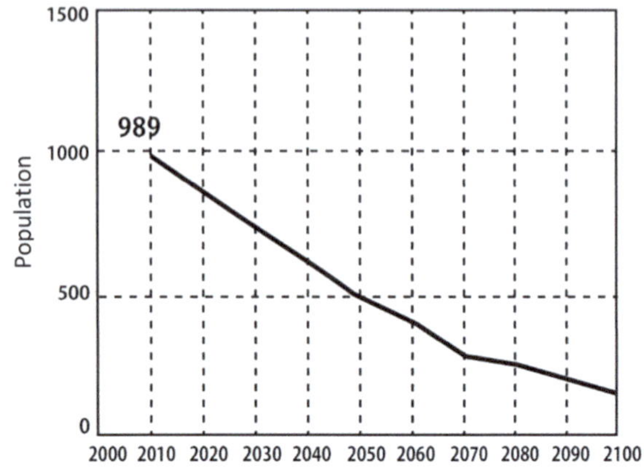

Fig. 14.14 Future population in Nyu district

For sustainable land management, the following four methodologies can be considered; (1) corresponding to the farmland by improving the efficiency of the fundamental labor accounts by relying on an association, (2) creating social systems to compensate for the outside economy for the fundamental labor accounts (collect the tax and volunteers of outside), (3) specifying the place to manage and the place to abandon if the total area management is impossible considering the balance of fundamental and profitable labor accounts, and (4) prioritizing the smaller management level in the same land use types such as fallow fields to biotope fields or forest under 30 years old to over 30 years old.

As one example, we calculated the future land use scenario. There is a tendency that the farmlands located on the upside area of Tachibai water channel are

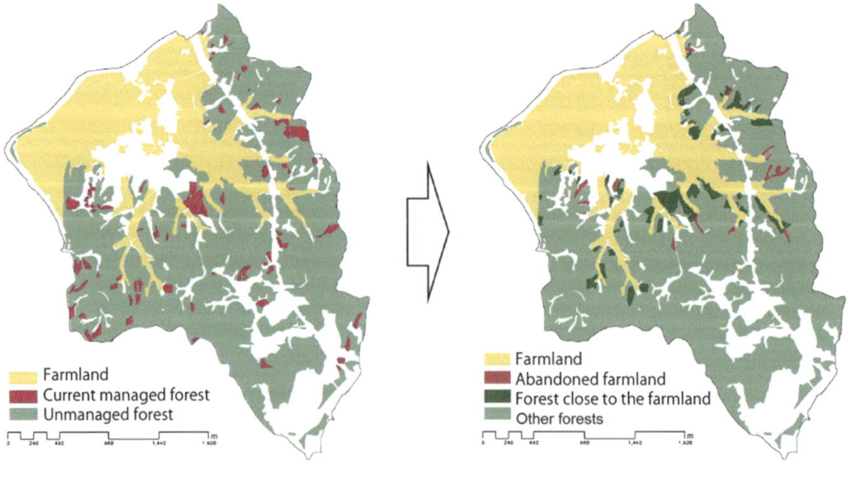

Fig. 14.15 Current and future scenarios in Nyu district

abandoned and it is said to cause harm from the wildlife. If the abandoned forest contacts farmland or the residential area, it also causes harm by the wildlife. Thus, we simulated two cases; one is to change all abandoned farmland into managed land (biotope type), and the other is to change the forest which is suitable for management close to the farmland into managed land. We then calculated how many labor accounts are needed in each case. The current labor account of forests is 9,225 h (Fig. 14.15). Case 1 needed 1,030 h and case 2 needed 7,668 h, so the total amount of case 1 and case 2 is under 9,225 h, and it is "ideally" and "quantitatively" possible. However, the land ownership and the stakeholders were not considered in this case study. In the future area planning in the Satochi-Satoyama area where the population is decreasing, we should consider where the spatially weak district is and how to manage the area.

References

Forestry Agency, Ministry of Agriculture, Forestry and Fisheries of Japan (2008) 2007 the forest and forestry paper

Ministry of Agriculture, Forestry and Fisheries of Japan (2012) Agriculture business management statistics investigation farm products life costs statistics, http://www.maff.go.jp/j/tokei/kouhyou/noukei/. Accessed 31 March 2016

Osawa K et al (2008) A new paradigm for conserving biodiversity at the fields of rural planning for the future. J Rural Plan Assoc 27(1):14–19

Terada T et al (2010) A citizen-based maintenance scheme for conservation of suburban Satoyama based on the Forest Working Plan. J Rural Plan Assoc 29:179–184

Uzawa H (2000) Social common capital. Iwanami Publisher

Chapter 15
Case Study of Other Field Satoyama Neighborhood Landscape Complex Units: Kayumi District

Nobuko Kawaguchi and Hiroyuki Shimizu

Abstract This chapter examines the case study of Kayumi District, Matsusaka City, Mie Prefecture, as landscape management labor accounts of an other field Satoyama type. Kayumi District is located in the mountainous countryside that can be commuted to easily from urban areas by car. The district shows a typical aged society in Japan, and systematized farmland and forest managements have been developed; for example, the corporate contracted farming system. Kayumi District has rich forest and farmland landscapes and a relatively large population; however, there are few people who manage the forests and farmlands. The development of further human resources to maintain the forest and agricultural land is required. This chapter describes the differences between the amount of landscape labor accounts of aging workers and corporative farming workers. The landscape management labor accounts are suggested to be used as base data for community discussions.

Keywords Other field satoyama neighborhood landscape complex units · Rural planning · Corporate farming · Tea plantation · Forestry

15.1 Introduction

This chapter considers landscape management labor accounts of an other field Satoyama type. The targeted district is Kayumi in the mountainous countryside of Matsusaka City (Fig. 15.1). The area is located midstream along the Kushida River and consists of river terraces surrounded by mountains.

N. Kawaguchi · H. Shimizu (✉)
Graduate School of Environmental Studies, Nagoya University,
Furo-cho, Chikusa-ku, Nagoya, Aichi 464-8601, Japan
e-mail: shimkankyou@msn.com

N. Kawaguchi
e-mail: nobqou@gmail.com

H. Shimizu et al. (eds.), *Labor Forces and Landscape Management*,
DOI 10.1007/978-981-10-2278-4_15

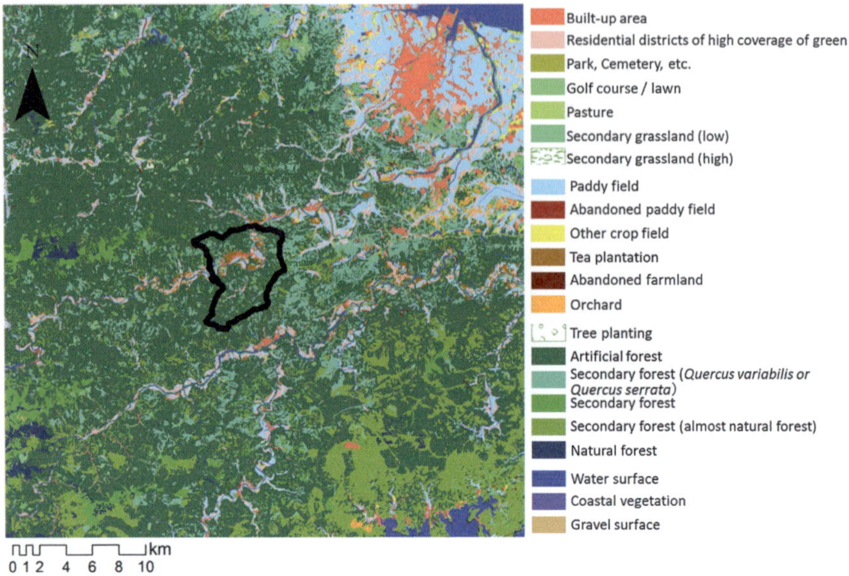

Fig. 15.1 Study area

Table 15.1 shows the size of the landscape area and its units and elements as well as management organizations within this district. The total size of the district area is 3,018 ha. Forests and farmlands occupy 83 % and 10 % of the total area, respectively, and developed areas occupy only 3 %.

Farmlands are developed along sloped areas between 8° and 15° as well as between 15° and 20°. Farmlands in these sloped areas consist mostly of tea plantations, due to the difficulty of developing paddy fields along such grades, as they require irrigated water. Some smaller paddy fields, however, are developed in the lower part of the terraces. Figure 15.2 shows some typical landscapes in the district.

According to the *2005 and 2010 Population Census* (Statistics Bureau 2005 and 2010), the population and number of households in Kayumi District were 2,200 and 844 in 2005, and 2,009 and 784 in 2010, respectively, with a continuing trend of decline. Residents older than 65 years occupied 33.5 % of the total population in 2010, revealing a typical aged society.

In Kayumi District, forests and farmlands are divided into small private sites. Farmlands of which are often managed by part-time farmers whose main income depends on activities other than farming. Most of the tea plantations and paddy fields in this district are cultivated by corporate, contracted farming. Artificial forests in the district are managed mostly by a forest cooperative, with few personal forest management enterprises.

The corporate contracted farming system, however, has a serious issue: the number of corporate farmers is limited and has nearly reached the maximum of their current cultivation abilities. Therefore, it is impossible to accept more offers from

Table 15.1 Landscape units/organizations

Landscape unit		Landscape element	Organization	A_l (managed) (a)	A_l (un-managed) (a)	A_l (total) (a)	S_l (a)	
Farmland	Tea plantation	Tea plantation	Self-employment and farm corporation	13,158	0	13,158	13,158	20,976
	Paddy field	Paddy field	Self-employment and Farm corporation	5,374	0	5,374	5,374	
	Other crop field	Other crop field	Self-employment	2,242	0	2,242	2,242	
	Mulberry field	Mulberry field	Self-employment	28	0	28	28	
	Orchard	Orchard	Self-employment	174	0	174	174	
Vacant land	Vacant land	Grassland and tree	Private owner	0	7,400	7,400	7,400	
Forest	Artificial forest	Tree	Self-employment	5,116	210,536	251,357	251,357	
			Forestry co-operative	35,705				
Built-up areas	Private building lots	Tree and grassland	Private owner	1,823	0	1,823	7,751	11,198
	Parking lots	Tree and grassland	Private owner	0	74	74	2,229	
	Public facilities	Tree and grassland	Private owner	105	0	105	989	
	Shrines and temples	Tree and grassland	Private owner	11	0	11	229	
Water surface	Rivers	–	Public	0	0	0	3,864	4,436
	Pond	–	Irrigation association	0	0	0	572	
Road	Road	–	Public	0	0	0	6,510	
Total				63,736	218,010	281,746	301,876	

Fig. 15.2 Typical landscapes in the district

aged farmers, who cannot cultivate their own fields any more, in future. Additionally, according to interviews with resident farmers, they are strongly concerned that abandonment of tea plantations might be increase rapidly in the future due to the recent decrease in tea prices (i.e., see Sect. 15.2).

The second characteristic of farming activities in this district is a simple water intake system for the paddy fields. Through the development of land readjustment projects, the water intake system in this area has changed from an irrigation pond system with open irrigation channels to a pumping-based system with intake pipelines. This improvement has reduced the burden of farmers to mow alongside the water channels. The irrigation ponds, however (which are not in use anymore), have been poorly maintained by the community. As such, in this district there is almost no voluntary collaborative management works, as shown in Asami District (Chap. 13) and Nyu District (Chap. 14) through "Deai." The reason for Kayumi District's lack of development of collaborative works is due to the development of tea plantations, which do not require collaborative management to maintain an irrigated water system.

These abandoned irrigation ponds and springs, however, can provide an important function for the cultural landscape by unifying the Satoyama landscape and also creating biodiversity and providing important habitats. Such a common consciousness has not yet taken hold in the community, even though some people are conscious about the benefits of these efforts. The landscape labor accounts proposed in this book can serve to incubate the awareness of residents to environmental management as base data for community discussions in future.

15.2 Change in Landscape Elements and Units

This section considers land use changes and (as this book also discusses) changes of landscape units in Kayumi District between 1966, 1975, and 2011. Figure 15.3 shows the spatial distribution changes of landscape units and Fig. 15.4 shows the details regarding the change of each landscape unit between 1966 and 2011 except for forests. Farmlands have decreased, but vacant lands, building sites, and roads have increased. Increase of vacant lands is a sign of population decline and an increase of developed areas is a sigh of urban sprawl. It is interesting to observe both changes in this same district. The population of farmers in 2010 decreased to one-third of the population compared to 1966 levels (Fig. 15.5). This means that the social change from a farmland society to an urban society is progressing under an aging population and declining resident rates.

Figure 15.6 shows the detailed changes of the different types of farmlands between 1966 and 2011. Paddy fields and other crop fields have decreased significantly; conversely, tea plantations and vacant spaces (e.g., abandoned farmlands) have increased. It can be said that the shrinking and deterioration of farmlands has occurred over this period in tandem with a reduced population of farmers. An increase in the number of tea plantations between 1966 and 2010 is likely due to farmers' intention to grow tea plants a cash crop. Recently, however, the price of tea has been decreasing, and the degradation of tea plantations are a serious concern. The abandonment of tea plantations will lead to a single-phase forest, which will not be beneficial to the district due to poor usage options and low biodiversity.

15.3 Landscape Units and Organizations

15.3.1 Overview

This section considers the landscape labor accounts and densities of each landscape unit. In Kayumi District, most paddy fields and tea plantations are managed by corporate organizations. That is, private land owners entrust their entire or partial farmland management to hired corporate organizations, only cultivating a very small part of their farmland for self-consumption, often done as kitchen garden style. Most forests in Kayumi District are possessed by private landowners, but their management is done mainly by the forestry cooperative of the area. There are few private forests for which management is done by small private forestry families.

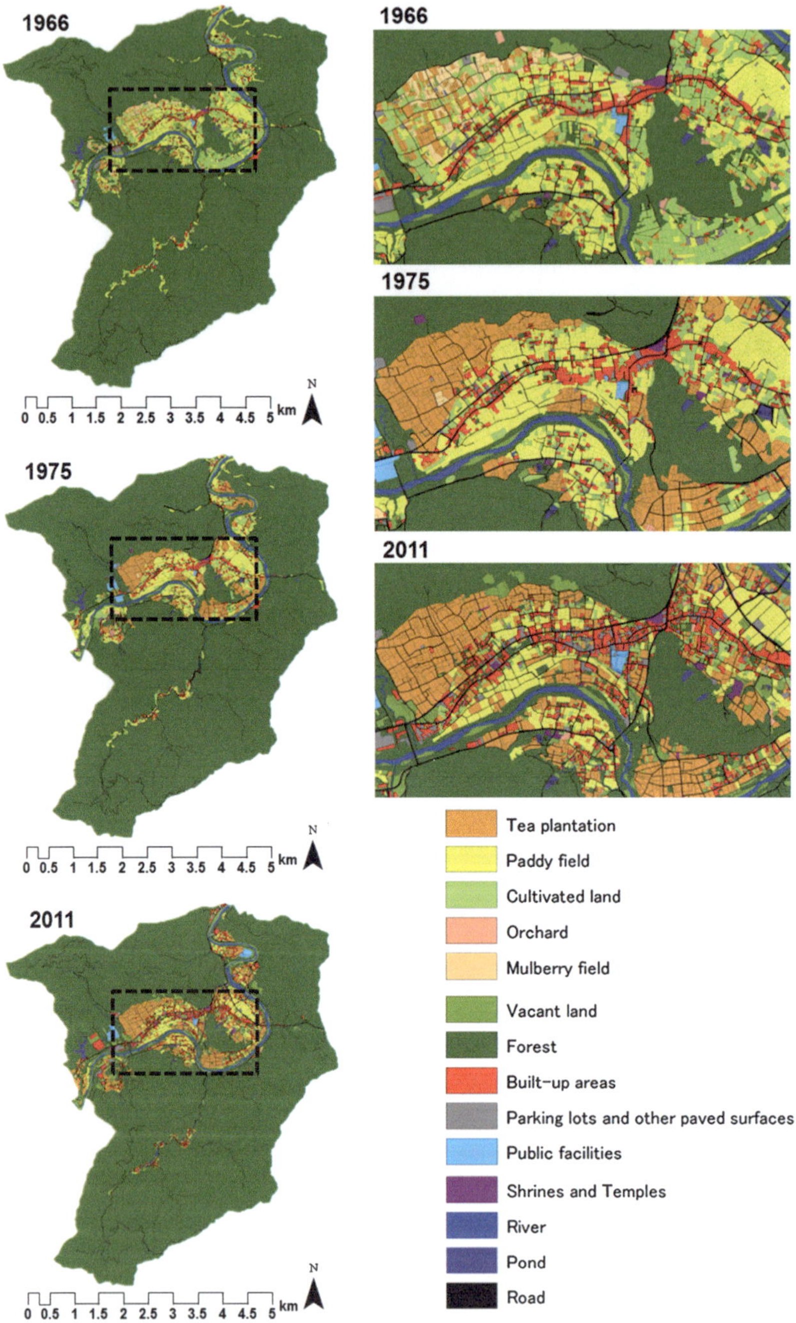

Fig. 15.3 Spatial distribution changes of landscape units

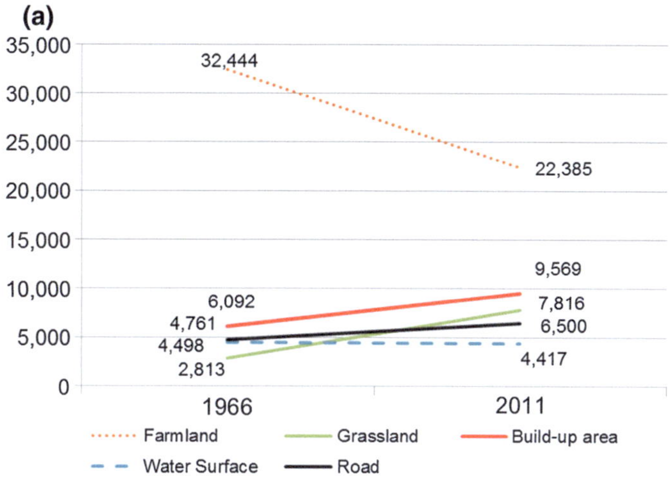

Fig. 15.4 Change of unit areas (except for artificial forest)

Fig. 15.5 Change in the number of farm households (1960–2010)

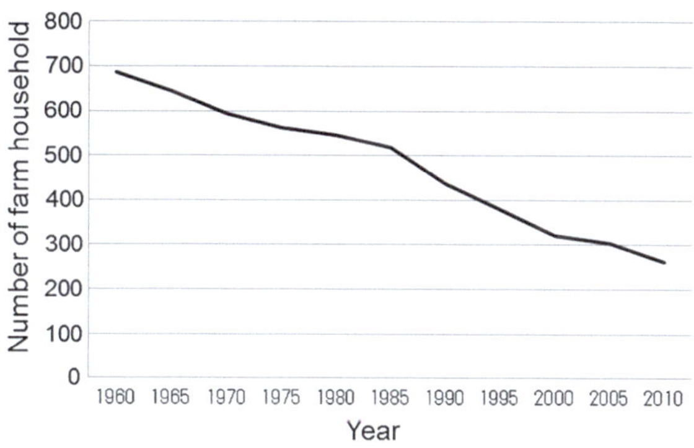

The interviews conducted for this book were done with six farmers or foresters, each with different management styles; for example, farmer A is a corporate managing paddy field farmer, B is a corporate managing tea plantation farmer, C is a private farmer managing both paddy fields and tea plantations, D is a corporate forester managing artificial forest, E is a private forester managing artificial forest, and F is a private forester managing both artificial forest and tea plantations.

15.3.2 *Paddy Fields*

Table 15.2 shows the landscape management labor accounts for paddy fields managed by a private farmer and a corporate farmer, respectively. In Kayumi District, unlike Nyu District in Chap. 14, only rice is cultivated during the summer

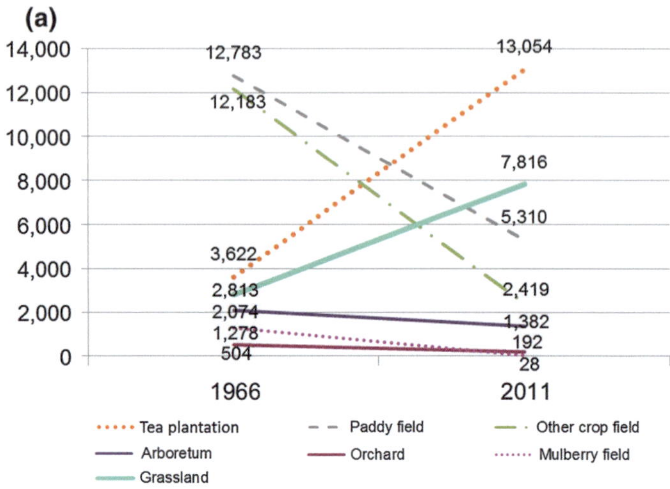

Fig. 15.6 Change of unit areas (agricultural land)

season, and the parallel cultivation of soybeans and winter cultivation of wheat are not introduced. In the case of corporate farmer A, the working farmlands are extended to the outer areas beyond Kayumi District, and labor accounts in the farmland of the outer areas are included. Individual land owners of paddy fields entrust management components such as nursery, planting, and harvest (which affect direct income) to corporate farmers, while basic management components such as plowing, mowing, and water management are done in many cases by the private landowners themselves. This is because management components such as planting and harvesting require a great deal of manpower and advanced machinery. The case of private paddy farmer C in Table 15.2 shows one example of such a case.

The labor densities of corporate farmer A and private farmer C are 2.02 h/a and 5.47 h/a, respectively. Private farmer C entrusts some work (e.g., raising seedlings, prevention of damage by wildlife, rice planting, manuring, harvesting, and drying) to corporate farmer A. The average value of the landscape labor density of paddy fields in Tokai area is 2.63 h/a (data from *Agriculture business management statistics investigation farm products life costs statistics* Ministry of Agriculture, Forestry and Fisheries of Japan 2005). The labor density of corporate farmer A is a little less than the average value of the labor density of paddy fields in Tokai area.

15.3.3 Tea Plantations

Table 15.3 shows the landscape management labor accounts for tea plantations managed by a private farmer and a corporate farmer, respectively. If a private farmer decides to entrust the management of his tea plantation to a corporate farmer,

Table 15.2 Landscape labor accounts and densities of paddy field

Operator	Labor content		Annual working days	Number of workers	Working hours per day	$TCLA_{jkl}$ (h)	S_l (a)	TLD_{jl} (h/a)
Operator A: corporate organization	Manuring		21	3	8	504	4,347	0.12
	Plowing and land preparation	Plowing 1	10	2	8	160		0.40
		Plowing 2	10	2	8	160		
		Soil crushing 1	14	2	8	224		
		Soil crushing 2	14	2	8	224		
		Land preparation	60	2	8	960		
	Raising seedling	Preprocessing 1	1	1	4	3		0.08
		Preprocessing 2	10	1	0.5	4		
		Plantation (house)	30	3	4	256		
		Raising seedling (house)	81	1	1	81		
	Rice-planting		60	2	8	960		0.22
	Weed killer	Border	5	1	8	40		0.01
	Water management and other	Prevention of damage by wildlife	14	1	8	112		0.15
		Water management	40	1	5.5	220		
		Mowing of border 1	11	1	8	144		
			7	1	8			
		Mowing of border 2	7	1	8	112		
			7	1	8			
		Water management	30	1	2	60		
	Harvest		60	6	8	2880		0.66
	Drying		60	1	23	1380		0.32
	Others (meeting, keep books, Repair agricultural implement)		18	2	8	288		0.07
	Total					8,771	4,347	2.02

(continued)

Table 15.2 (continued)

Operator	Labor content		Annual working days	Number of workers	Working hours per day	$TCLA_{jkl}$ (h)	S_l (a)	TLD_{jl} (h/a)
Operator C: self-employment	Manuring		1	1	1	1	16.3	0.06
	Plowing and land preparation	Plowing 1	1	1	2	2		0.67
		Plowing 2	1	1	1	1		
		Soil crushing	1	1	2	2		
		Land preparation	2	1	3	6		
	Raising seedling[a]		–	–	–	–		0.00
	Water management and other	Repair border	1	1	4	4		4.60
		Mowing of border 1	1	1	2	2		
		Mowing of border 2	1	1	2	2		
		Mowing of border 3	1	1	2	2		
		Water management	130	1	0.5	65		
		Prevention of damage by wildlife[a]	–	–	–	–		
	Weed killer	Paddy field (water surface)	1	1	0.16	0.16		0.01
	Manuring[a]		–	–	–	–		0.00
	Pest control	Pest control 1	1	2	0.5	1		0.12
		Pest control 2	1	2	0.5	1		
	Harvest[a]		–	–	–	–		0.00
	Drying[a]		–	–	–	–		0.00
	Total					8,771	16.3	5.47

[a]Farming by contractor (corporate organization)

Table 15.3 Landscape labor accounts and densities of tea plantation

Operator	Labor content		Annual working days	Number of workers	Working hours per day	$TCLA_{jkl}$ (h)	S_l (a)	TLD_{jl} (h/a)
Operator B: corporate organization	Plowing and manuring			10		3,260	3,100	1.05
	Pruning			10		4,720		1.52
	Weeding, pest control	Weeding		13		8,640		3.06
		Pest control		10		840		
	Harvest	Harvest		13		6,800		3.35
		Other		10		3,600		
	Others (meeting, keep books, repair agricultural implement)			10		390		0.13
	Total					28,250	3,100	9.11
Operator C: self-employment	Plowing and manuring	Manuring	1	1	6	6	65	0.09
			1	1	6	6		0.09
			1	1	6	6		0.09
			1	1	4	4		0.06
			1	1	4	4		0.06
			1	1	6	6		0.09
		Plowing	1	1	6	6		0.09
	Pruning	Pruning leaf buds	2	2	3	12		0.18
		Pruning 1	3	2	4	24		0.37
			3	2	4	24		0.37
		Pruning 2	1	1	4	4		0.06

(continued)

Table 15.3 (continued)

Operator	Labor content		Annual working days	Number of workers	Working hours per day	$TCLA_{jkl}$ (h)	S_l (a)	TLD_{jl} (h/a)
	Soil disinfection		1	2	5	10		0.15
			1	2	5	10		0.15
			1	2	5	10		0.15
			1	2	5	10		0.15
			1	2	5	10		0.15
	Harvest	First-grade tea	1	2	5	10		0.49
					2	4		
					5	10		
					4	8		
		Second-grade tea	1	2	5	10		0.55
					5	10		
					4	8		
					4	8		
	Weeding, pest control	Weed killer	1	1	5	5		0.08
			1	1	5	5		0.08
			1	1	5	5		0.08
			1	1	5	5		0.08
		Weeding and patrol	150	1	2	300		4.62
	Total					540	65	8.31
Average of Mie prefecture (MAFF 2005) (except for clerical work)								9.65

he will entrust all components, unlike in the case of paddy fields. The harvesting of tea leaves is done three times per year, in spring, summer, and autumn; in winter, the plants are monitored. In summer, weeding is also done. The management works of pruning and fertilizing are done according to the harvesting times. Herbicides are used on the outskirts of the plantation, but the weeding inside the plantation is done by hand. The labor densities of farmer C and corporate farmer B are 8.31 h/a and 8.98 h/a, respectively. The average labor density of tea plantations according to the data from Mie Prefecture (*Agriculture business management statistics investigation farm products life costs statistics* Ministry of Agriculture, Forestry and Fisheries of Japan 2005) is 12.54 h/a. (The total value of the work item, except for clerical work is 9.65 h/a.)

The labor density of tea plantations is four times higher than that of paddy fields. In general, corporate farmer A, who manages only paddy fields, and corporate farmer B, who manages only tea plantations, both work more than 5 days per week; on the contrary, private farmer C, who cultivates both paddy fields and tea plantations, works a little more than once per week, and farmer D, who cultivates only paddy fields, works once per week. The corporate farmers work as a profession, but the private farmers work on the land only part-time or on weekends. In reality, the working load differs greatly across the seasons, with some seasons that are very busy and some that are now. In busy seasons, private farmers often must work every day.

According to an interview with corporate farmer A, the toughest labor load is closed up. He usually works 8 h per day but during a busy agricultural season, he will work more than 20 h per day and has almost no holidays throughout the year. Under these conditions, there is no additional room to be entrusted with more fields; in addition, due to a lack of potential farmers, there is a shortage of professionals that can be entrusted the extra fields that are likely to appear in future, due to the progressing demographic changes as well as the depopulation and aging of the society.

15.3.4 Irrigation Equipment

Some sub-districts along Kayumi's irrigation channels use water from the Kushida River, but some other sub-districts do not have tributaries with rich water, and must use irrigation ponds, which collect smaller water flows. In Kayumi District, there are some irrigation associations based on such kinds of water resources. Figure 15.7 shows the distribution of irrigation ponds and managing organizations.

Because the current irrigation system pumps water from Kushida River and along pipeline networks, most of the irrigation ponds are no longer in use. The management of these ponds is instead done by the leaders of irrigation associations, and the collaborative working system called "Deai" by residents. This can be seen in the cases of Asami District in Chap. 13 and Nyu District in Chap. 14.

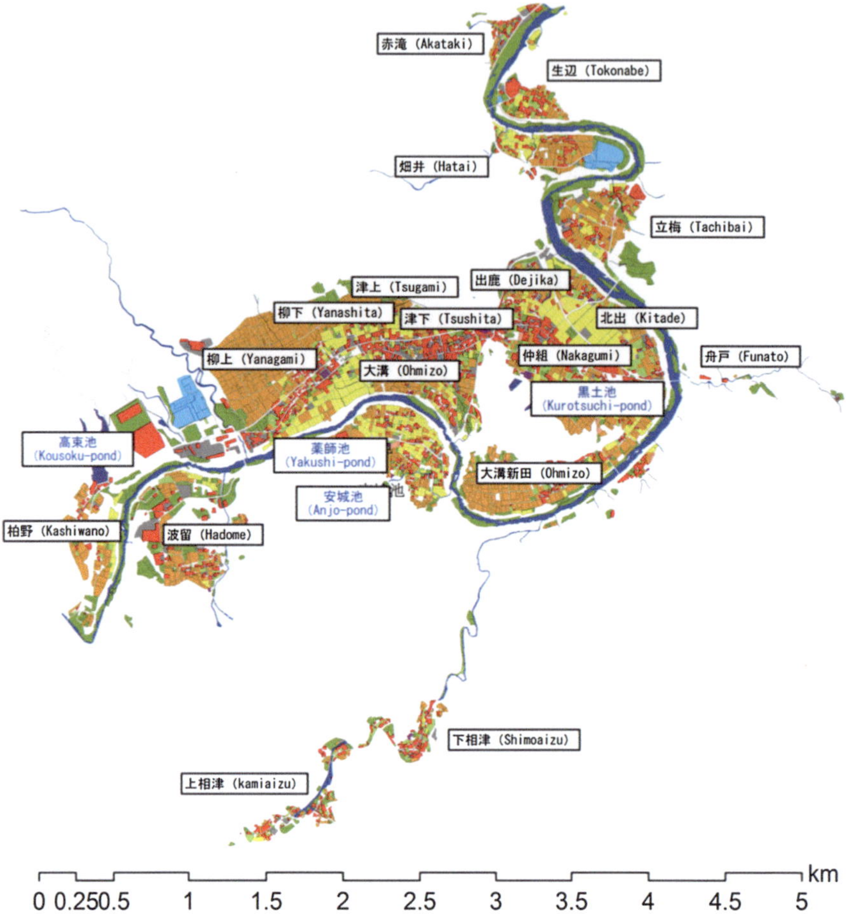

Fig. 15.7 Distribution of irrigation ponds and managing organizations

15.3.5 Artificial Forests

Figure 15.8 and Table 15.4 show the zoning of forests according to their functions (i.e., sustainable productive forest, ecological conservation forest, and landscape and cultural-service conservation forest) as designated by forest planning as well as zoning according to the owner types. The managing areas of the forests (based on the *Forest Register* Mie Prefecture 2013) comprise 13 % of productive forests, 4 % of environmental forests, and 17 % of the total forest area. Figure 15.9 shows pictures of each forest type.

Sustainable productive forests are those that aim to produce timbers and other products in a stable and structured manner with periodic, scheduled management (i.e., artificial planting, weeding, thinning, and felling with intervals of 10 years, 25 years, 35 years, and 60 years and 90 years, respectively) (Miyata et al. 2011).

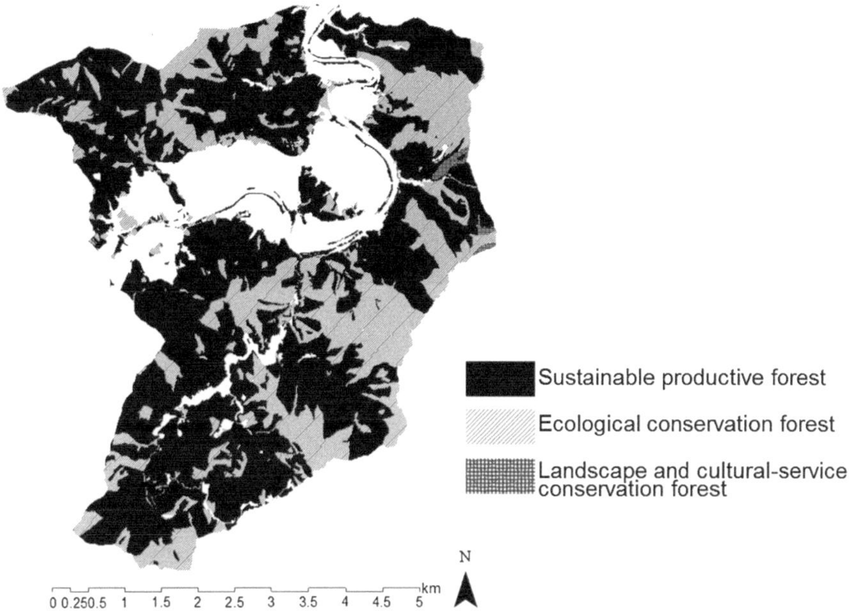

Fig. 15.8 Distribution of forest zoning (Mie Prefecture)

Ecological conservation forests are those that aim to increase public function, such as recharging the water supply and preventing disasters. Conservation forests in Kayumi District are a type of environmental forest, where artificial forests are cut down and left to undergo a natural transition to broadleaf forests.

Landscape and cultural-service conservation forests are those that aims to conserve and recreate the forest environment by residents. In Kayumi District, landscape and cultural-service conservation forests do not even account for 1 % of the total forest area, as related projects to encourage these forests has not been carried out.

Almost of all the management of forests is dependent upon the type of projects that relate to governmental grants (i.e., forest conservation projects, afforestation projects, and forest-environment creation projects). Most of the forest management is done by the hand by those in forest cooperatives, though there are some individual forest owners who entrust only the operational applications to the government (e.g., deforestation), but the management itself is done by the operators of the entrusted forestry enterprise. These operators, however, are mostly aged persons more than 65 years old. In the forest industry, it can be difficult to work in agriculture during post-retirement age, as forest industry management requires a high level of professional skill and physical ability.

Figure 15.10 summarizes the landscape labor densities of artificial forest. The amount of the landscape labor density of corporate forester D is rather low, clearly showing that this corporate forester works effectively. Conversely, the amount of

Table 15.4 Area size of managed/unmanaged forest in each zoning type

Zoning type		Unmanaged		Managed		Total	
		Area (a)	Percentage (%)	Area (a)	Percentage (%)	Area (a)	Percentage (%)
Sustainable productive forest		134,251	57	32,540	14	166,792	71
Environmental forest	Ecological conservation forest	67,505	29	8,739	4	67,505	29
	Landscape and cultural-service conservation forest	1,195	1	0	0	1,195	1
Total		67,505	29	41,279	18	235,491	100

Fig. 15.9 Pictures of each forest type

the labor density of private forester E, who manages artificial forest and tea plantations, is rather high. In the hilly and mountainous areas of Japan, there are many private foresters that engage in part-time agriculture and forestry.

Fig. 15.10 Summary of landscape labor density of artificial forest

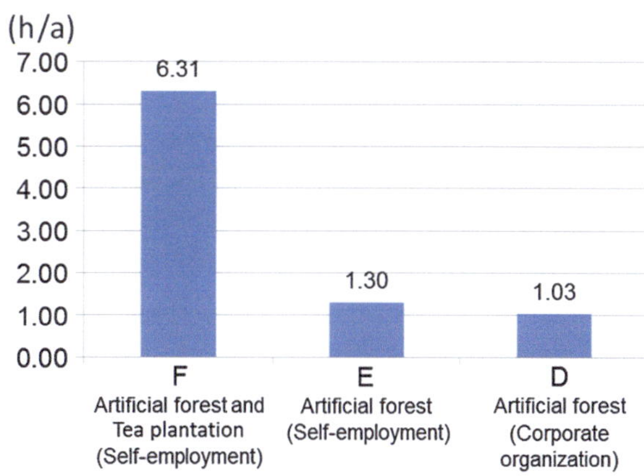

15.4 Summary of Landscape Labor Accounts

Figure 15.11 shows that the labor densities of farmlands and forests managed by corporate organizations are relatively low and have high efficiency. In general, tea plantations require high labor densities compared to paddy fields and forest management.

Figure 15.12 shows personal annual labor accounts and densities of different types of landscape units and management styles. The annual labor accounts of people in corporate organizations are very high—more than 2,100 h per year, which means 7 h work days, 6 days per week, 50 weeks per year. The workers have almost no vacation time. On the contrary, private farmers have relatively high labor densities, but low annual labor accounts. According to interviews, part-time farmers often change to full-time farmers after their retirement, but they begin to feel their physical limitations after just a few years, and then tend to entrust the management of their farms to a corporate organization. These aged individual farmers also own kitchen gardens for self-consumption, and they work in these gardens about two hours every day, in addition to the work they do on their productive farms.

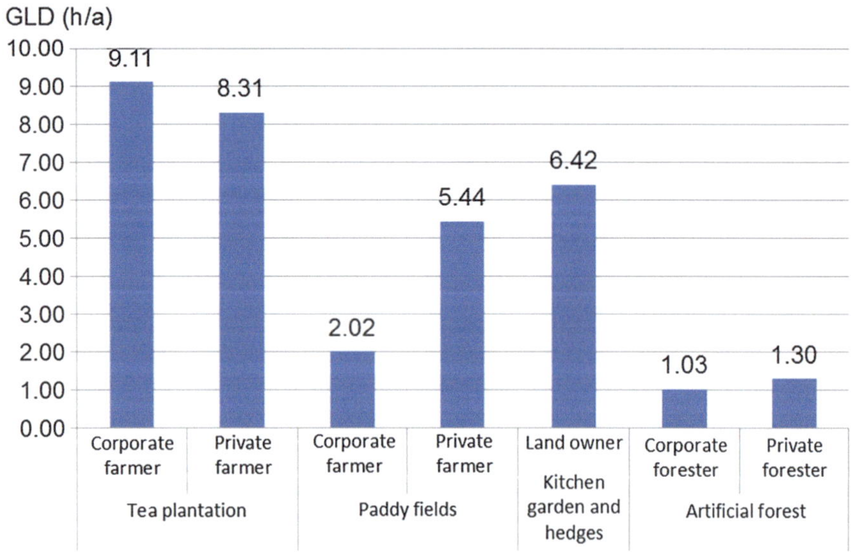

Fig. 15.11 Labor density of different landscape units and management organizations

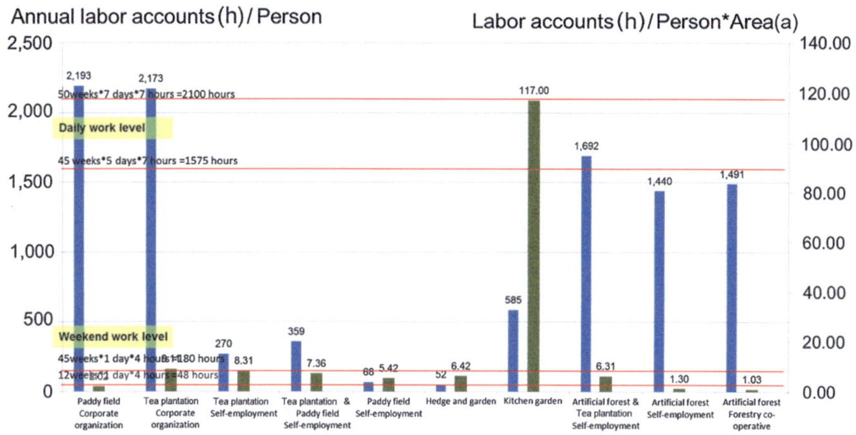

Fig. 15.12 Personal annual labor accounts and density

Table 15.5 shows the resulting estimation of the landscape labor accounts for all of Kayumi District. The total amount of the landscape labor accounts for the district is 319,207 h. Among the farmland units, tea plantations require the highest amount of labor. When comparing paddy fields and other crop fields, the landscape labor account in other crop fields is higher. Within the forest units, the labor density of the total area is lower than the labor density of the managed area. This is due to the fact that the managed area of forests comprises only 17 % of the total forest area. This district is occupied by almost all artificial forest, but the area of managed forest is small.

15.5 Consideration of Future Landscape Management

Kayumi District is considered an other field Satoyama landscape type originally shown in Chap. 2, though it has relatively wide farmlands. While most of these farmlands are efficiently managed by corporate farming organizations, the workloads of the professionals of these organization are already very high, and there is no room to accept additional entrusted works, now or in the future, particularly as the aging and depopulation of society is expected to progress.

Figure 15.13 shows the estimated future population of Kayumi District. In 2035, the population is expected to decrease by half of what it was in 2010. If the decreasing trends of the tea plantations and fields also continue, then enormous increase of undesired forests and abandoned grasslands can be expected.

In the future, it can be assumed that the population of farmers and foresters will decline in tandem with the general population decline. In addition, the aging society in this district will coincide with a decreased number of the residents who engage in agriculture and forestry. Consequently, new market entrants of farmers and foresters

Table 15.5 Overview of landscape labor accounts and densities in Kayumi district

Landscape unit		Organization	A_l (managed) (a)	A_l (total) (a)	S_l (a)	TLD_{jl} (h/a)		TLA_l (h)		NLD_l (managed) (h/a)		GLD_l (h/a)	SLI
Farmland	Tea plantation	S&Fa	13,158	13,158	20,976	12.54		165,000		12.54		12.54	128
	Paddy field	S&Fa	5,374	5,374		2.63		14,134		2.63		2.63	
	Other crop field	S	2,242	2,242		29.77		66,752		29.77		29.77	
	Mulberry field	S	28	28		28.26		789		28.26		28.26	
	Orchard	S	174	174		29.26		5,087		29.26		29.26	
Vacant land	Vacant land	Pr	0	7,400	7,400	0.00		0		0.00		0.00	
Forest	Artificial	S	5,116	251,357	251,357	1.30	1.07	6,653	43,573		1.07	0.17	
	Forest	Fo	35,705			1.03		36,920					
Built-up areas	Private building lots	Pr	1,823	1,823	11,198	12.38		22,575		12.38	12.31	2.13	
	Parking lots	Pr	0	74		0.00		0		0.00			
	Public facilities	Pr	105	105		12.38		1,295		12.38			
	Shrines and Temples	Pr	11	11		0.23		2		0.23			
Water	Rivers	Pu	0	0	4,436	0.00		0		0.00	0.00	0.00	
Surface	Pond	I	0	0		0.00		0		0.00		0.00	
Road	Road	Pu	0	0	6,510	0.00		0		0.00	0.00	0.00	
Total			63,736	281,746	301,876			319,207		5.01		1.06	128

Fa Farm corporation, *Fo* Forestry co-operative, *I* Irrigation association, *Pr* Private owner, *Pu* Public, *S* Self-employment

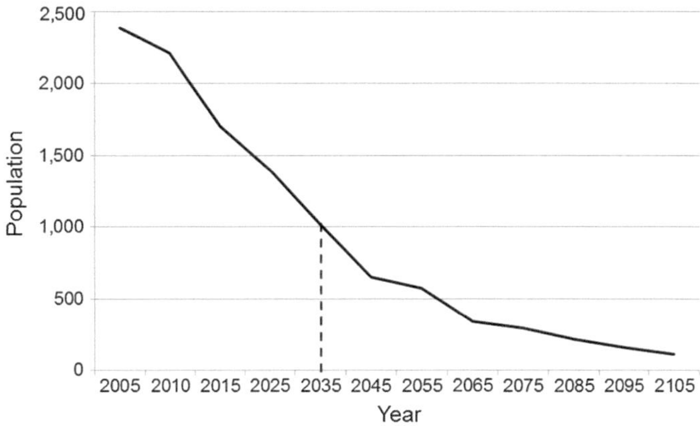

Fig. 15.13 Future population in Kayumi District

as cooperative and corporate association is necessary to conserve the forest and farmland in Kayumi District. The landscape management labor accounts should be considered material for investigating the directionality of methodology regarding who will fill up the vacancies in the labor force.

References

Ministry of Agriculture, Forestry and Fisheries of Japan (2005) Agriculture business management statistics investigation farm products life costs statistics. http://www.maff.go.jp/j/tokei/kouhyou/noukei/. Accessed 1 July 2015

Mie Prefecture (2013) Forest Register, Mie Prefecture

Miyata M, Kamegai K, Kato H, Kawase Y, Hayashi Y (2011) A proposal of the social necessity indicator for artificial forests in hilly and mountainous areas. J Jpn Soc Civil Eng 67(5). 67_I_389 67_I_398

Statistics Bureau (2005) 2005 Population census. Director-General for Policy Planning (statistical standards) and Statistical Research and Training Institute. http://e-stat.go.jp/SG2/eStatGIS/page/download.html. Accessed 1 Mar 2015

Statistics Bureau (2010) 2010 Population census. Director-General for Policy Planning (statistical standards) and Statistical Research and Training Institute. http://e-stat.go.jp/SG2/eStatGIS/page/download.html. Accessed 1 Mar 2015

Chapter 16
Case Studies in Coastal Neighborhood Landscape Complex Units: Saki-Shima Peninsula

Taito Yasufuku and Hiroyuki Shimizu

Abstract This chapter shows the case study of Saki-Shima Peninsula, Shima City, Mie Prefecture, as a coastal neighborhood landscape complex unit type. Saki-Shima includes five districts—Katada, Fuseda, Wagu, Koshiga, and Goza—which have a total population of 12,000, with 10 % of the population living in fishing villages and employed as fishermen. Because of the low international economic competitiveness of marine products and changes in the natural environment, the decrease in number and aging of fishermen has progressed in this district. The demographic, industrial, and social changes over the past 50 years in Saki-Shima Peninsula are analyzed in this chapter. Then, three scenarios are assumed according to different future land uses, and the sustainability of land management and employment is considered.

Keywords Coastal neighborhood landscape complex units · Demographic change · Land use change · Fishing and agricultural village

16.1 Introduction

The survey on the landscape management of coastal areas is also important because Japan is surrounded by oceans and special lifestyles are developed in such areas. The targeted neighborhood landscape complex units of the survey are five districts (Katada, Fuseda, Wagu, Koshiga, and Goza) of Saki-Shima peninsula in Shima city, Mie prefecture shown in Figs. 16.1 and 16.2.

The total population of the districts is 12,000 and 10 % of the population in fishing villages are employed fishermen. Because of the low international economic competitiveness of marine products and the change in the natural environment, the decrease in the number and aging of fishermen has progressed in these districts.

T. Yasufuku · H. Shimizu (✉)
Graduate School of Environmental Studies, Nagoya University,
Furo-cho, Chikusa-ku, Nagoya, Aichi 464-8601, Japan
e-mail: shimkankyou@msn.com

H. Shimizu et al. (eds.), *Labor Forces and Landscape Management*,
DOI 10.1007/978-981-10-2278-4_16

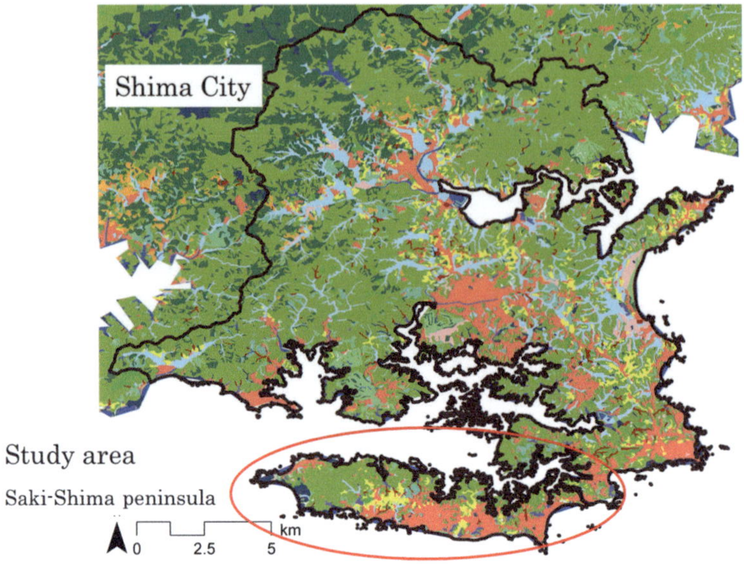

Fig. 16.1 Location of Saki-Shima peninsula

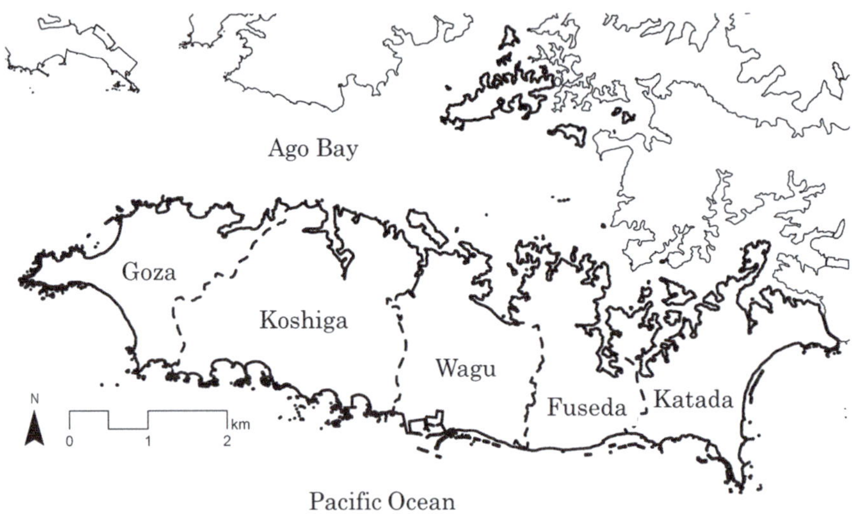

Fig. 16.2 Landscape complex units in study area

Furthermore, the ratio of fishermen is expected to decrease more in the future. The corporatization of fishermen will also progress because of productive efficiency.

In the period between 1972 and 1993, in which Japanese economic growth was the highest, tourism flourished and the socio-economic structure of the districts greatly changed. But after 1993, which was the end of Japanese economic growth, the social-economic structure was changed again and aging and shrinking tendencies accelerated.

These changes in employment and working styles will affect land use and land management in the future if the demographic changes, aging and depopulation, continue. There might not be enough population for the appropriate management of the land for environmental conservation and disaster prevention in the future.

16.2 The Landscape Units and Organizations

Table 16.1 shows the area sizes of landscape units and the dominant management organizations in the five districts.

Forests, which are privately owned, are mostly secondary forests and occupy half of the area of the districts. Until the middle of the 20th century, these forests were used for gathering fuel materials and fertilizer, but afterwards, such usage disappeared. Consequently, management of the forests is not sufficiently done any

Table 16.1 Landscape units and management organizations in Saki-Shima peninsula

Landscape unit		Landscape element	Organization	A_l (a)		S_l (a)	
Built-up area		Tree and grassland	Private owner	1,894		25,761	
Forest		Tree	Private owner	103,229		103,229	
Farmland	Paddy field	Paddy field	Private owner	18	8,202	18	8,202
	Other crop field	Other crop field	Private owner	7,812		7,812	
	Orchard	Orchard	Private owner	155		155	
	Tea plantation	Tea plantation	Private owner	63		63	
	Arboretum	Arboretum	Private owner	154		154	
Road	National highway	Tree and grassland	Construction office in Shima city	1,308	1,446	1,308	9,516
	Prefectural road	Trees and grasslands	Government	138		8,208	
Beach	Beach	Volunteer, fishermen cooperatives		1,605		5,561	
Water surface		Non-element	Private owner	0		867	
Grassland		Grasslands	Private owner	7,594		7,595	
Others		Non-element	Private owner	0		10,448	
Total				123,970		171,178	
Coast	Seaweed forest	Seaweed forest	Women divers	6,000		201,403	

more. Also in early times, farmlands were developed widely and cultivated for the self-use of residents, but due to the change in the occupational structure of residents and in the commercial flow of fresh goods, these farmlands have shrunk and remain partially as fields for sweet potatoes which are a special product of these districts, and partially in the style of kitchen gardens. The management body of farmlands and forests depends on the sense of duty of individuals.

Beaches are managed by volunteers and fishermen cooperatives independently. Seaweed is managed by Ama divers, i.e., women divers, to protect them from sea urchins and starfish, especially in the Koshiga and Goza districts. The management of coastal landscape elements is done by the fishery organizations to protect their income. But, because of the recent degradation of the natural environments, more people and time are needed to manage them appropriately.

16.3 Analysis of Demographic, Industrial and Social Changes in 50 Years in Saki-Shima Peninsula

By examining the regional documents (Shima Town 2009) and the interviews of residents and the municipality, the relationship between demographic, industrial and social changes in the last 50 years are considered in this section.

Three typical periods are extracted as shown in Fig. 16.3.

(1) The period of self-help agro-fishery community (1920–1972)
(2) The period of employment conversion to tourism (1972–1993)
(3) The period of fishermen and population reduction (1993–2015)

Figure 16.4 shows the typical life and working style of the period of self-help agro-fishery community (Yokohama Shinema Shokai (Yokohama Cinema Commercial Company 1941)). Figure 16.5 shows the typical life and working style of the period of employment conversion to tourism. Figure 16.6 shows the typical life and working style of the period of fishermen and population reduction.

(1) The period of self-help agro-fishery community

The activities of women divers and pearl farming flourished until around 1972. Residents lived in a self-sufficient society by using their regional resources and there was little inflow of external capital. According to the statistics (*Miekentoukeisho* (the Statistics of Mie prefecture) (Mie Prefecture 1960–2010), the *Census of Fishery* (Ministry of Agriculture, Forestry and Fisheries of Japan 1963–2013) and the *Census of agriculture and forestry* (Ministry of Agriculture, Forestry and Fisheries of Japan 1960–2010)), about 35 % of the population were fishermen and half of the individual fishery families were engaged in agriculture in 1965. In spite of harsh land conditions, most of the households owned and managed their farmland. As for forests, the residents went to the Goza district and bought branches to use as fuel wood and fishing implements. Thinning and pruning of forests were done by the forest owners.

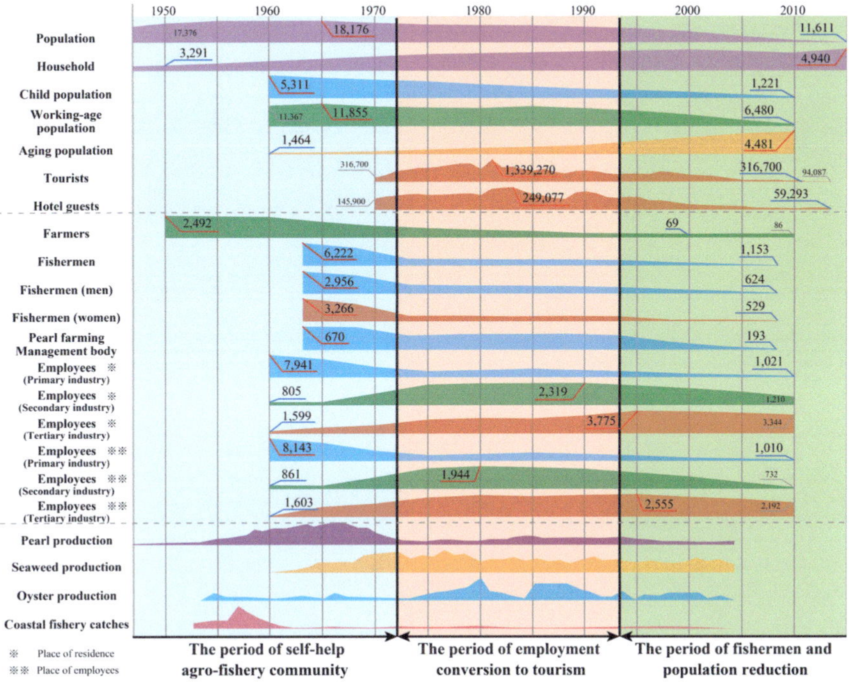

Fig. 16.3 Classification of periods by demographic, industrial and social changes

Fig. 16.4 Typical life and working style of the period of self-help agro-fishery community (Wagu no Ama 1941)

(2) The period of employment conversion to tourism

In the beginning of the 70's the pearl industry began to decrease, and the employment structure converted to tourism. The pearl production changed to oyster and seaweed production. This period continued until the collapse of the bubble economy in 1993.

Fig. 16.5 Typical life and working style of the period of employment conversion to tourism (provided by Shima City)

Fig. 16.6 Typical life and working style of the period of fishermen and population reduction

The reasons of the change were, (1) the lowering of international economic competitiveness of the marine products, (2) the change in the natural environment, (3) the decrease of population and the tourist increase of 2 times due to the development of the automobile society. In particular, employment in the Goza district increased according to the development of tourism contents such as sea bathing and camping. Residents didn't need to live in a self-sufficient condition any more. As a consequence, residents didn't use forests as fuel wood and fertilizer any more, and more forests began to be left uncared for.

(3) The period of fishermen and population reduction

After the bubble collapse in 1993, the decrease of the fishermen's population and also total population began to progress at a high speed. The aging rate reached 38 %, and the number of fishermen decreased to 10 % of the population. Also the farmers' population in 2010 decreased to 4 % of the population. A decline of the amount of fish caught and the selling price and environmental changes due to global warming collapsed the primary industry. Utilization of local products by residents was reduced because of the development of interregional commercial markets. The number of tourists also decreased.

This integrated decrease in economic, social and environmental aspects changed the employment environment in the tertiary industry, and accelerated the outflow of the population. The society of the districts is going to depend more and more on the extra-regional economy.

16.4 Analysis of Land Use Changes from 1968 to 2011

Figures 16.7, 16.8 and 16.9 show the spatial distributions of the landscape units in three periods, 1968, 1975 and 2011, respectively. The increase of forests and decrease of farmlands are easily recognized. Through the interviews of the stakeholders, it was clarified that the changes in the landscape were caused by the change in the usage of regional resources.

Figure 16.10 shows the typical landscape units in 2014. The figure in the upper left shows the abandoned farmlands among a small valley. Such abandoned farmlands change firstly to grasslands and then to forests. The figure in the lower left is a vacant space in a residential area used for a parking lot. The figure in the upper right shows the mixed area of houses and farmlands in the hilly side. Residential houses dot the sunny farmlands in a distributed manner. The figure in the lower right shows the seaweed in the sea. The floor bed of seaweed deteriorates because of environmental changes, such as changes in nutrients, seawater temperatures and so on.

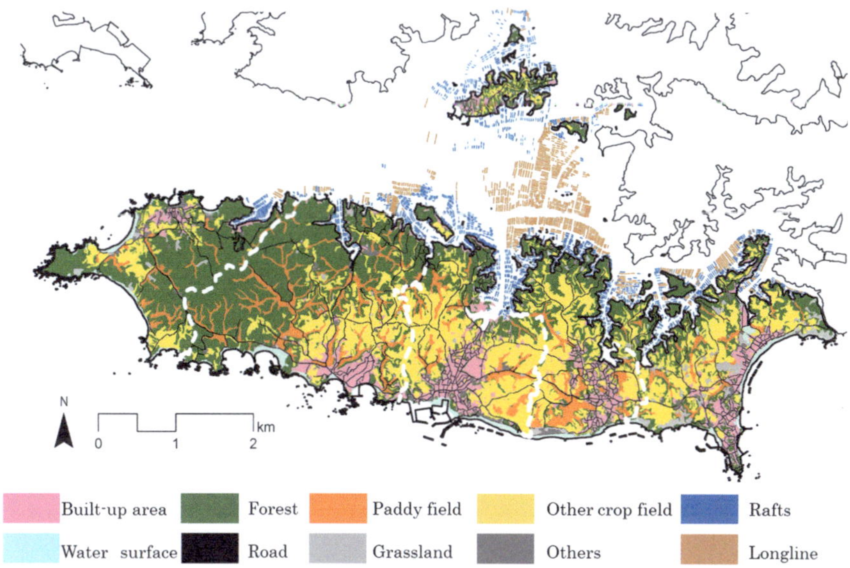

Fig. 16.7 Spatial distribution of landscape units in 1968

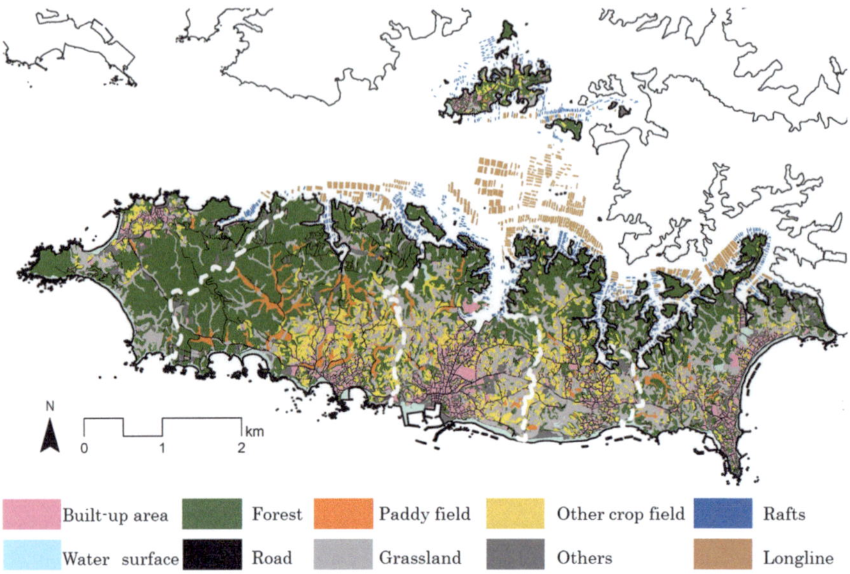

Fig. 16.8 Spatial distribution of landscape units in 1975

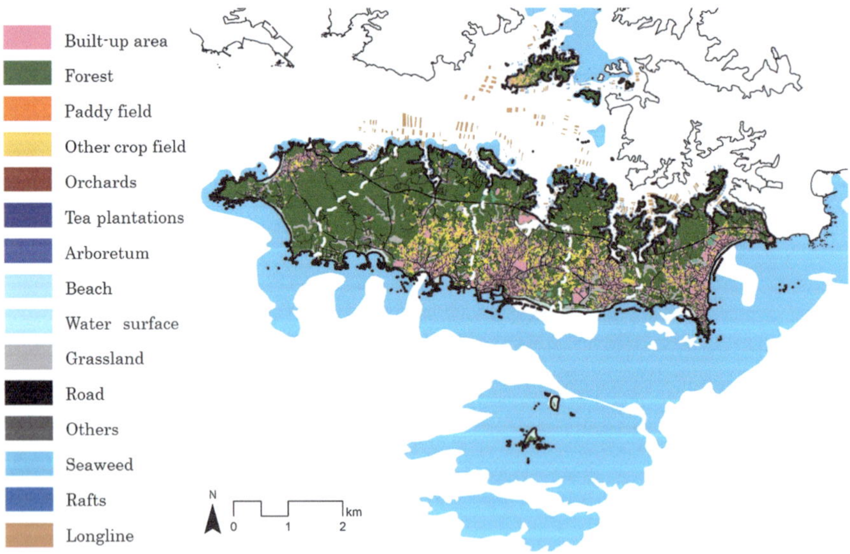

Fig. 16.9 Spatial distribution of landscape units in 2011

Fig. 16.10 Landscape units in 2014

Figure 16.11 shows the coastal landscape change between 1960 and 2014. The figure in the left shows the coastal landscape in 1960. There are many floating farms in the ocean in 1960, but in 2014 they become much fewer. On the contrary, trees grow and the biomass of forests increases.

Figure 16.12 shows the change in the ratio of landscape units in each district. In each district the increase of forests and decrease of farmland are observed. Figure 16.13 shows the change in landscape units' ratio according to altitudes. Built-up areas in lower elevations decrease and built-up areas in hilly areas increase. Reforestation progresses in all altitudes. Farmlands, which mainly exist in the altitudes between 7 and 37 m, also decrease. Figure 16.14 shows the change in landscape units' ratio according to average land gradients. Reforestation can be observed in all land gradient categories.

Figure 16.15 shows the change in landscape units' ratio in three periods. In 1965, 30 % of the population engaged in fishing and pearl farming. In the fishing

Fig. 16.11 Coastal landscape change between 1960 and 2014 (provided by Shima City)

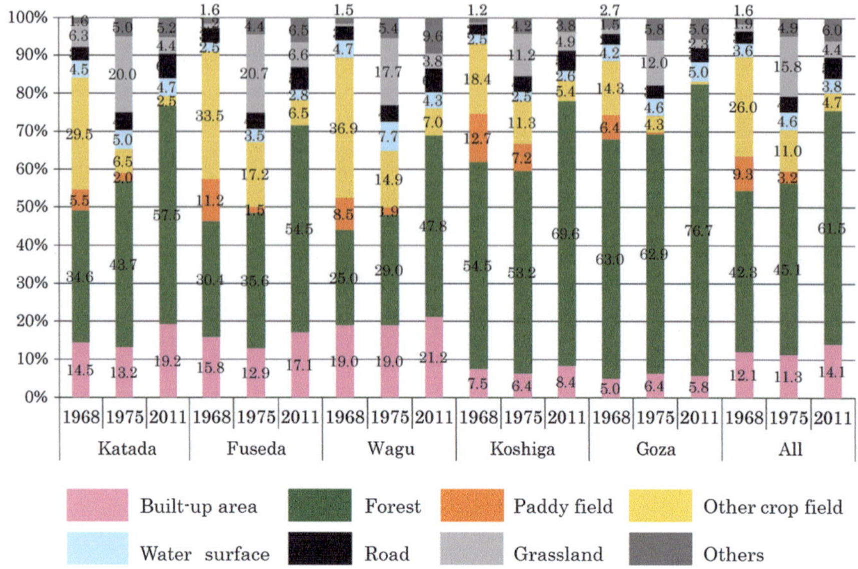

Fig. 16.12 Change in landscape units' ratio in each district

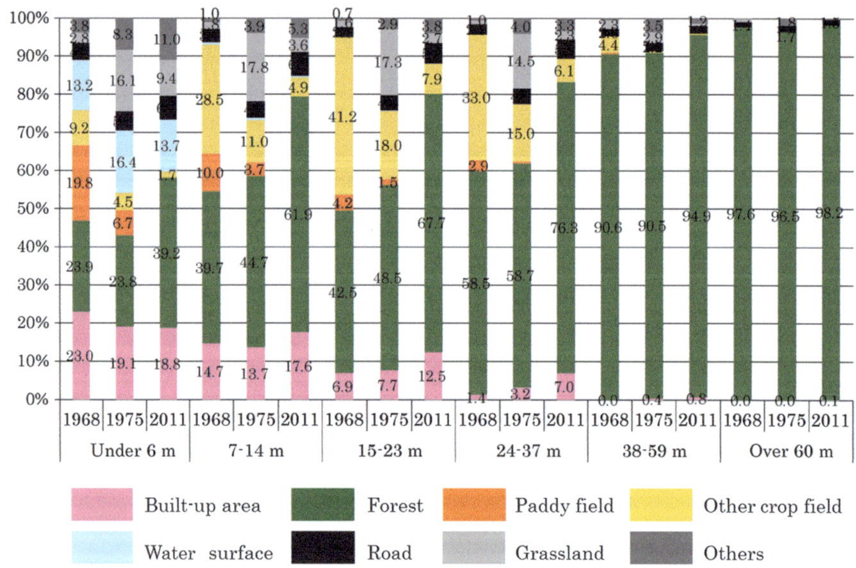

Fig. 16.13 Change in landscape units' ratio according to altitudes

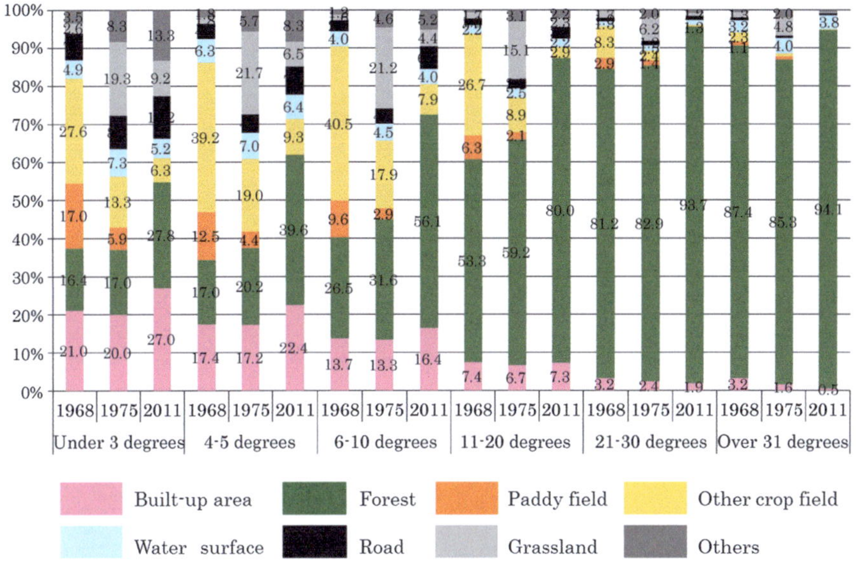

Fig. 16.14 Change in landscape units' ratio according to average land gradients

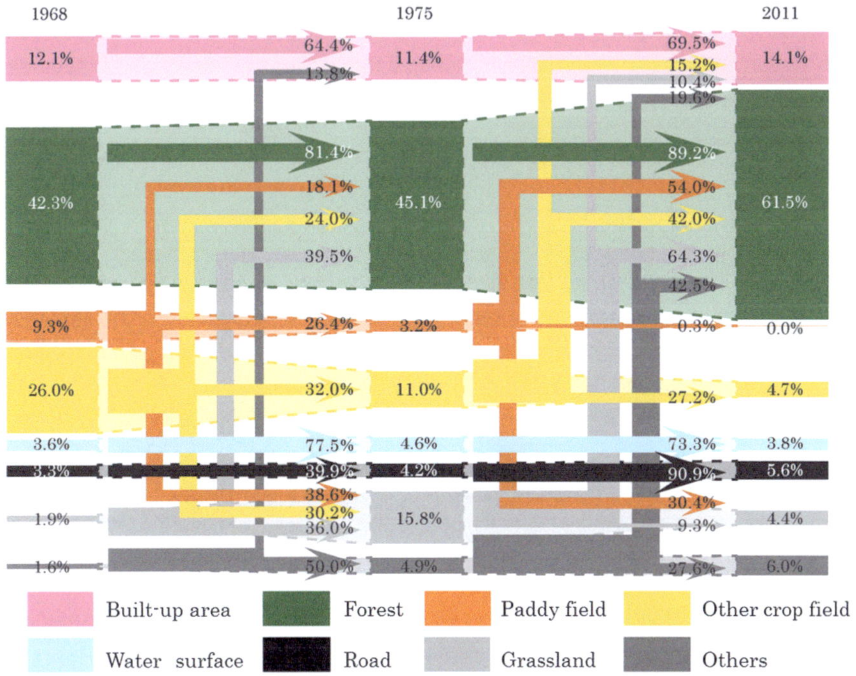

Fig. 16.15 Change in landscape units' ratio in three periods

villages, residential sites were located near the ports and the farmlands extended around the residential sites. Every district had farmlands, of which the areas were 3 times wider than residential sites, and the agro-fishery lifestyle was established. Most of all residents in the five districts used timber for fuel, which was gathered in the Goza district. Many residents worked as foresters, with their households reaching 443 in 1960.

In 1975, some parts of the residential sites near the coastal area became vacant because cheaper pearl products were imported, and many pearl farmers lost their jobs. Residential areas decreased in most parts of the districts, but only in the Goza district did they increase, because of the increase of immigrants from outside due to the expanded use of private cars. The use of private cars became common, roads improved, and many people began to visit from outside the district. As the consequence, hotels and other tourist facilities were constructed, and residents changed to engage in tourism in the Goza district.

At the same time, cultivated lands didn't need to be used any more, especially in harsh land conditions, because of the increase of inflows of agricultural products through the markets. Farmlands decreased drastically by about 20 %, and grasslands, which were abandoned farmland, increased about 15 %.

Observing the statistics and the distribution pattern of landscape units in 2011, the residential sites extended toward the north, shifting from coastal areas to the hilly areas. One reason for the extension was the increase in nuclear families. In the Katada and Fuseda districts, the residential areas increased 5 % compared to the areas in 1975 because of the development of road networks. In the inside hilly areas, residential sites began to spread, while the vacant houses and lands appeared in the lowlands of the old villages.

Reforestation began in the grasslands, and the forest areas increased about 20 %, especially in the places that are hard to access. On the other hand, forests were cut down and developed to become residential sites alongside new roads because of the easy access. In the inner coastal areas of Ago Bay, the numbers of rafts and longlines declined, and trees became densely grown, which changed the water surface and coastal landscape. Seaweed production decreased during the last 10 years because of the deterioration of the sea environment and the damage by sea urchins and starfish at the outside coasts along the Pacific Ocean.

16.5 Calculation of Landscape Labor Accounts and Changing Regional Resources Use Through Interviews

Labor contents and their labor accounts in the current and past conditions are collected by interviews with residents, fishery cooperatives and the government. Table 16.2 shows the sizes of each landscape unit and the amount of their labor accounts, which can be obtained through the interviews.

Table 16.2 Size of landscape units and amount of labor accounts

Landscape units		Organizations	Number	Days	Hour (h)	Area (a)	TLA_l (h)	TLD_{jl} (h/a)
Built-up area		Private owner	–	–	–	–	–	–
Forest		Private owner	–	–	–	–	–	–
Farmland	Paddy field	Private owner	–	–	–	–	–	–
	Other crop field	Private owner	1	279	4	13	1,116	85.8
	Orchard	Private owner	–	–	–	–	–	–
	Tea plantation	Private owner	–	–	–	–	–	–
	Arboretum	Private owner	–	–	–	–	–	–
Road	National highway	Construction office in Shima city	4–5	2	8	298	72	0.24
	Prefectural road	City hall	7	2	8	138	336	0.00
Beach	Volunteer		90	3	1	1,605	102	0.21
	Fishery cooperative		200	1	5	10	1,000	1.00
Water surface		Private owner	–	–	–	–	–	–
Grassland		Private owner	–	–	–	–	–	–
Others		Private owner	–	–	–	–	–	–
Total								
Coast	Seaweed forest	Women divers (Koshiga)	28	1	1.5		42	
		Women divers (Goza)	10–15	10–15	1–2	3,000	234	0.08

Other crop fields and seaweed are managed by residents. Some crop fields are managed by elderly women and are used for self-consumption. This work is also for health maintenance, and involves a small number of labor hours but a high number of management days. And also they show inefficient high labor densities because of the small allocation of working sites.

In seaweed beds, removal of organisms that cause feeding damage is done by women divers. But enough management is not done because of the aging of women divers and difficulty of the work in the sea.

Pruning of roadside trees is done along the roadside of the national road by the municipality, and securing of the sea turtles spawning ground is done voluntarily in the coastal area. Such work is only a few times in a year. For the management of the hotel's garden, professionals are hired and use many workers.

Figure 16.16 shows the results of the net labor densities of each managed landscape element. These values are the sum of the labor accounts of all participants per the net worked area in the targeted landscape elements.

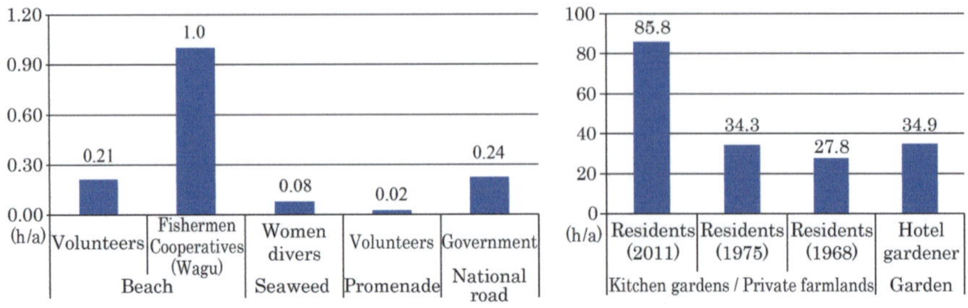

Fig. 16.16 Net labor densities of each managed landscape element

Labor densities of kitchen gardens/private farmlands managed by residents and that of a hotel garden are relatively higher than those of the beach, seaweed, promenade and national road, which are managed by volunteers, cooperatives and the government. Labor densities of kitchen gardens/private farmlands had a trend of increasing from 1968 to 2011. The data of kitchen gardens/private farmlands in 1968 through 1975 are introduced according to the statistics (*Noka Keiei Chosa Hokoku* (Survey of farmer economy) (Ministry of Agriculture, Forestry and Fisheries 1968, 1975)), and the data in 2011 are introduced by the interview of residents.

It seems to be against the progress of agro-mechanization, but the reason might be that recent private farmers tend to work mainly as their hobby or for health maintenance rather than for food or money. They might spend their time in the farmland even though they have little work to do.

The management of hotel gardens is entrusted to professional gardeners. The survey was made in the garden of a resort hotel, which is located outside of, but near to Saki-Shima districts. The main labor contents done by three gardeners in a day are mowing and pruning of forests and mowing of grasses. The annual labor account is high, because they are employees and they manage the garden throughout a year. They have high working ability, but only a limited area (about 10 %) of all gardens is managed by them. The middle area of the forest is left with no care. As a consequence, the labor densities of the managed elements are high, but the gross labor density of the total garden decreases to a lower value.

Roads are classified into national, prefectural and municipal roads. Green elements of municipal roads aren't managed, but the roadside trees and hedges of the national road, which stretches in the central area of the peninsula, are managed. Pruning and fertilizing are conducted by subcontractors annually. Promenades specified by the Ministry of the Environment are managed by resident volunteers.

Coastal management areas are divided and allocated to each fishery organization according to the fishery rights. As the landscapes of beaches are damaged by illegal dumping and flowing garbage, volunteers and fishermen cooperatives clean and maintain them. Recently, abandoned materials used by pearl farmers are carried out to the sea and fishermen must collect them in order not to damage their fishery work.

Seaweed tends to disappear because of global warming and damage by sea urchins and starfish. Women divers try to disinfect such vermin in order to protect sea products. Compared to the past, more management labor is needed due to the degradation of the sea environment, however, aging and declining of labor forces makes it difficult to manage the needed areas appropriately.

As a summery, kitchen garden labor density is the highest of all. The other landscape elements do not need regular management, and are managed by the cooperatives or government. Therefore the number of hours and days to be able to be provided is limited. Especially, seaweed management depends on the personal labor of women divers who work individually, and is extremely limited compared to the labor which can be provided by the organizations. As the management of the coastal area links directly to the income of fishermen, the municipality and fishermen cooperatives will promote it as much as they can in Saki-Shima peninsula. Hotel gardens are managed by professional employees and they work efficiently. But such a gardenized forest area with dense management is rare and most forests are abandoned.

16.6 Summary of Landscape Labor Accounts

Table 16.3 shows the overview of landscape labor accounts and densities in Saki-Shima districts. As our interviews in the districts were limited, all labor densities of landscape elements are not clarified. Especially labor densities of elements in built-up areas and in farmlands are different and have a large diversity. Therefore the results of Minamoto (2015) and integrated labor densities shown in Chap. 17 are used.

The labor density of farmlands was 29.8 as the average value of other crop fields. The average value of interviewed farmers in this district is 85.8, which is higher than the above mentioned value. This high value is due to the kitchen garden style management of the farmland. In this district, productive farmlands develop still to some extent, and using the high value of kitchen gardens might lead to overestimation of labor density of the farmland.

As shown in Table 16.3, most labor accounts come from the management of farmlands. Even though in Saki-Shima districts, where more than half of the land is covered with forests, labor forces are not generated because the forests are not used any more. Net labor density, gross labor density and social labor intensity are 2.0, 1.5 and 21.7 respectively.

Table 16.4 shows total labor accounts (TLA), gross labor densities (GLD) and social labor intensity (SLI) in each district. In the districts, which have much farmland, GLDs tend to be higher. SLI of the Koshiga district is very high because of the high value of the farmland area per population.

Table 16.3 Each landscape unit of element, net, gross of labor accounts

Landscape unit		Landscape element	Organization	A_l (a)		S_l (a)		TLD_{jl} (h/a)	TLA_l (h)	NLD_l (h/a)	GLD_l (h/a)	NLD (h/a)	GLD (h/a)	SLI
Built-up area		Tree and grassland	Private owner	1,894		25,761		6.4	12,161	6.4	0.5	2.0	1.5	21.7
Forest		Tree	Private owner	103,229		103,229		0.0	0.0	0.0	0.0			
Farmland	Paddy field	Paddy field	Private owner	18	8,202	18	8,202	2.6	48	0.0	2.6			
	Other crop field	Other crop field	Private owner	7,812		7,812		29.8	232,557	29.8	29.8			
	Orchard	Orchard	Private owner	155		155		28.3	4,386	0.0	28.3			
	Tea plantation	Tea plantation	Private owner	63		63		12.5	784	0.0	12.5			
	Arboretum	Arboretum	Private owner	154		154		0.0	0.0	0.0	0.0			
Road	National highway	Tree and grassland	Construction office in Shima city	1,308	1,446	1,308	9,516	0.7	72	0.1	0.1			
	Prefectural road	Tree and grassland	Government	138		8,208		2.4	336	2.4	0.0			
Beach		Beach	Volunteer, Fishermen cooperatives	1,605		5,561		1.2	1,102	0.7	0.2			
Water surface		Non-element	Private owner	0		867		0.0	0	0.0	0.0			
Grassland		Grassland	Private owner	7,594		7,595		0.0	0	0.0	0.0			
Others		Non-element	Private owner	0		10,448		0.0	0	0.0	0.0			
Total				123,970		171,178			252,347	2.0	1.5			
Coast	Seaweed forest	Seaweed forest	Women divers	6,000		201,403		0.1	276	0.0	0.0			

SLI population: 11,611

For the calculation of TLA, NLD and SLI, the amount of TLD are used as the results of Minamoto (2015) and integrated labor densities shown in Chap. 17

Table 16.4 Gross labor accounts, densities and per population in each district

	TLA (h)			GLD (h/a)			SLI			Population
	Land area	Coastal area	Sum	Land area	Coastal area	Sum	Land area	Coastal area	Sum	
Katada	26,660	0	26,560	0.8	0.0	0.8	11.4	0.0	11.4	2,337
Fuseda	52,400	0	52,400	2.0	0.0	2.0	24.7	0.0	24.7	2,121
Wagu	82,211	0	82,211	2.3	0.0	2.3	16.7	0.0	16.7	4,929
Koshiga	83,958	42	84,000	1.6	0.002	1.6	50.2	0.025	50.2	1,674
Goza	7,093	234	7,328	0.3	0.014	0.3	1.4	0.046	1.4	550
Total	252,561	276	252,837	1.5	0.001	1.5	21.7	0.024	21.8	11,611

16.7 Personal Annual Labor Accounts and Densities and Working Style Types

Figure 16.17 shows the net personal labor accounts and labor densities in each landscape element. A professional gardener has higher labor accounts and densities than that of the volunteer workers. The labor densities of kitchen gardens are extremely high, because workers do not need efficiency in their work.

Figure 16.18 shows the typical working style types in the districts. There are three types of working styles, i.e. the professional working type in the hotel garden, the self-help management of kitchen gardens and the voluntary or governmental management of coasts, seaweed and roads. And also in the past period a working style of personal farmers are distinguished as the fourth category of working types.

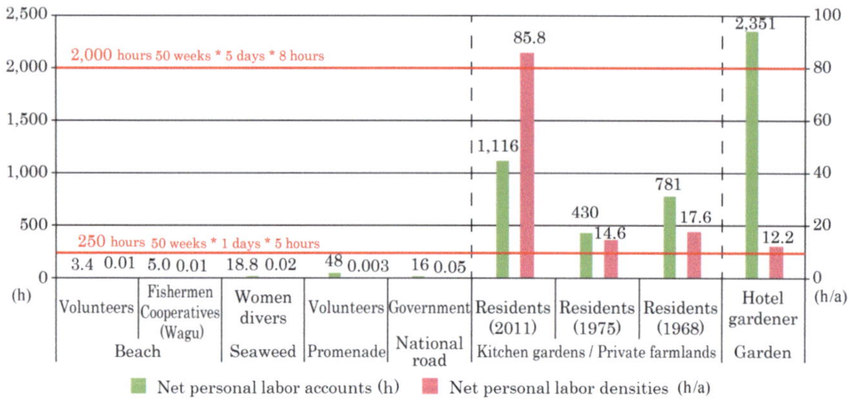

Fig. 16.17 Net personal labor accounts and labor densities in each landscape element

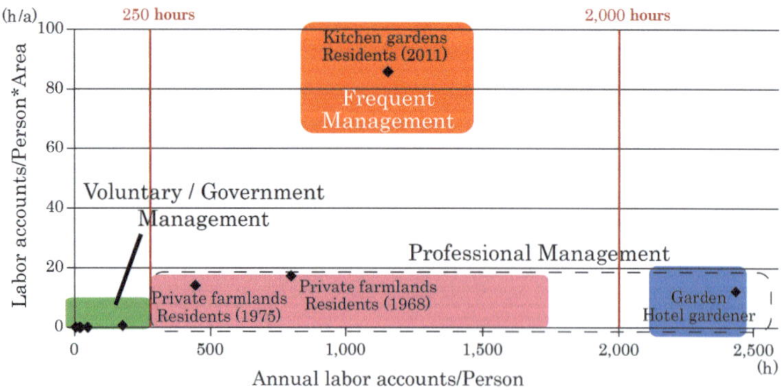

Fig. 16.18 Working style types of Saki-Shima districts

16.8 Consideration of Landscape Management in the Future

As shown above, the recent situation of landscape management of Saki-Shima peninsula has many issues which seem to be difficult to solve. The demographic, industrial and social changes in the last 50 years emerge, leaving many uncared for spaces in farmlands, forests and coastal areas. The revitalization of landscape management is needed, but the labor forces necessary for the activities are lacking at the moment. How can it be achieved? Figure 16.19 shows the estimation of the future population. In 2055 the population will reduce to 19 % of the current population. Appropriate land management will be impossible with such a drastic population decrease. This district is designated as a part of Ise-Shima National Park, and the protection and conservation of the ecological environment and landscapes must be considered. Therefore, too many introductions of industries will not be welcomed. But the inflow or stopping the outflow of population is needed. One option is the revitalization of tourism. To overcome the competition among tourist destinations, extra charm for the district is needed. Perhaps an increase of the charm of the landscape might be the best possible attempt. But to accelerate such a trend, economical possibility or benefits which can be obtained by land management must be considered.

Then, three scenarios are assumed according to the different future land use, and the sustainability of land management and employment will be considered.

(1) Scenario that the current trend of land use and population changes will continue (Type of extension of today's trend)

Figure 16.20 shows the expected land uses in the scenario of the extension of today's trend, in which the scattering of residential sites, abandonment of farmlands, and reforestation will continue as in today. In many villages, the vacancy of houses and housing lots will increase in the flatlands, and the land use change from farmlands to residential areas will progress in the highlands. These changes will lead to the fragmentation of landscapes and require continuous upgrading of infrastructures. On the other hand, the reforestation of abandoned farmland will also progress.

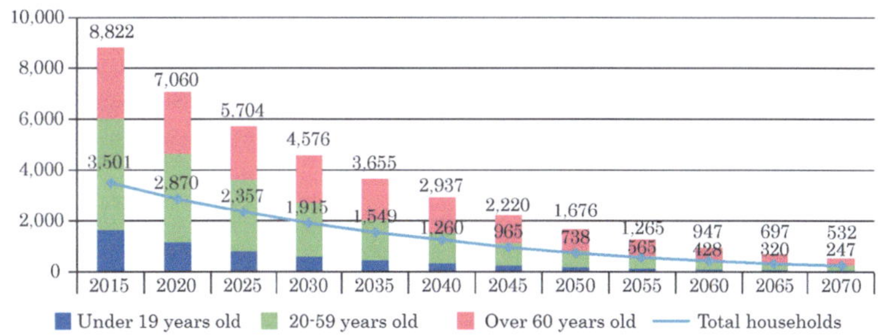

Fig. 16.19 Estimation of future population

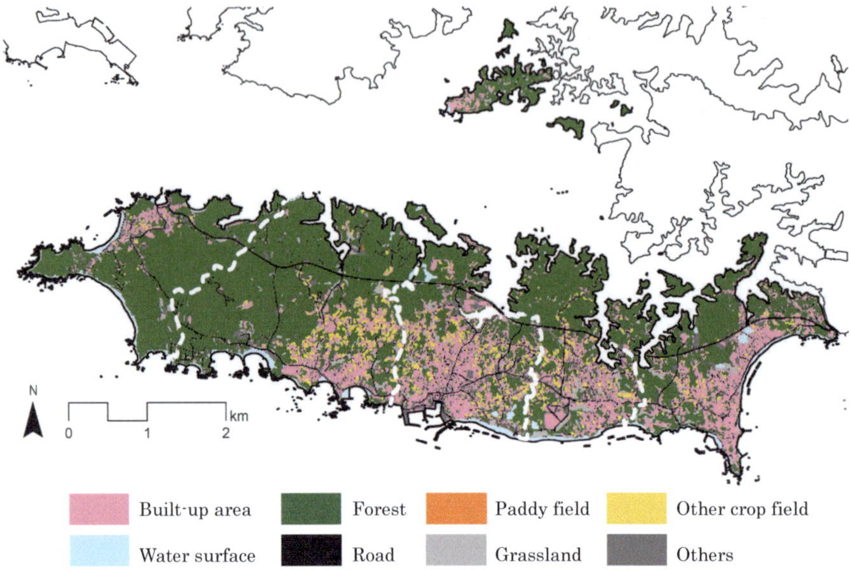

Fig. 16.20 Spatial distribution of landscape units in the scenario that the current trend of land use and population changes will continue

(2) Scenario according to the development vision of the municipality (Type of government promotion)

Figure 16.21 shows the scenario of the government promotion, in which the future land uses designated in the development plans, i.e. Urban master plan, Green basic plan and Landscape plan, of Shima city are considered. In the urban masterplan, residential areas of all villages should be agglomerated in the southern port areas. In the green basic plan, green coverage ratios in residential areas should be increased. In the landscape plan, landscape conservation by the appropriate management of forests should be promoted.

(3) Scenario in the assumption that the farmland will recover as in the past (Type of farmland restoration)

Figure 16.22 shows the land uses of the farmland restoration, in which land management and ensuring of employment will be pursued by the revitalization of farming. In the past many farmlands were cultivated. In this scenario, abandoned grasslands and afforested farmlands will be reactivated as farmlands, and the creation of employment and the reduction of harm from insects and animals by the intensive management of land will be aimed for.

In Fig. 16.23 the land use area sizes of each scenario are compared. In the scenario of the extension of today's trend, the scattering of residential areas, increase of grasslands, decrease of farmlands and the trend of afforestation are confirmed. By the scenario of the governmental promotion, the increase of

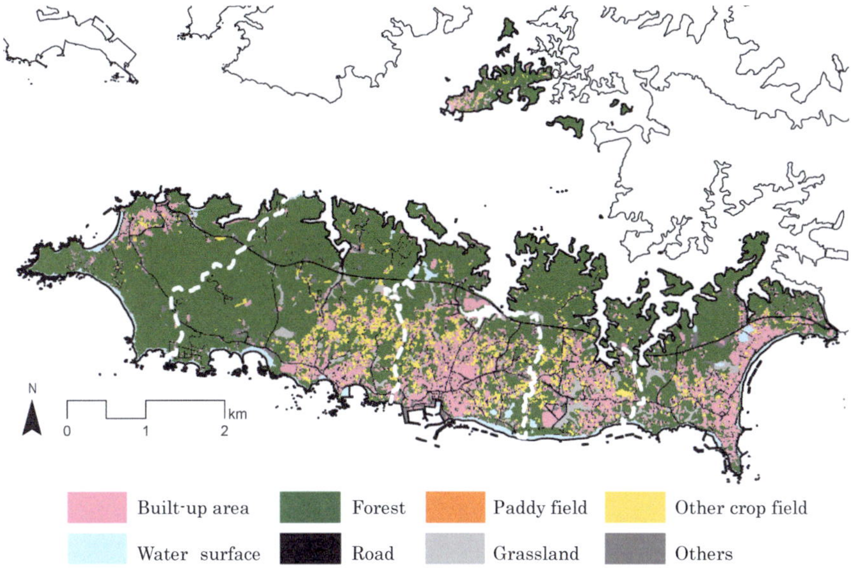

Fig. 16.21 Spatial distribution of landscape units in the scenario according to the development vision of the municipality

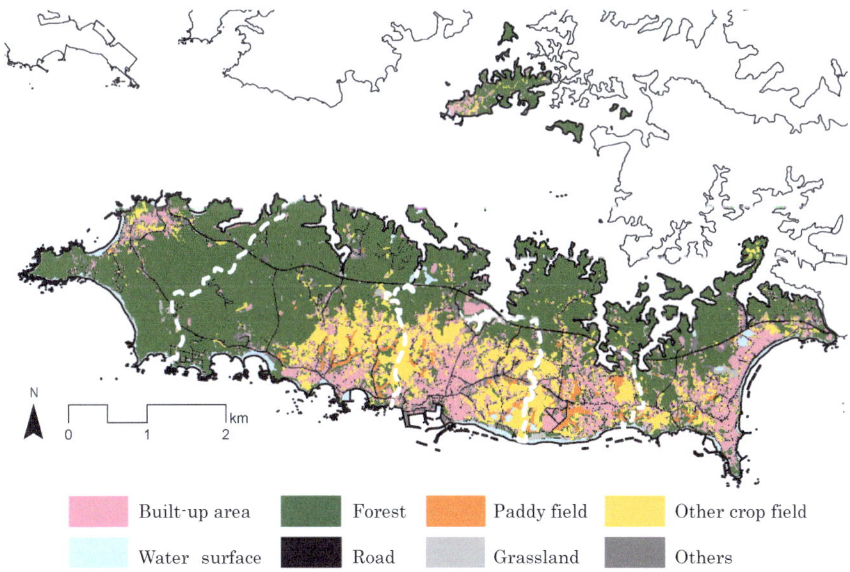

Fig. 16.22 Spatial distribution of landscape units in the scenario in the assumption that the farmland will be recovered as in the past

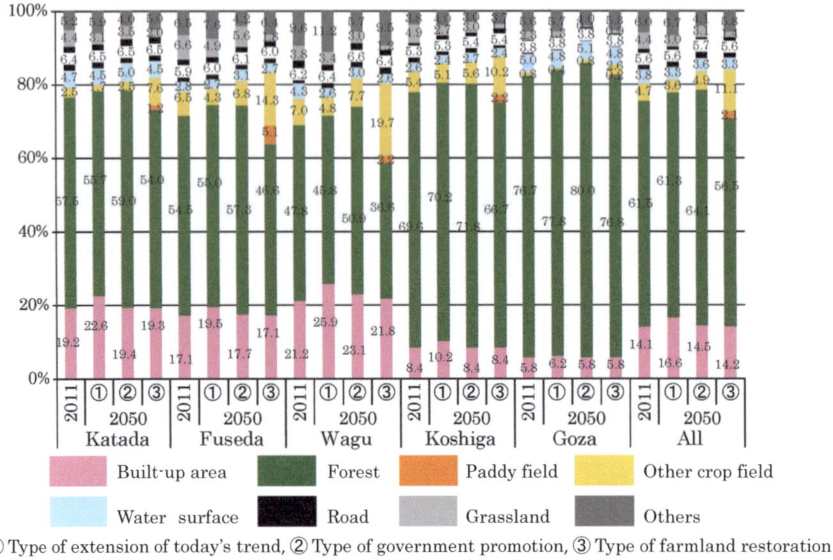

Fig. 16.23 Change in landscape units' ratio according to each scenario

residential areas in Wagu is confirmed, and the zones of prime farmlands in Wagu and Koshiga are preserved. In the residential areas, greening of vacant spaces is promoted in order to increase green coverage ratios.

In the scenario of farmland restoration, good farmlands surrounding the villages are aimed to be agglomerated, and the farmlands in the mountainous areas are aimed to be returned to forests. As a result, the reduction of forests is 5 % and increase of farmlands is 9 %. The land use of this scenario is regarded as almost similar to the past land use.

Figure 16.24 shows the landscape management labor accounts needed when the three scenarios are calculated. In the scenario of the extension of today's trend, 37 % of today's labor accounts will be reduced because of the decrease of farmlands. This decrease is not surprising, but are the natural results of population decrease expected in the future. By the scenario of the government promotion, labor accounts will increase because of the emerging of new management works or expanding of working areas according to the promotion of the preservation of landscapes and ecological environment in the forests and coasts as written in the development plans. Also in the scenario of the farmland restoration, the labor accounts will increase to 2.8 times of today, because of the expansion of farmland areas.

The increase of needed labor forces of the scenarios of the governmental promotion and the farmland restoration shown in Fig. 16.24 must be afforded extra, because the labor forces needed in the situation of the shrinking population shown in the scenario of the extension of today's trend are limited. The extra needed labor forces in the two scenarios are counted in Table 16.5.

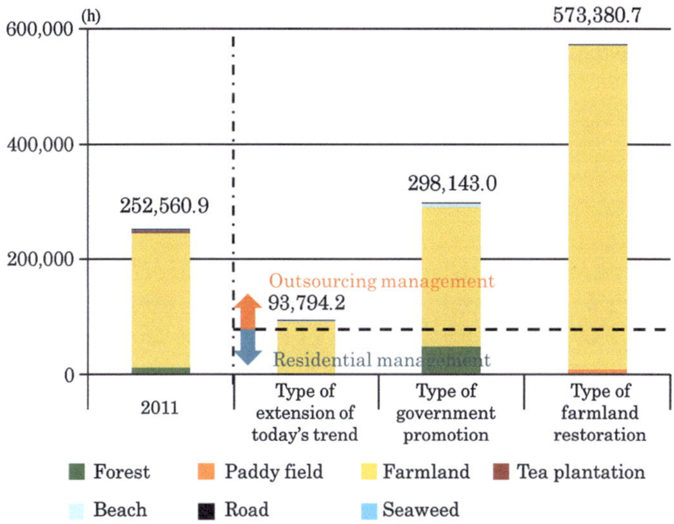

Fig. 16.24 Labor accounts of each scenario

Table 16.5 Management population in each landscape unit

	Type of government promotion		Type of farmland restoration
	Volunteer	Professional	Professional
Forest	0	24	0
Paddy field	0	0	5
Farmland	0	75	235
Beach	1,655	0	0
Total	1,655	99	240

Volunteer/government management type: 3.4 h/year
Professional management type: 2,000 h/year

The main working style types in the governmental promotion should be considered as the weekend working type by volunteers and daily working style type by professionals.

The number needed to manage the beach is 1,655, and this will reach to 5.7 times of today's account. If the scenario of an increase of farmland is included, labor forces should be provided as employees by corporate farming, because there are few private farmers in the future. The needed number of employees is estimated at 75 persons. Also the management of forests needs 24 professionals extra in order to ensure the safety of the work.

It should be noted that works such as beach and forest management are usually economically externalized, in other words, there are very few economic returns. In the consequence, without any subsidy, the sustainable management of this scenario will not be possible. This is the most difficult point for the realization of the scenario of governmental promotion.

The main working style type in the scenario of the farmland restoration is assumed as the professional type, because the increase of the number of private farmers might not be possible because of the population shrinking. If economic efficiency is pursued, the introduction of corporate farming will be needed. In the case of all needed labor forces being provided as professional workers, then 240 workers must be hired for the cultivation of all additional farmlands. The issue of this scenario is that enough economic returns of the additional farming are available for the sustainable employment of the 240 professional workers.

All of the above mentioned estimations based on the labor accounts are ideals and not realistic enough, but such thinkings might be a base of public debates for the creation of the future vision and of the beginning of actions towards the future.

References

Mie Prefecture (1960–2010) Miekentoukeisho (Statistics of Mie prefecture)

Minamoto K (2015) Estimation of landscape management labor accounts based on the land use classification map in the Chubu area, proceedings

Ministry of Agriculture, Forestry and Fisheries of Japan (1960–2010) Census of agriculture and forestry

Ministry of Agriculture, Forestry and Fisheries of Japan (1963–2013) Census of fishery

Ministry of Agriculture, Forestry and Fisheries of Japan (1968, 1975) Noka Keiei Chosa Hokoku (survey of farmer economy)

Shima Town (2009) Shima-cho town history

Yokohama Shinema Shokai (Yokohama Cinema Commercial Company) (1941) Wagu no Ama (women divers in Wagu)

Chapter 17
Integration of Landscape Management Labor Accounts

Hiroyuki Shimizu, Chika Takatori, Nobuko Kawaguchi and Keidai Minamoto

Abstract This chapter summarizes the results of the case studies in previous chapters and then classifies the working style types by analyzing the working time and working area of all neighborhood landscape complex units. Then, the total labor accounts of neighborhood landscape complex units are compared to one another. In addition, the integration of landscape labor accounts from small scale to large scale is considered. On a city scale, the targeted area is Nagoya City, the total labor accounts of which are calculated and future prospects for sustainable landscape management are proposed. On a regional scale, the targeted area is Chubu Metropolitan District, the total labor accounts of which are calculated in 1 km mesh and the future situations, particularly the shortage of landscape management plans, are examined through comparing the current and forecasted populations.

Keywords Working style types · Comprehensive spatial planning · City scale · Region scale · Population simulation

17.1 Introduction

In this chapter, a summary is presented of the results of the case studies described in the previous chapters. In addition, integration of the small- to large-scale landscape management labor accounts will be considered.

In the previous chapters, labor accounts and labor densities for different landscape types, i.e., landscape management complex units, were calculated and described. In addition, it was shown that these indicators could be employed to determine the prevailing conditions relevant to the labor available for landscape management of the different landscape types. Furthermore, in view of an aging and shrinking population, these indicators could be useful for estimating the likely

H. Shimizu · C. Takatori (✉) · N. Kawaguchi · K. Minamoto
Graduate School of Environmental Studies, Nagoya University, Furo-cho, Chikusa-ku, Nagoya, Aichi 464-8601, Japan
e-mail: takatori@corot.nuac.nagoya-u.ac.jp

H. Shimizu et al. (eds.), *Labor Forces and Landscape Management*,
DOI 10.1007/978-981-10-2278-4_17

challenges that would have to be addressed, and for pointing out the prospective solutions to these challenges. In particular, by using the personal annual labor account and personal labor density indicators, the categories pertaining to the working style types could indicate the quality of the management labor, as well as the attitudes of the residents and the professionals to their work. The prevailing opinion is that an appropriate mix of working style types in a local community could create a moderate working environment and a sustainable social system.

Currently, the integration of the small- to large-scale landscape management labor accounts has to be considered. As regards this integration, each labor density factor has to be identified according to the landscape elements and units that were presented in Chap. 9. Although considerable effort has been made to identify each labor density factor, the samples were not sufficient for identifying all the types of landscape elements and units. In view of this constraint, various ambiguous values, obtained from the limited samples or from the literature, had to be used. If sophisticated and comprehensive research could be conducted, data that are more appropriate could be obtained and the values could be modified accordingly. The results of this chapter are considered tentative; even so, the information could be considered sufficient for assessing the various managed units at a regional scale.

17.2 Summary of Case Studies

17.2.1 Working Style Types

The working style types were classified by calculating the annual personal labor accounts (hour:h), set on the horizontal axis, and the annual personal labor densities (h/a), set on the vertical axis. Figure 17.1 shows the integrated results of the field surveys.

The following working styles types were extracted from the information contained in the results of the case studies (see Part 3):

(1) "Professional management type" indicate people who manage the landscape units as professional or daily workers. Two types of professional management were determined, namely, daily, i.e., done every day by corporate or cooperative organizations that manage farmlands and artificial forests, and that done 2–3 days per week by self-employed farmers.
(2) "Self-support management type" pertains to the management of typical kitchen gardens. Two types of self-support management were determined, namely, that done 2–3 days per week ("frequent management type") for health reasons or as a hobby, and that done once per week ("weekend management type"). Both types manage small areas of land at a low efficiency level.
(3) "Voluntary and government management type" indicates volunteers or government-appointed managers of landscape units, e.g., for the maintenance of secondary forests in urban parks or for the upkeep (mowing) of roadside

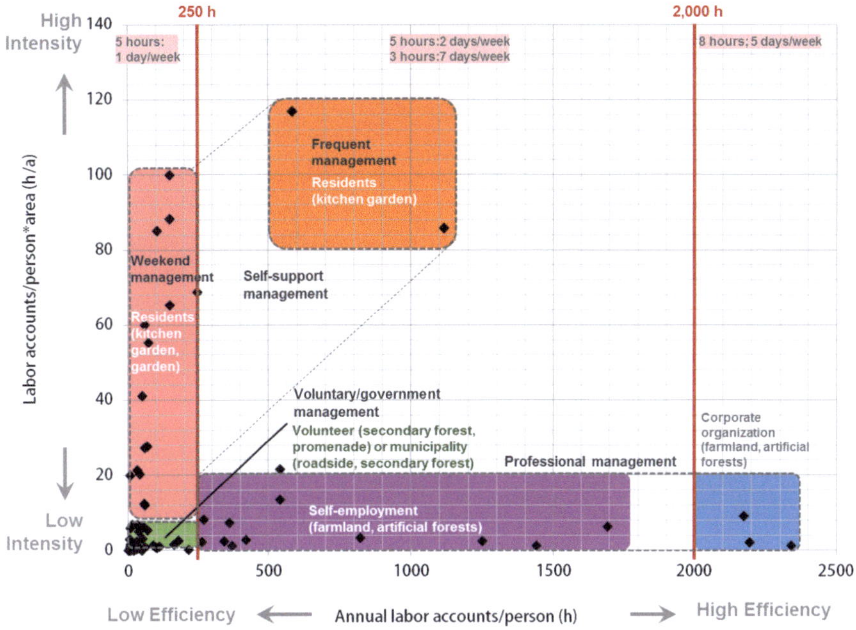

Fig. 17.1 Classification of working style types

grasslands. The frequency of this type is low, namely, a few times per year, or once per month to twice weekly. However, the efficiency of this management type is relatively high.

As indicated in Chap. 12, relevant to the management of urban forests, the management force of the public sector is expected to shrink in future because of expected budget constraints. Therefore, the introduction of a volunteer labor force is a welcome development. However, such a personal labor account is generally low; consequently, effective organization and citizen networks would be required to muster the necessary management forces for the appropriate management of all the urban forests.

In a paddy field community, such as that of the Asami district, shown in Chap. 13, the labor force of farmers is expected to shrink in future because of the aging and declining population. Therefore, utilizing cooperative organizations or corporate farming would have to be considered to compensate for this decline. On the other hand, such a type of working system could lead to the collapse of the well-maintained collaborative system of the farming society, such as that of the "Deai." Moreover, a working style in pursuit of economic efficiency could detrimentally affect the ecosystem balance or the biodiversity of an area. Therefore, an appropriate arrangement has to be pursued of combining the economically efficient working styles with the working style of self-employed farmers, i.e., high environmental affinity coupled with a sense of pride. Such a combined management style could also apply to the management of the Satoyama land areas, such as the Nyu district, discussed in Chap. 14, and the Kayumi district, discussed in Chap. 15. In addition, the kitchen garden management type is an important feature relevant to the health and self-sufficiency of elderly people. Consequently, governmental policy that places disproportionate emphasis on efficiency by indiscriminate promotion of land-improvement projects could jeopardize the incentives of the residents for healthy living.

In addition, an effective measure could be the introduction of volunteer workers from urban areas to supplement the shrinking farming labor force of the countryside. Such interaction has been introduced already in various regions. In addition, systems that allow urban residents to purchase the rights of the owners to fruit trees or to the terraced paddy fields have been started already. Kieninger et al. (2011) has stated "The Ownership System can be regarded as an urban–rural coalition, where non-farmers (predominantly city dwellers) engage in farming activities. Against the payment of a participation fee, they rent their own agricultural land, cultivate it, and thus contribute to the revitalization of the rural society and preservation of the traditional cultural landscape." As the interest in preserving the rural environments of the urban residents would probably grows in future, such intimate interaction between urban and countryside residents is expected to increase.

Numerous plantation forests exist in Japan, of which many parts are not managed, as shown in Chaps. 14 and 15. This lack is ascribed to such forest management requiring the services of a professional labor force. The economic value of forest resources is expected to rise if these resources could be managed and utilized appropriately. In recent years, proposals have been put forward to utilize biomass or timber in urban areas. This implies that active collaboration has to be developed between the forestry and the electric power plant industries or the building companies.

17.3 Integration at a City Scale (Case Studies in Nagoya City)

17.3.1 Methodology

In this section, the estimation of labor accounts and densities at a city scale is discussed. The targeted area is Nagoya city, and Fig. 17.2 shows the spatial distribution of the landscape elements and units in Nagoya city. This section is based on the paper by Kawaguchi et al. (2016).

To calculate the landscape labor account at a city scale, the landscape units and elements will be classified in accordance with the maintenance method of each vegetation type. The total area size of each classified type of landscape element will

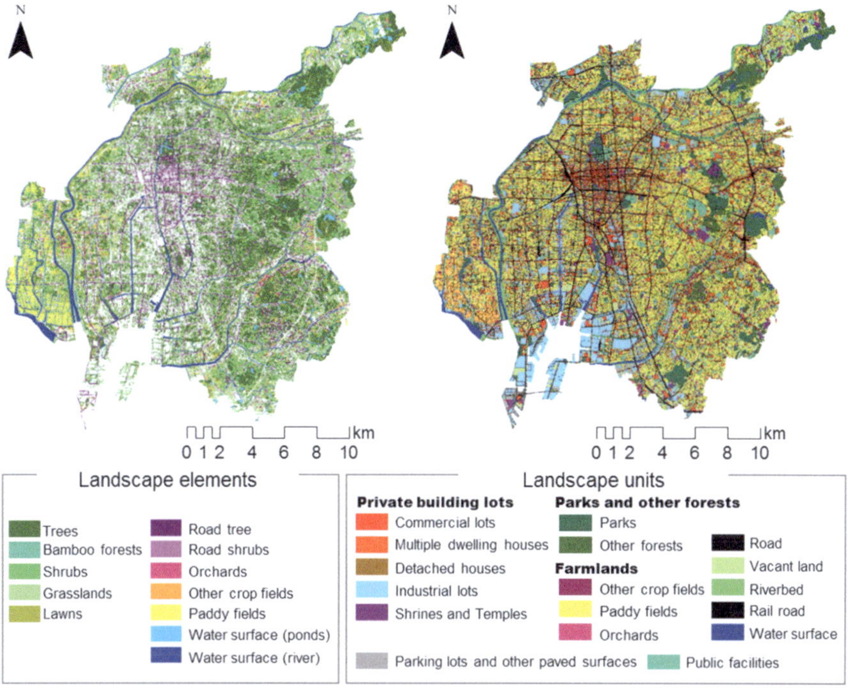

Fig. 17.2 The spatial distribution of landscape elements and units (urban area)

be calculated. The managed area, which differs for each type of land use, will be considered, and the total managed area size of each type of landscape element will be estimated.

Finally, the total annual labor account of each unit type (TLA_l) will be calculated by multiplying the total annual labor density (TLD_{jl}, extracted from the case studies) of each element type in a landscape unit by the managed area. Table 17.1 shows the definition of the allocated value of TLD_{jl}, which was used for the calculation of the TLA_l for each type of landscape unit, as obtained from the case studies.

The land use data were mainly used for identifying the spatial distribution and the area size of the units and elements. These data were obtained from the *Basic Survey of City Planning (*Nagoya city 2012), whereas the green coverage data were obtained from the *Green Status Survey* (Nagoya city 2010), issued by the municipality of Nagoya.

Figure 17.3 and Table 17.2 show the classification method of the urban green units and elements. Urban green elements can be divided into two major groups, namely, (1) the large-scale urban forests, large-scale grasslands, and roadside trees (hereafter called group A); and (2) the small-scale green units (hereafter called group B). Group A has various functions as urban green networks and green hubs. The landscape units and elements will be further classified into several groups in the current subsection.

Table 17.1 Definition of set value TLD_{jl}

Landscape unit		TLD_{jl} (h/a)	
Private building lot	Commercial lot	8.45	*1
	Industrial lot	8.45	
	Multiple dwelling house	5.77	*2
	Detached house	38.30	*3
	Shrine and temple	4.88	*4
	Parking lots and other paved unit	0.00	*5
Public facility		8.45	*1
Farmlands	Orchard	28.26	*6
	Other crop field	29.77	*7
	Paddy field	2.63	*8
Park		16.43	*9
Road		9.55	*11
Vacant land		1.96	*10
Riverbed		1.96	
Rail road		1.96	
Parks and forests in private lot	Park	8.09	*12
	Forests in private lot	7.00	*13
Private building lots	Commercial lot	7.00	
	Industrial lot	7.00	
	Multiple dwelling house	7.00	
	Detached house	7.00	
	Shrines and temple	7.00	
	Parking lots and other paved unit	7.00	
Public facility		7.00	
Farmlands	Orchard	7.00	
	Other crop field	7.00	
	Paddy field	7.00	
Vacant land		0.00	*14
Riverbed		0.00	
Edge of other water body		0.00	

Referent for TLD_{jl} ([ ] worker type); **1* The value of two commercial lots in Nagoya city [professional gardener], **2* The value of twelve multiple dwelling houses in Nagoya city [professional gardener], **3* The value of six detached houses in Nagoya city [landowner], **4* The value of two shrines and temples in Matsusaka Chuou district, [landowner], **5* No work, **6* The value of TLD (orchard) is identified according to the statistical data of Census of Agriculture and Forestry, **7* The value of TLD (other crop fields) is identified according to the statistical data of Census of Agriculture and Forestry, **8* The value of TLD (paddy fields) is identified according to the statistical data of Census of Agriculture and Forestry, **9* Nagoya city, (2015) 'DOBOKU KOUJI SEKISAN KIJUNSHO' (Guidelines for cost estimating for public works) [gardener], **10* Nagoya city, (2015) 'DOBOKU KOUJI SEKISAN KIJUNSHO' (Guidelines for cost estimating for public works) [gardener], **12* Nagoya city, (2015) 'DOBOKU KOUJI SEKISAN KIJUNSHO' (Guidelines for cost estimating for public works) [gardener], **13* The value of private forest TLD_{ij} of surrounding area of road in Nagoya city [gardener], **14* No work

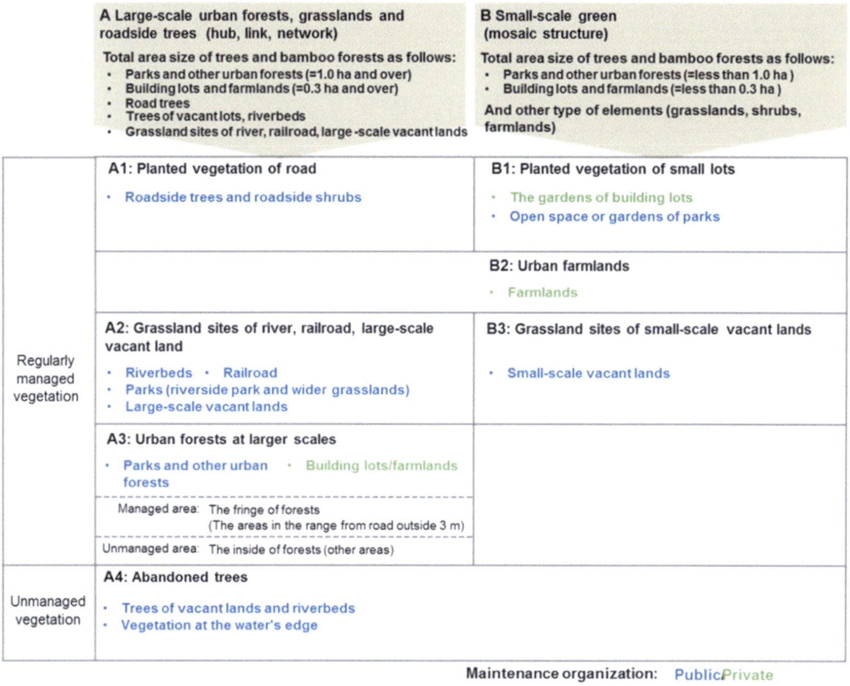

Fig. 17.3 The classification method of unit and element in urban green

Group A was subdivided into several subgroups (A1–A4). Roadside trees and shrubs were allocated to group A1, whereas the grassland banks of rivers, railroads, and large-scale vacant land areas were allocated to group A2. Group A3 consists of large-scale urban forests, such as tree and bamboo forests larger than 1.0 ha in parks and other private forests, as well as areas larger than 0.3 ha on building lots and on farmlands. In group A3, the fringe of the forest is assumed to be managed, whereas the inner area of the fringe is assumed unmanaged. Consequently, the forest located at the inner area of the roadside (inner fringe), up to 3 m away from the road, would be considered unmanaged area, whereas the forest located further than 3 m from the road would be considered the managed area. Group A4 relates to trees located on vacant land, at riverbeds, and at the edge of water bodies, all of which are considered unmanaged vegetation. Therefore, group A4 is classified as an "abandoned landscape" element.

The green sites of groups A1, A2, and A3 are considered areas of regularly managed vegetation and those of groups A1 and A3 are managed by public hands ("public hands" means the employees of the government sector or urban green conservation groups for public green sites; see Chap. 10). Furthermore, the parks of Group A3 are managed by the employees of the government sector and groups of volunteers (e.g., Higashiyama Park and Fujimaki district).

Table 17.2 Classification of unit and element and calculation of TLA based on management levels

Element type			Landscape unit		Tree + bamboo forest	Other elements	A_l (total) (a)	Managed area ratio (%)	A_l (managed) (a)	TLD_{jl} (h/a)		TLA_l (h)	
B Small-scale green (mosaic structure)	B1	Planted vegetation of small lot	Park		Less than 1.0 ha	All areas	63,258	100.0	63,258	16.43	*9	1,039,326	6,825,910
			Private building lot	Commercial lot	Less than 0.3 ha	All areas	9,491	100.0	9,491	8.45	*1	80,195	
				Industrial lot			15,469	100.0	15,469	8.45		130,715	
				Multiple dwelling house			15,395	100.0	15,395	5.77	*2	88,832	
				Detached house			86,920	100.0	86,920	38.30	*3	3,328,913	
				Shrines and temples			11,909	100.0	11,909	4.88	*4	58,117	
				Parking lots and other paved unit			5,353	100.0	5,353	0.00	*5	0	
			Public facility				18,818	100.0	18,818	8.45	*1	159,016	
	B2	Urban farmland	Farmland	Orchard			6,746	100.0	6,746	28.26	*6	190,632	
				Other crop field			53,588	100.0	53,588	29.77	*7	1,595,325	
				Paddy field			58,875	100.0	58,875	2.63	*8	154,840	
	A2 + B3	Grassland vegetation	Park (riverside park and wide grassland)		–	All areas (except for shrubs)	19,464	100.0	19,464	1.96	*10	38,149	193,288
			Vacant land				38,875	100.0	38,875	1.96		76,195	
A Large-scale urban forest, grassland, and road tree (hub, link, network)			Riverbed				36,712	100.0	36,712	1.96		71,956	
			Railroad		All areas		3,565	100.0	3,565	1.96		6,988	
	A1	Planted vegetation of road	Road		All areas		22,831	100.0	22,831	9.55	*11	218,037	321,999,
	A3*	Urban forests at larger scale	Park		1.0 ha and over	–	36,983	1.6	591	8.09	*12	4,784	
			Forests in private lot		All areas		87,809	14.3	12,517	7.50	*13	93,877	
			Private building lot	Commercial lot	0.3 ha and over	–	59	9.5	6	7.50		42	
				Industrial lot			1,150	8.9	102	7.50		766	
				Multiple dwelling house			373	7.9	30	7.50		221	
				Detached house			320	8.0	26	7.50		191	
				Shrines and Temple			5,281	5.9	309	7.50		2,321	
				Parking lots and other paved unit			22	11.5	3	7.50		19	
			Public facility				3,952	5.8	231	7.50		1,731	
			Farmland	Orchard			0	0.0	0	7.50		0	
				Other crop field			109	1.2	1	7.50		10	
				Paddy field			1	0.0	0	7.50		0	
	A5	Abandoned tree	Park (riverside park)		All areas	Shrubs	4,034	0.0	0	0.00	*14	0	
			Vacant land				14,758	0.0	0	0.00		0	
			Riverbed				4,730	0.0	0	0.00		0	
			Other water's edge			All area	3,966	0.0	0	0.00		0	

In a further classification step, the remaining elements of group A, namely, the small-scale green areas were allocated to group B. These green areas are the small-scale plots of vegetation near buildings, in parks, and on farmland, as well as the grasslands on small-scale vacant land areas.

17.3.2 Calculation of Landscape Management Labor Accounts and Densities

Table 17.2 also shows the calculated results of TLA_l, which indicate that the TLA value of the large-scale urban forests, roadside trees, and grassland vegetation amounts to only 4.4 % of the TLA value of the entire city. As the inner areas of almost all urban forests are unmanaged, the amount of TLA in large-scale urban forests is assumed low. In the green areas that belong to Group A3, the large-scale forests of parks are managed by municipality staff and volunteer groups. The calculation does not include the value of TLD_{jl} contributed by voluntary work. Consequently, if all the volunteer activities at the city scale were taken into account, the total amount of TLA would be higher.

Table 17.3 shows the results of the TLA, GLD, and the social labor intensity (SLI) of each unit type in Nagoya city. The TLA amounts to 7,341,196 h and the GLD to 2.26 (h/a). The area size of the landscape unit of the private work accounts for 58 % of the area size of all the units; therefore, the area size of private work and public work is almost equal. In contrast, when the two work types are compared according to the amount of TLA, the TLA of the private work is 78 % and that of the public work is 22 %. Accordingly, the results indicate that the labor account of private work is significantly higher.

Figure 17.4 shows the values of the TLA and GLD for each type of landscape unit. Figure 17.5 shows the spatial distribution of the total labor accounts of each unit type in Nagoya city. The graph illustrates that the detached houses, farmlands, and the parks have a high TLA value, whereas the orchards and other crop fields have a high GLD value.

17.3.3 Future Prospects for Sustainable Land Management at a City Scale

(1) Large-scale Urban Green Sites

Large-scale urban green sites, such as parks, private forests, and the forests of temples and shrines are considered a function of the urban green space networks. The vegetation succession and the growth of the biomass in these forests have progressed undisturbed by human actions, such as cutting timber for fuel. In addition to the vegetation succession being beneficial to nature preservation, urban

Table 17.3 Calculation of landscape labor account and density at city scale

Landscape unit		S_l (a)	A_l (a)	Managed area size (a)	Managed area ratio (%)	TLA_l (h)	NLD_l (h/a)	GLD_l (h/a)	GLD (h/a)	SLI (population: 2263,894)
Private building lot	Commercial lot	225,973	9,550	9,496	99.4	80,237	8.40	0.36	0.02	
	Industrial lot	284,603	16,619	15,571	93.7	131,481	7.91	0.46	0.04	
	Multiple dwelling house	178,881	15,768	15,425	97.8	89,053	5.65	0.50	0.03	
	Detached house	791,317	87,240	86,946	99.7	3,329,104	38.16	4.21	1.02	
	Shrines and temples	62,352	17,191	12,219	71.1	60,438	3.52	0.97	0.02	
	Parking lots and other paved unit	118,468	5,375	5,356	99.6	19	0.00	0.00	0.00	
Farmland	Orchard	6,747	6,746	6,746	100.0	190,632	28.26	28.25	0.06	
	Other crop field	65,026	53,697	53,590	99.8	1,595,334	29.71	24.53	0.49	
	Paddy field	61,579	58,876	58,875	100.0	154,840	2.63	2.51	0.05	
Forests in private lot		92,507	87,809	12,517	14.3	93,877	1.07	1.01	0.03	
Total (private)		1,887,453	358,870	276,739	77.1	5,725,015	20.69	15.95	1.76	2.5
Public facility		155,036	22,771	19,049	83.7	160,747	7.06	1.04	0.05	
Park		191,137	123,738	83,313	67.3	1,082,259	8.75	5.66	0.33	
Road		632,171	22,831	22,831	100.0	218,037	9.55	0.34	0.07	
Vacant land		201,403	53,633	38,875	72.5	76,195	1.42	0.38	0.02	
Railroad		26,843	3,565	3,565	100.0	6,988	1.96	0.26	0.00	
Riverbed		62,316	41,443	36,712	88.6	71,956	1.74	1.15	0.02	
Water surface		95,990	3,966	0	0.0	0	0.00	0.00	0.00	
Total (public)		1,364,898	271,947	204,346	75.1	1,616,181	7.91	5.94	0.50	0.7
Total		3,252,351	630,817	481,085	76.3	7,341,196	11.64	2.26	2.26	3.2

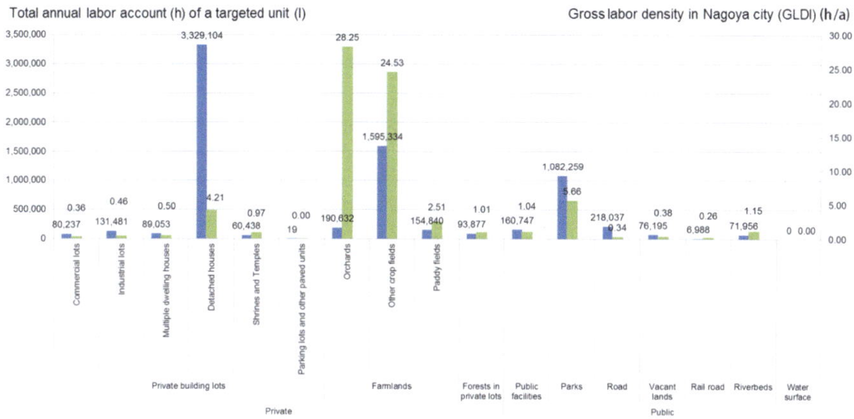

Fig. 17.4 Calculation of labor accounts and densities in units (urban area)

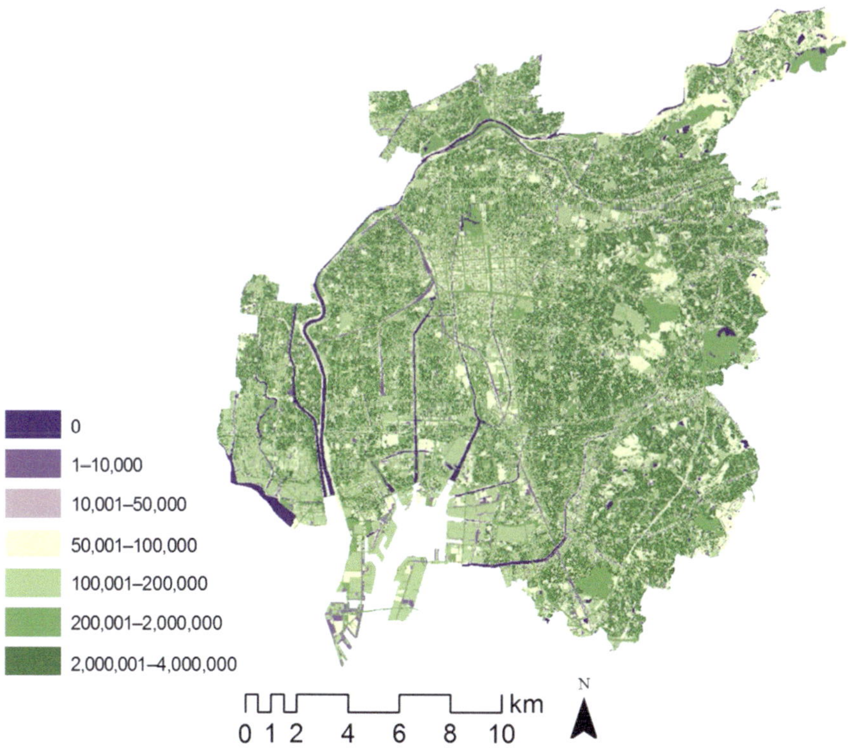

Fig. 17.5 The spatial distribution of total labor accounts (TLA_l)

green spaces have another function, namely, a cultural function. This function is maximized when forests play an important role in the daily activities of people, such as walking, playing, sports, bird watching, and similar activities. As regards the green management for such activities, various challenges have to be addressed, including an increase in fallen trees and removing the fallen leaves adjacent to residential areas, paths, and roads, and the illegal dumping of waste in the forests.

Municipalities are charged with managing fallen trees and branches next to roads and at the edge of public green sites. On the other hand, because of the heavy burden on owners, many secondary forests on private green sites are nearly abandoned.

As mentioned above, large-scale green sites in urban areas allow many people to come into closer contact with nature. The contribution of the private volunteer organisations is therefore essential to manage such areas of cultural value when the amount of public money available for management is limited. For instance, in the Association of Nagoya Forest Management Partnership Network system, the Nagoya Higashiyama Forest Management Association engages in green management in collaboration with the local municipalities.

The large-scale urban green areas can be regarded as a common good. As regards the large-scale private forests, the ownership is often fragmented and complicated and it is therefore difficult to obtain the required permission from all the owners to enter the sites for appropriate management. Consequently, a legal system has to be developed to facilitate the management of these sites by the public or by commissioned public bodies. Fujimaki town is an illustration of the possibilities of such a system.

The management of roadside trees is the responsibility of municipalities. Roadside trees are multi-functional elements of the ecosystem by providing shadow in high summer, compensating for gas emissions, enhancing urban spaces, and others. Therefore, the budget for maintaining these trees has to be maintained; however, public awareness of the cost of such maintenance must be increased. Accordingly, the concept of landscape management labor accounts could contribute to increasing such public awareness.

(2) Small-scale Private Green Sites

As private building lots dominate urban areas, the green spaces in these areas are under threat. Various factors contribute to the decline of green areas, namely, reducing garden spaces or changing to more manageable plant species, dividing of sites, and changing green spaces into parking lots.

This trend is noticed even in residential areas with abundant green spaces, such as the Arako district and the Matsusaka Chuo district. In these districts, green hedges are being replaced by artificial fences, and vacant spaces are being changed into parking spaces. In this process, the landscape of the districts is being disfigured. In addition, in the Arako district, small farmlands managed by small landowners are under pressure of urbanization.

In order to maintain or increase private small green spaces, certain measures were proposed, as follows.

In districts with aging residents, the gardens of the detached houses could become a burden to these elderly people, as managing the gardens could be labor intensive. However, such a green environment constitutes a common good; therefore, a system is required that would allow public funds to be used for the professional management and maintenance of these gardens. By implementing the district planning system or the landscape ordinances that are already in place as planning instruments, the green landscapes of residential areas could be managed as a regional resource. In addition, adjacent small residential sites and vacant land could be integrated into one area that could be developed as a large-scale garden. However, currently, the system is not well developed.

As regards landscape management labor accounts, it could be possible to create green spaces more efficiently on land occupied by multiple dwelling residences, as more residents could share the expense associated with managing the green spaces.

In areas where detached houses are being changed into multiple dwelling residences because of the high housing demand, it could be necessary to introduce regulations to increase the green spaces by controlling the green ratio of the area.

In a landscape management complex unit, a system could be developed to manage temporary vacant plots as temporary parks or allotment gardens by employing the help of cooperative organizations.

The management of green spaces on private land is the responsibility of the landowners. However, these landowners often neglect their responsibility because of the associated costs. Even though these green sites are privately owned, they function as healthy biodiverse environments in cities. However, these functions are economically externalized.

As regards the policy relevant to farm fields (green spaces on farms), a direct payment system was introduced to compensate this type of externalization. However, various public fund allocations are utilized to maintain the multiple functions of forests. In addition, support systems, such as public grants or preferential tax treatment as incentives to maintaining the green areas, should be considered as compensation for the externalized values that constitute the common good. The concept of landscape management labor accounts could provide an integrated perspective for developing a consistent policy toward public support for the different landscape types that are managed in accordance with different legal systems.

17.4 Integration at a Regional Scale (Case Studies in Chubu Metropolitan Area)

17.4.1 Methodology

In this section, the integration of landscape management labor accounts at a regional scale is discussed. The site of focus is the Chubu metropolitan area, which includes eight prefectures, namely, Shizuoka, Aichi, Gifu, Mie, Nagano, Toyama, Ishikawa, and Fukui. The labor account totals were calculated relevant to each standard 1 km mesh (grid), as provided by the data from the *National Land Numerical Information, Land utilization tertiary mesh Data* (National land Information Division, National Spatial Planning and Regional Policy Bureau, Ministry of Land, Infrastructure, Transportation and Tourism (MLIT) of Japan 2009). The totals were compared with the basic landscape types, as shown in Chap. 2, as well as the current and the future population to gain a perspective on sustainable landscape management at a regional scale. The following steps were conducted to calculate the landscape management labor accounts.

The first step in the calculation was to identify the gross labor density of the targeted landscape unit (GLD_l). The landscape elements and units were classified in accordance with the *Vegetation survey, Natural Environmental Information GIS* (Ministry of the Environment of Japan 2005), as described in Chap. 9.

Where the landscape units were nearly equal to the landscape elements, the total labor density of the landscape unit was calculated as follows:

The total labor density in each landscape unit (TLD_i) was identified in accordance with the statistical data of the *Agriculture business management statistics investigation farm products life costs statistics* (Ministry of Agriculture, Forestry and Fisheries of Japan 2005) or the data available from the field surveys, as presented in the preceding chapters. However, the labor densities in some landscape units could not be identified clearly because of the constraints of the investigation. In such instances, the values of similar landscape units were employed. A total labor density of zero indicates there is no management in the landscape unit. Subsequently, the total labor accounts of each landscape element (TLA_l) were calculated by multiplying the TLD_l and the targeted element size. Where there was no management of any landscape element, relevant to units such as plantation forests, where most of the area is left unmanaged, the management area ratio was multiplied to calculate the net managed landscape size of the element. This size value was used to calculate the TLA_l of the unit.

Where the size of a landscape unit, such as that in a built-up area, included a non-landscape area and was not equal to the area size of the landscape element, the element coverage ratio, such as the green coverage ratio, was additionally multiplied with the total unit size to calculate the TLA_l.

The gross labor density of a landscape unit (GLD_l) was calculated by dividing the TLA_l by the size of the total unit area. This GLD_l value was used as a base factor for integration at a regional scale.

The detail description of landscape elements and units is as follows:

(1) "Built-up Areas"
According to the *Green Status Survey* (Nagoya city 2010), built-up areas include various land use types, such as commercial, industrial, residential, religious and public facilities, parks, and others. It was assumed that these areas would have a variety of labor densities. Therefore, the *Basic Survey of City Planning* (Nagoya city 2012) was used to classify built-up areas into detailed landscape elements and

units. Nagoya city was selected as a case study site in this book, as it is a typical example of a built-up area. The labor densities in each detailed landscape element and unit were surveyed, and the total labor density of the built-up areas was calculated, based on the value of the labor density of Nagoya city. Moreover, this value was considered the typical value of built-up areas and it was introduced for the calculation at a regional scale.

(2) "Golf course", "Residential districts with high green coverage", "Paddy field", "Tea plantation", "Other crop fields", and "Orchard"
The data on the labor density in "golf courses/lawns" were obtained from a survey conducted among the landscaping contractors who manage the golf courses in Nagoya city. The data on the "residential districts with high green coverage" were obtained from interviews with the owners of the detached houses in Nagoya city. The data on the labor density in "paddy field" and "tea plantation" were based on the results of a statistical survey on the national average of labor time in each 1,000 m^2 (are) according to the *Agriculture business management statistics investigation farm products life costs statistics* (Ministry of Agriculture, Forestry and Fisheries of Japan 2005). The data sets of labor density in "other crop fields" and "orchard" vary according to the crops; therefore, the average labor density of the three major crops in each prefecture was calculated by using the data of the *Agriculture business management statistics investigation farm products life costs statistics* (Ministry of Agriculture, Forestry and Fisheries of Japan 2005).

(3) "Abandoned farmland", "Secondary Grassland", "Artificial Forest", "Tree Plantings"
The labor density of "abandoned farmland" and "secondary grassland" was set at zero, as it was assumed there was no management of these land use types. The labor densities of "artificial forest" were obtained from interviews with the forest owners in the Kayumi districts. The *Forest Register* (Mie prefecture 2013) of the Mie prefecture was utilized to obtain the ratio of the management areas for "artificial forest".

(4) "Secondary Forest", "Park, Cemetery, and Other Areas that have a Remaining Plant Community"
"Secondary forest" can be divided into three types, namely, "secondary forest (*Quercus*)", "secondary forest", and "secondary forest (almost natural forest)", according to the extent of the anthropogenic impact. The labor density of "secondary forest (*Quercus*)" was based on interviews with the volunteer groups in Fujimaki town and Higashiyama Park in Nagoya city. The interview data of two towns were utilized as the ratio of the management area in "secondary forest (*Quercus*)". In this book, "park, cemetery, and others that have a remaining plant community" were assumed to have almost the same labor density and ratio of management area as "secondary forest (*Quercus*)". The labor densities of "secondary forest" and "secondary forest (almost natural forest)" were set at zero because it was assumed there was no management of these land use types.

(5) "Natural Forest," "Coast, Riparian Vegetation," "Natural Vacant Land," and "Water Surface"
The labor densities of "natural forest," "coast, riparian vegetation," "natural vacant land," and "water surface" were set at zero because it was assumed that there was generally no management of these land use types.

Table 17.4 shows the total labor accounts and densities in each landscape unit in the Chubu metropolitan area.

The second step is the integration of landscape management labor accounts at a regional scale. All the GLD_l of the existing landscape units in the targeted 1 km mesh were multiplied with the corresponding unit area sizes, and all the results were summarized as the TLA of the targeted 1 km mesh. In addition, the GLD and the SLI of each 1 km mesh were calculated. By considering these three indicators, the characteristics of landscape management labor accounts at a regional scale were determined.

17.4.2 Distribution of Landscape Management Labor Accounts in Chubu Metropolitan Area

By mapping the total number of labor accounts in each 1 km mesh, it was possible to distinguish easily where the labor accounts were high or low at a regional scale. As shown in Figs. 17.6 and 17.7, the suburban area with "paddy field," "cultivated land," and "residential district with high green coverage" have relatively larger labor accounts. However, the annual personal labor density was relatively high in "residential district with high green coverage" that is managed by the people of the "self-support management type."

Table 17.5 shows the calculation gaps in the data of TLA, NLD, GLD, and SLI, which were calculated from the district-scale data mentioned in the previous chapters and the regional-scale data calculated by employing these statistics. Tables 17.6 and 17.7 indicate a comparison of the TLD_{ij} of each landscape unit. The results indicate that various factors could cause the TLA calculation gaps.

The first of these causative factors is a mismatch in the land areas (S_l, A_l). Since the definitions of the landscape units at each scale differ, these cannot be matched exactly, which causes differences in the unit area size. This situation mainly pertains to farmlands, for example, those of Asami and Shima. The element area size, whether managed or unmanaged, such as the artificial forest in Kayumi, also causes differences in the TLA.

The differences in the TLD_{ij} constitute another causative factor. In some instances, according to Tables 17.6 and 17.7, the TLD_{ij} of a landscape unit in the district-scale data was higher than was that in the regional-scale data. This is also relevant to the built-up areas in all the case studies, the traditional Jorisei paddy

Table 17.4 GLD lists for integration of landscape labor accounts at regional scale

Landscape unit		S_l (a)	Road ratio	LER		Managed area ratio	A_l (a)	TLD_{jl} (h/a)	TLA_l (h)	GLD (h/a)
				City block	Road					
Built-up area	Urbanization promotion area	18,434,121	0.16	0.29	0.03	1.00	9,992,395	2.26	22,582,812	0.57
	Other area	21,131,407	0.11		0.00					
Golf courses/lawns		2,935,308	0.01	1.00	0.00	1.00	2,908,139	12.53	36,438,983	12.41
Residential districts with high green coverage		20,527,968	0.11	0.30	0.00	1.00	5,459,978	4.21	22,986,508	1.12
Park, cemetery, etc.		591,187	0.07	1.00	0.00	1.00	548,949	1.34	735,592	1.24
Pastures		1,823,472	0.09	1.00	0.00	1.00	1,658,284	3.84	6,367,812	3.49
Other crop fields		14,419,419	0.07	1.00	0.00	1.00	13,480,951	29.77	401,327,923	27.83
Paddy fields		52,622,334	0.06	1.00	0.00	1.00	49,346,273	2.63	129,780,699	2.47
Tea plantations		3,476,886	0.07	1.00	0.00	1.00	3,236,940	12.54	40,591,233	11.67
Orchards		6,668,653	0.06	1.00	0.00	1.00	6,249,672	28.26	176,615,732	26.48
Tree plantings		75,065	0.11	1.00	0.00	0.00	0	0	0	0.00
Abandoned farmlands		206,871	0.02	1.00	0.00	0.00	0	3.84	0	0.00
Abandoned paddy fields		946,703	0.03	1.00	0.00	0.00	0	3.84	0	0.00
Secondary grasslands (low)		10,799,871	0.05	1.00	0.00	0.00	0	0	0	0.00
Secondary grasslands (high)		3,320,168	0.03	1.00	0.00	1.00	3,228,552	0	0	0.00
Artificial forests		176,167,852	0.01	1.00	0.00	0.19	32,436,538	1.07	34,623,289	0.20
Secondary forests (Quercus variabilis or Quercus serrata)		79,037,614	0.01	1.00	0.00	0.00	0	1.34	0	0.00
Secondary forests		81,148,123	0.01	1.00	0.00	0.00	0	0	0	0.00

(continued)

Table 17.4 (continued)

Landscape unit	S_l (a)	Road ratio	LER		Managed area ratio	A_l (a)	TLD_{jl} (h/a)	TLA_l (h)	GLD (h/a)
			City block	Road					
Secondary forests (almost natural forest)	20,297,140	0.01	1.00	0.00	0.00	0	0	0	0.00
Natural forests	67,269,538	0.00	1.00	0.00	0.00	0	0	0	0.00
Coastal vegetation	5,457,807	0.04	1.00	0.00	0.00	0	0	0	0.00
Gravel surface	3,850,999	0.01	0.00	0.00	0.00	0	0	0	0.00
Water surface	8,567,100	0.00	0.00	0.00	0.00	0	0	0	0.00

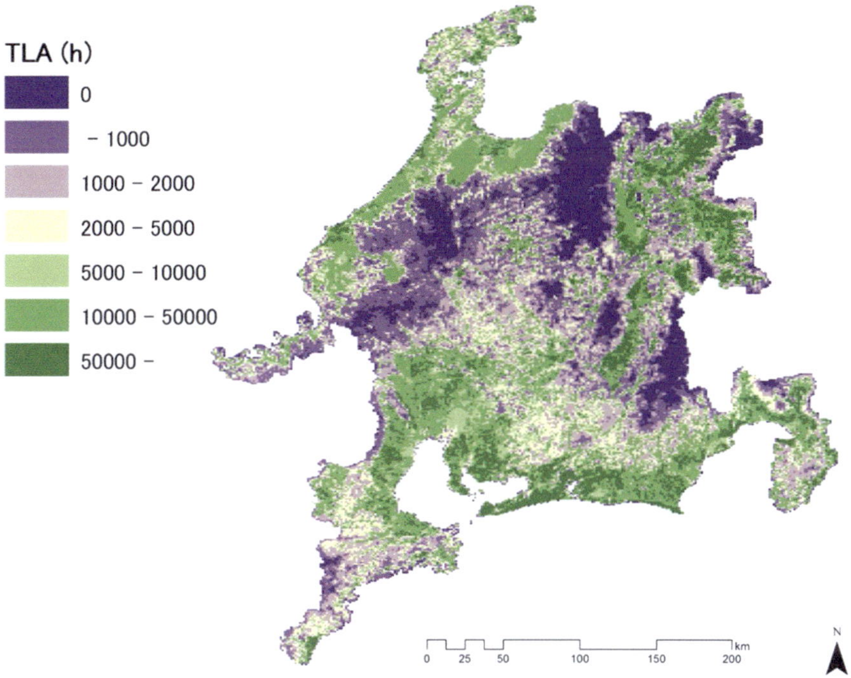

Fig. 17.6 Labor accounts in Chubu metropolitan area (1 km mesh)

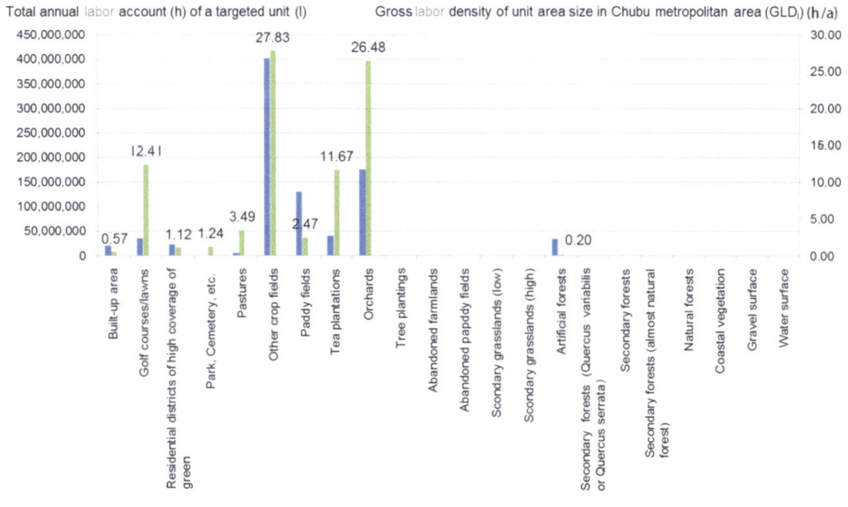

Fig. 17.7 Labor density and labor accounts of each landscape unit in Chubu metropolitan area

Table 17.5 The gaps between calculation by using regional-scale data and district-scale data: (district)—(region)

Case study	S_l (a)	A_l (a)	A_l (a)	TLA_l (h)	NLD_l (h/a)	GLD_l (h/a)	SLI_l
Arako	0	−1141	−918	13,393	11.17	1.52	1.85
Matsusaka	−229	−1458	−1219	17,509	12.50	1.52	4.82
Fujimaki	−28	169	441	14,759	5.88	4.88	33.62
Asami	−127	−2015	−35	68,245	1.28	0.93	31.80
Kayumi	0	1341	−174	34,262	0.12	0.11	13.72
Shima	356	15,601	97,022	−72,838	−0.98	−0.43	−6.27

field in Asami, a secondary forest in Fujimaki (landscape management only at the district scale), an artificial forest managed by "self-employment" in Nyu, and the coastal vegetation in Shima (landscape management only at the district scale). Well-managed landscape units exist in each district that was selected as a case study, giving rise to higher TLD_{ij} than that of the regional-scale data. Consequently, the district-scale TLA could be larger than that of the regional-scale TLA. In Arako and Matsusaka, each NLD has a large gap, which was caused mainly by the TLD_{ij} of the built-up area. In the built-up area, the value of the labor density elements included in the landscape unit differed between the district-scale and the regional-scale data.

In addition, Tables 17.6 and 17.7 indicate differences in perception between the two scales of data relevant to what constitutes a landscape unit. The cells in grey in Tables 17.6 and 17.7 indicate the absence of data on the landscape unit. In the farmlands of Matsusaka, for example, only paddy fields at the regional scale are included, although these are actually "other crop fields." A large gap exists in the data on the TLD_{ij} of these two landscape types, namely, "paddy fields" and "other crop fields" that affects the calculation of TLA. The regional-scale data were produced from the information obtained from satellite photographs. The district-scale and the regional-scale data were created at different scales, with that of the district data being 1:2,500 and that of the regional data being at 1:25,000. Consequently, an error occurs in the area when regional data are employed that are based on the detail information of the district data.

Figure 17.8 shows the ratio of the total land area of the basic landscape types in the Chubu metropolitan area. The "nature type" occupies more than half of the total area. The second-largest landscape type is the "paddy field Satoyama type", and the third largest is the "other field Satoyama type" The farmland landscape types, including the "paddy field type" and the "other field type," occupy 10.6 % of the area. The "urban type" and the "urban paddy field mixed type" occupy 11.2 % in total.

Figure 17.9 shows the total landscape management labor accounts for the seven categories of basic landscape types, as indicated in Chap. 2. The total labor account of the "other field type" is the highest, even though it occupies only 3.1 % of the total land area. This is ascribed to the high value of the labor density of the "other field type". In addition, the total labor account of the "paddy field type" is high.

Table 17.6 The comparison of TLD_{ij} between the district- and the regional-scale data (Arako, Matsusaka, Fujimaki, Kayumi)

	(A) District-scale data (1:2,500)		(B) Regional-scale data (1:25,000)		TLD_{ij} (h/a)	
	Land scape unit	Landscape element	Land scape unit		(A)	(B)
Arako	Built-up areas	Public facilities	Built-up areas	Urbanization Promotion Area	0.32	2.14
		Shrines and temples			4.88	
		Parking lots			0.00	
		Railroad			1.96	
		Private building lots		Residential districts of high coverage of green	21.94	4.21
	Parks		Park, Cemetery, etc.		4.00	1.34
	Other crop field and orchard		Other crop field and orchard		29.77	–
	Secondary grasslands (low)		Secondary grasslands (low)		–	0.00
	Vacant land				1.96	
	Forests		Secondary forests (almost natural forest)		0.00	0.00
	Road		Road		0.00	0.00
Matsusaka	Built-up areas	Public facilities	Built-up areas	Urbanization Promotion Area	0.32	2.14
		Cultural heritage building			9.65	
		Shrines and temples			4.88	
		Parking lots			0.00	
		Vacant land			1.97	
		Railroad			1.97	
		Private building lots		Residential districts of high coverage of green	38.30	4.21
	Parks		Park, Cemetery, etc.		0.81	
	Paddy fields		Paddy fields		–	2.63
	Other crop fields		Other crop fields		29.77	–
	Secondary forests (almost natural forest)		Secondary forests (almost natural forest)		–	0.00
	Water surface	Rivers (water surface)	Water surface		0.00	0.00
		Dry riverbed			1.97	
		Water channel			47.12	
	Road		Road		0.00	0.00
Fujimaki	Built-up areas	Private building lots	Built-up areas	Urbanization Promotion Area	30.73	2.14
	Residential districts of high green coverage			Residential districts of high coverage of green	–	4.21
	Parks		Park, Cemetery, etc.		0.81	1.34
	Paddy fields		Paddy fields		–	2.63
	Secondary grasslands (low)		Secondary grasslands (low)		–	0.00
	Forests	Secondary forest (*Quercus*)	Secondary forests (*Quercus variabilis* or *Quercus serrata*)		3.42	0.00
		Bamboo forest	Secondary forests		2.24	0.00
	Road		Road		0.00	0.00

(continued)

Table 17.6 (continued)

	(A) District-scale data (1:2,500)		(B) Regional-scale data (1:25,000)		TLD_{ij} (h/a)	
	Land scape unit	Landscape element	Land scape unit		(A)	(B)
Kayumi	Built-up area	Private building lots	Built-up areas	Other area	12.38	2.14
		Parking lots			0.00	
		Public facilities			12.38	
		Shrines and temples			0.23	
	Residential districts of high coverage of green			Residential districts of high coverage of green	–	4.21
	Other crop fields		Farmland	Other crop fields	29.77	29.77
	Paddy fields			Paddy fields	2.63	2.63
	Mulberry fields			Orchard	28.26	–
	Orchards				29.26	–
	Tea plantations			Tea plantations	12.54	12.54
	Tree plantings		Tree plantings		–	0.00
	Secondary grasslands (low)		Secondary grasslands (low)		0.00	0.00
	Secondary grasslands (high)		Secondary grasslands (high)		0.00	0.00
	Artificial forests		Artificial forests		1.07	1.07
	Secondary forests (*Quercus variabilis* or *Quercus serrata*)		Secondary forests (*Quercus variabilis* or *Quercus serrata*)		–	1.34
	Secondary forests		Secondary forests		–	0.00
	Secondary forests (almost natural forest)		Secondary forests (almost natural forest)		–	0.00
	Natural forests		Natural forests		–	0.00
	Coastal vegetation		Coastal vegetation		–	0.00
	Gravel surface		Gravel surface		–	0.00
	Water surface	Rivers	Water surface		0.00	0.00
		Pond			0.00	
	Road		Road		0.00	0.00

Figure 17.10 shows the relation between the GLD and SLI of each basic landscape type in 2010. The SLI of the other field Satoyama type is the highest. In addition, the SLIs of the countryside are higher than are those of the urban and urban periphery areas.

Figure 17.11 shows the relation between the total land area ratio, population ratio, and the mesh TLA ratio. As the population ratio is high in urban type and low in Satoyama and nature types, the total land area ratios show contrasting behaviors. The mesh TLAs are high in the farmland types.

Table 17.7 The comparison of TLD_{ij} between the district and the regional scale data (Asami, Shima)

	(A) District-scale data (1:2,500)		(B) Regional-scale data (1:25,000)		TLD_{jl} (h/a)	
	Land scape unit	Landscape element	Land scape unit		(A)	(B)
Asami	Built-up area	Private building lots	Built-up area	Urbanization promotion area	41.10	2.14
		Shrines and Temples			28.10	
		Public facilities			0.32	
		Parking lots and other paved surface			0.00	
		Rail road			0.00	
	Residential districts of high coverage of green		Residential districts of high green coverage		–	4.21
	Other crop fields	Other crop fields	Farmland	Other crop fields	29.77	29.77
		Tree			0.00	
	Paddy fields	Rice (type1)		Paddy fields	3.33	2.63
		Rice (type1)			1.56	
		Rice (type2)			2.46	
		Rice (type2)			1.46	
		Rice (type3)			2.23	
		Rice (type3)			1.31	
		Wheat (type1)			1.02	
		Wheat (type2)			1.02	
		Wheat (type3)			1.02	
		Soybean (type1)			1.02	
		Soybean (type3)			1.02	
		Grassland			0.00	
	Tea plantations		Tea plantations		–	12.54
	Orchard		Orchard		28.26	–
	Abandoned fields		Abandoned farmlands		0.00	–
			Abandoned paddy fields		0.00	–
	Vacant land	Tree	Secondary grasslands (low)		0.00	0.00
		Grasslands	Secondary grasslands (high)		0.00	0.00

(continued)

Table 17.7 (continued)

	(A) District-scale data (1:2,500)		(B) Regional-scale data (1:25,000)		TLD_{jl} (h/a)	
	Land scape unit	Landscape element	Land scape unit		(A)	(B)
	Artificial forests		Artificial forests		–	1.07
	Secondary forests (*Quercus variabilis* or *Quercus serrata*)		Secondary forests (*Quercus variabilis* or *Quercus serrata*)		–	1.34
	Natural forests		Natural forests		–	0.00
	Coastal vegetation	Riparian forest	Coastal vegetation		0.00	0.00
	Water surface	Ponds and canals*	Water surface		0.00	0.00
		River			0.00	
	Road		Road		0.00	0.00
Shima	Built-up area	Built-up area	Other area		12.38	2.14
	Residential districts of high green coverage		Residential districts of high green coverage		–	12.53
	Golf course/lawn		Golf course/lawn		–	4.21
	Other crop field		Farmland	Other crop field	29.77	29.77
	Paddy field			Paddy field	2.63	2.63
	Orchard			Orchard	28.26	28.26
	Tea plantation			Tea plantation	12.54	
	Tree planting	Arboretum	Tree planting		0.00	0.00
	Secondary grasslands (low)	Grassland	Secondary grasslands (low)		0.00	0.00
	Secondary grasslands (high)		Secondary grasslands (high)		0.00	0.00
	Artificial forests		Forest	Artificial forests	–	1.07
	Secondary forests (*Quercus variabilis* or *Quercus serrata*)			Secondary forests (*Quercus variabilis* or *Quercus serrata*)	0.00	1.34
	Secondary forests			Secondary forests	0.00	0.00
	Secondary forests (almost natural forest)			Secondary forests (almost natural forest)	0.00	0.00

(continued)

Table 17.7 (continued)

	(A) District-scale data (1:2,500)		(B) Regional-scale data (1:25,000)	TLD_{jl} (h/a)	
	Land scape unit	Landscape element	Land scape unit	(A)	(B)
	Natural forests		Natural forests	0.00	0.00
	Coastal vegetation	Beach	Coastal vegetation	1.21	0.00
	Gravel surface	Others	Gravel surface	0.00	0.00
	Water surface	Non-element	Water surface	0.00	0.00
	Road	National highway	Road	0.74	0.00
		Prefectural road		2.40	

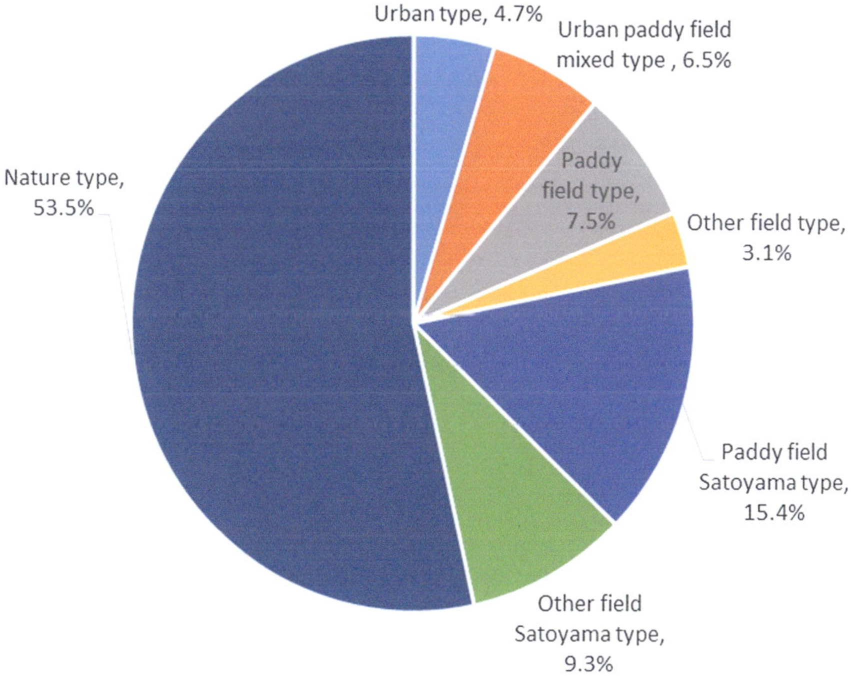

Fig. 17.8 Total land ratio of basic landscape types in Chubu metropolitan area

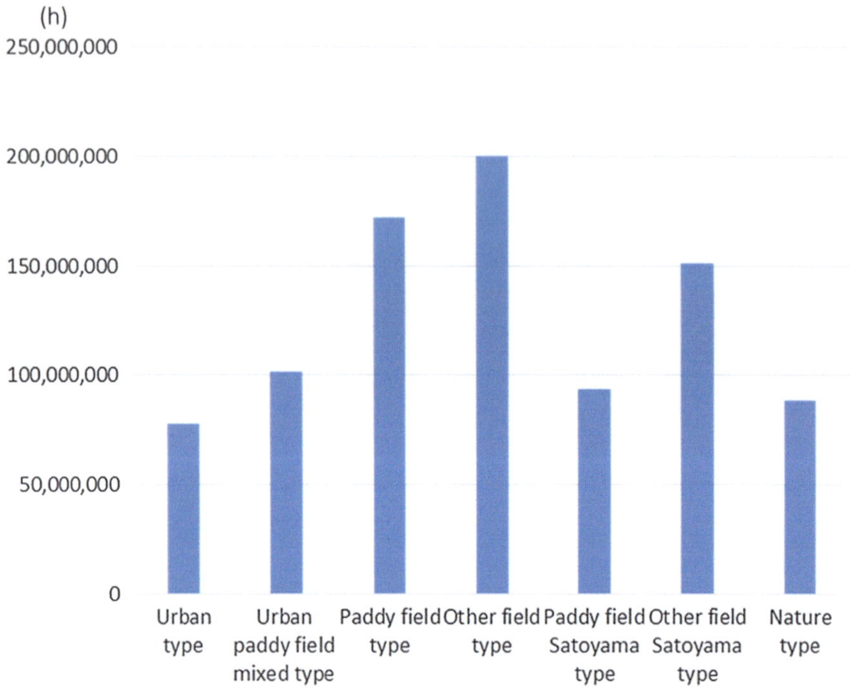

Fig. 17.9 Total labor accounts in 2010 of basic landscape types in Chubu metropolitan area

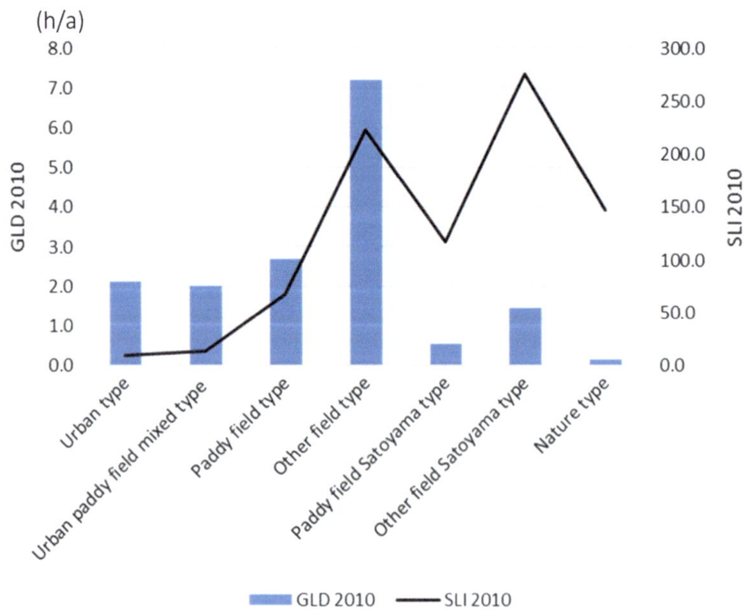

Fig. 17.10 Gross labor densities and social labor intensities in 2010 of basic landscape types of Chubu metropolitan area

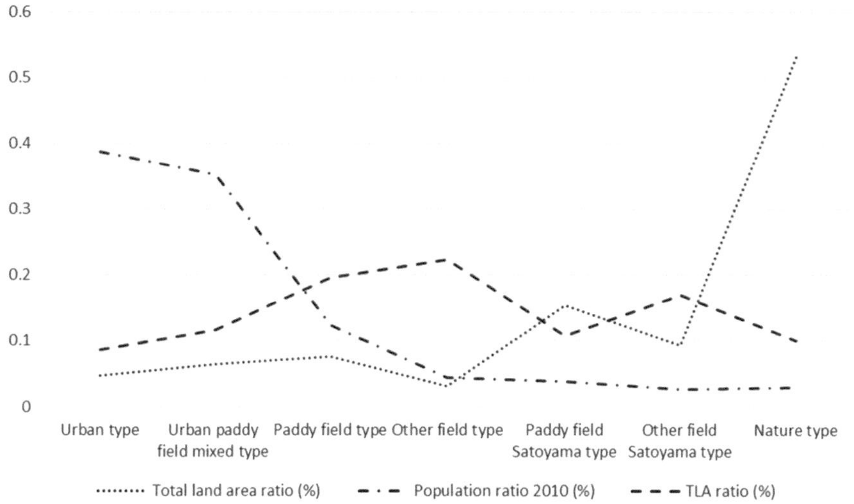

Fig. 17.11 Total land area ratio, population ratio, and TLA ratio in 2010 of basic landscape types of Chubu metropolitan area

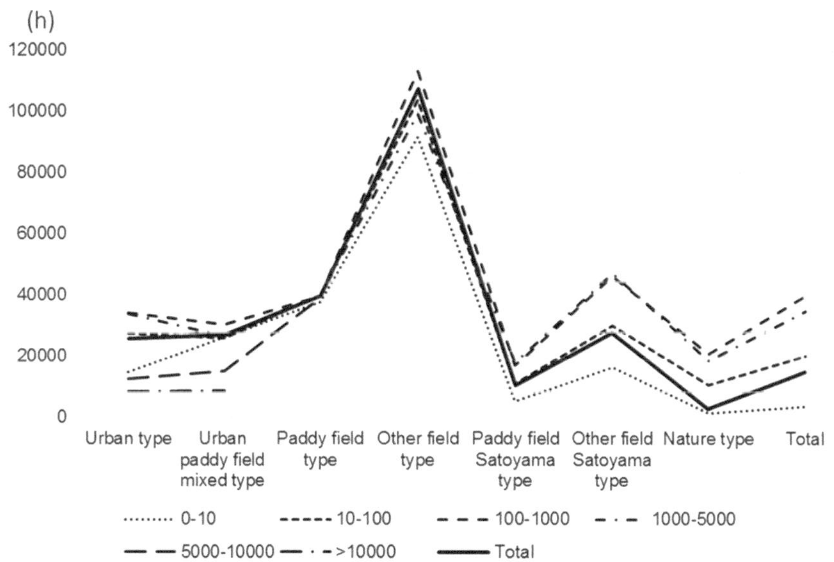

Fig. 17.12 TLA per mesh in 2010 of basic landscape types and mesh population classes in Chubu metropolitan area

Figure 17.12 shows the TLA per mesh in 2010 of the basic landscape types and the mesh population classes in the Chubu metropolitan area. The value in the other field type is the highest, even though the land ratio of the other field type is not high. This is ascribed to the labor density of the other field type being higher than is that of the rest of the landscape unit types.

17.5 Toward an Affordable Landscape Labor Account and Management

17.5.1 Population and Labor Accounts Simulation at a Region Scale

Subsequently, the total labor accounts relevant to the 1 km mesh in the Chubu metropolitan area were compared with the corresponding future population at the same scale mesh for 2050. The population data for 2050 were published in the *National Land Numerical Information, Land utilization tertiary mesh Data* (National land Information Division, National Spatial Planning and Regional Policy Bureau, Ministry of Land, Infrastructure, Transportation and Tourism (MLIT) of Japan 2014). The ratios of the population in 2010 to the population in 2050, relevant to the basic landscape types, and the mesh population classes in the Chubu metropolitan area are shown in Fig. 17.13. The ratios in the paddy field Satoyama type, other field Satoyama type, and the nature types are higher compared with the

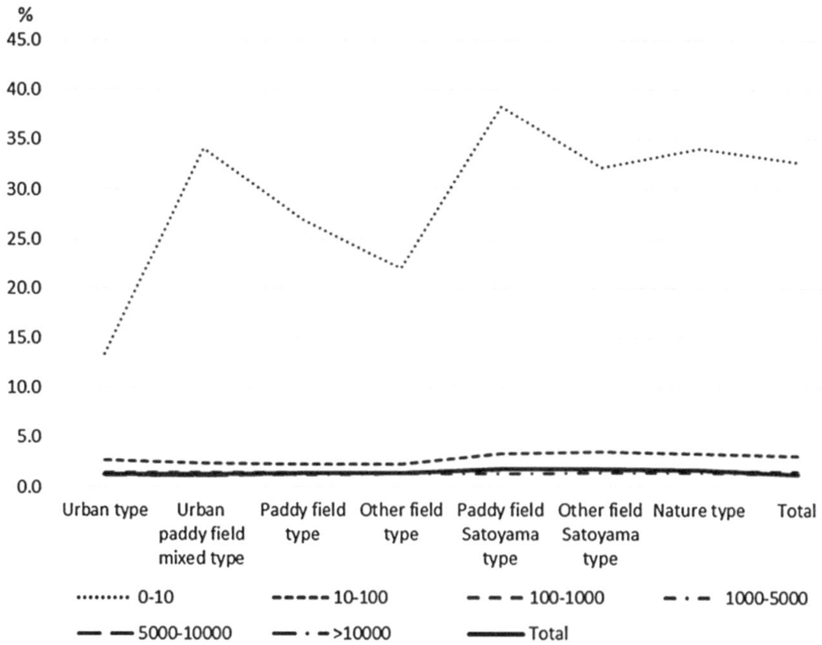

Fig. 17.13 Population ratio of 2010/2050 of basic landscape types and mesh population classes in Chubu metropolitan area

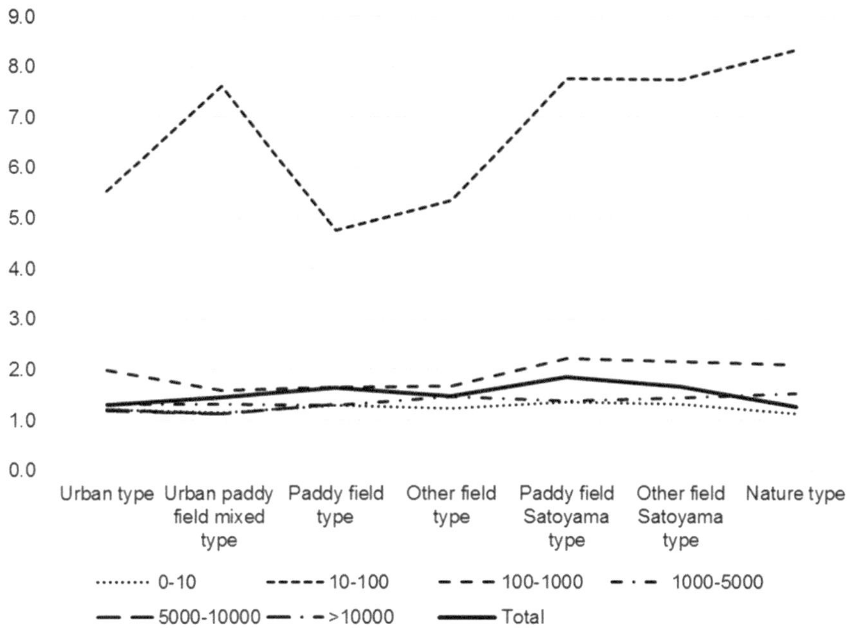

Fig. 17.14 SLI ratio of 2050/2010 of basic landscape types and mesh population classes in Chubu metropolitan area

rest of the landscape types. This means that the population decline in these landscape types is higher than in the rest of the types. Particularly, in the areas where the population per one mesh is less than 100 persons, the population decline is higher than is the average.

Figure 17.14 shows the social labor intensity ratio (SLI:TLA per population) of the basic landscape types and the mesh population classes in the Chubu metropolitan area for 2050/2010. In this instance, the assumption is that the TLA of 2010 would be maintained in 2050, even with a shrinking population. In the population classes between 10 and 1,000 persons, and especially between 10 and 100 persons, the SLIs are higher than the average SLIs for each basic landscape type.

Figure 17.15 shows the spatial distribution of SLI in the Chubu metropolitan area. Compared with the spatial distribution of the basic landscape types, the SLI in 2050 is higher than is that of 2010, especially for the paddy field Satoyama type and the other field Satoyama type.

Figure 17.16 shows the GLD ratio for 2050/2010 of the basic landscape types and the mesh population classes in the Chubu metropolitan area, on the assumption that the TLA in 2050 would decrease according to the ratio of the population decline. In the population classes for less than 100 persons, the decrease of GLD is stronger than are the averages of each basic landscape type. In addition, in the urban type, urban paddy field mixed type, paddy field type and other field type, the

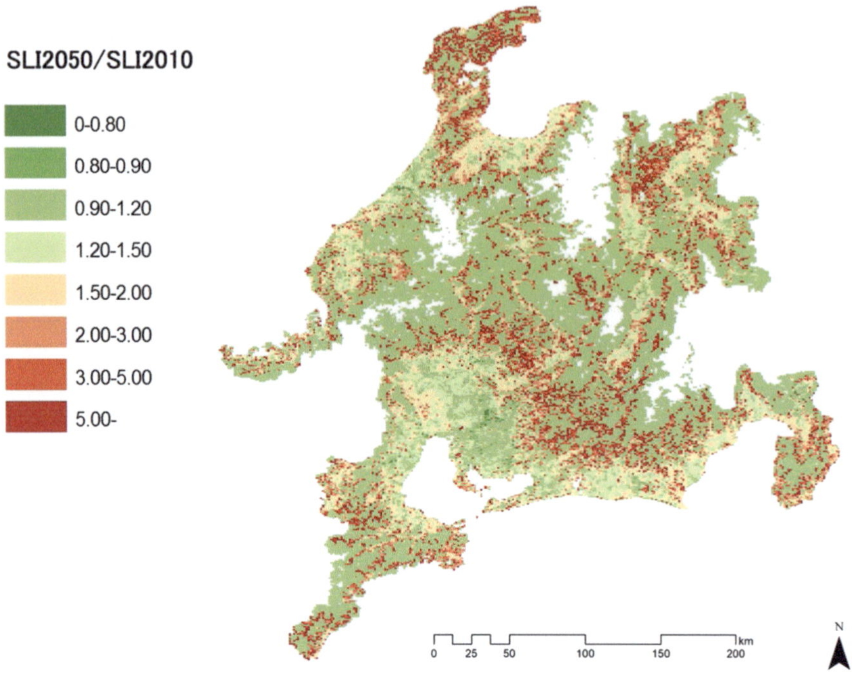

Fig. 17.15 Spatial distribution of SLI ratio of 2050/2010 in Chubu metropolitan area

decrease of GLD is stronger than are the averages in the population classes of 100 to 1,000 persons. It is estimated that in these areas there would be a significant shortage of the labor required for appropriate management of the landscapes. Figure 17.17 shows the spatial distribution of the GLD ratio for 2050/2010 in the Chubu metropolitan area. The indication is that the GLD in the Satoyama landscape types tends to decline from 2010 to 2050.

Figure 17.18 shows the spatial distribution of the gap in the total labor accounts between 2050 and 2010, on the assumption that the total labor accounts would decrease in accordance with the ratio of the population decline. It appears that the labor accounts would decrease almost everywhere in Chubu, which would exacerbate the problems associated with landscape management.

17.5.2 Concerns About Landscape Management Labor Accounts in Demographic Change

In the preceding subsection, the total labor accounts in the Chubu metropolitan area were calculated and were compared with the current and future population, and the areas where landscape management was estimated to be lacking were specified. In

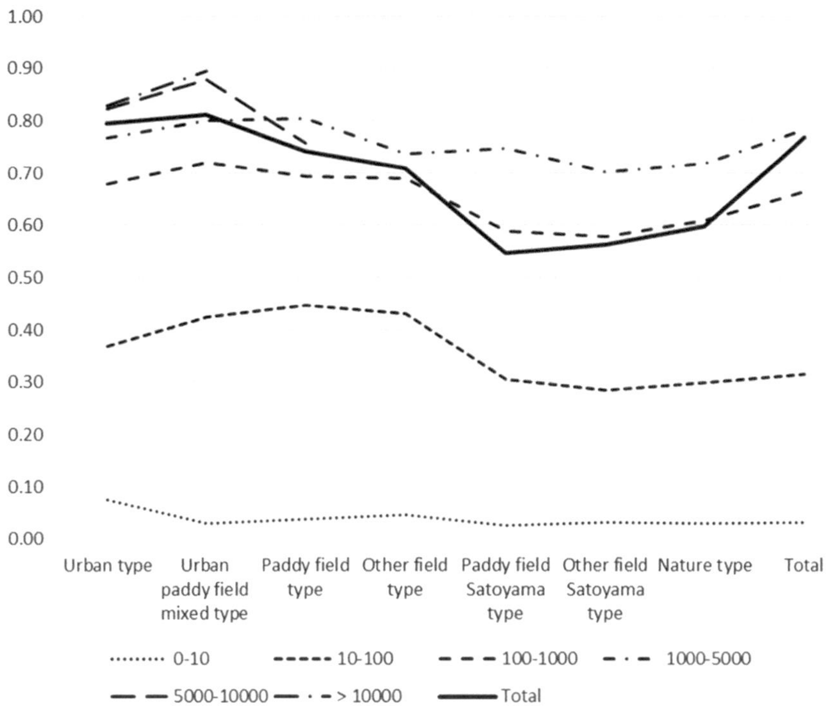

Fig. 17.16 GLD ratio of 2050/2010 of basic landscape types and mesh population classes in Chubu metropolitan area

view of this regional-scale analysis, it could be inferred that in future almost every area would have to contend with a lack of labor for landscape management.

As regards the sustainable and appropriate management of landscapes at a regional scale in future, the environmental management policy should make provision for effectively matching those individuals willing to manage the landscape with the particular locations suffering from a lack of management. The shrinking population is especially of concern in the Satoyama area and the coastal areas. In these areas, the value of the cultural landscapes and ecosystem services is estimated to be high; however, there is considerable concern about the lack of a labor force for landscape management. Therefore, appropriate policies and measures have to be developed and implemented to promote migration from the urban areas to such areas. The landscape management labor accounts could provide a basis for considering such incentives.

The imbalance between the landscape management needs and the labor force required to meet such needs is ascribed partly to an underestimation of the importance of the labor for landscape management. Moreover, the necessary investigations are not being conducted and financial support is not forthcoming, as most of the labor for the maintenance of valuable landscapes is externalized. To

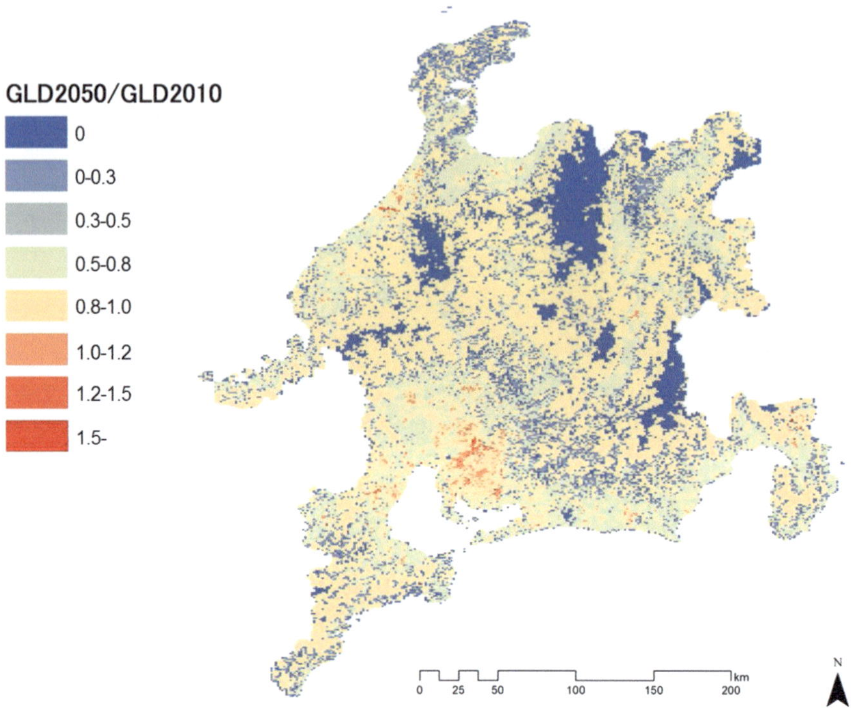

Fig. 17.17 Spatial distribution of GLD ratio of 2050/2010 in Chubu metropolitan area

eliminate such an imbalance, local efforts have to be considered to develop the domestic economy by cooperation between the public and the private sectors. On the other hand, economic valencing systems, such as direct payments relevant to fragmented agricultural, forestry, and fishery fields, attributable to the governmental policy of sectionalism, have to be coordinated and integrated in future. In addition, the information relevant to the landscape management labor accounts could serve as a vehicle for such collaboration and for the integration of various governmental sections.

The following four points, introducing the concept of neighborhood landscape complex units, should be considered in future management plans to compensate for economic externalities.

(1) To distinguish between a place that should be managed and that which should not be managed.
    In view of the lack of management labor, a distinction should be made between landscapes that require management and those that do not, according to the ecological and economic values of the areas concerned. Negotiation with and the consensus of the landowners in the communities are needed to determine such particular areas.

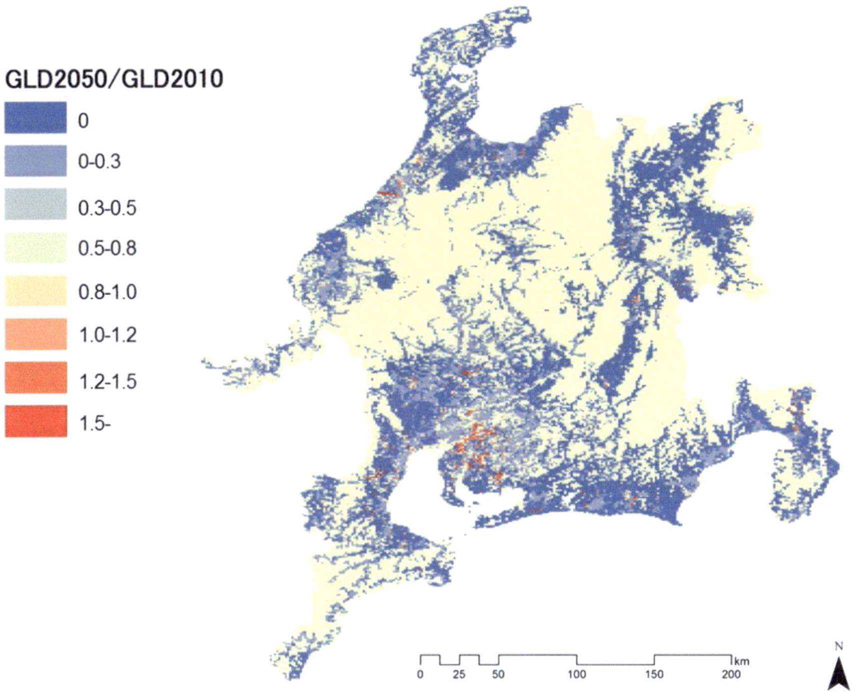

Fig. 17.18 Gap of labor accounts between 2050 and 2010

(2) To change the same land use types from high-management types to low-management types.
For example, the biotope type "abandoned farmland" requires significantly less management than a fallow field does. Moreover, an artificial forest older than 30 years requires significantly less management than a forest younger than 30 years does. To change the same land use types from a high-management level to a low- management level is another solution.

(3) To increase the efficiency of landscape management in cooperative forms.
Particularly, the management efficiency of the "farmland" and "artificial forest" sites was low because of the small scale of the land and the scattered pockets of land belonging to the landowners. A possible solution would be to combine these scattered fields into one larger unit, which would facilitate management efficiency.

(4) To establish or enhance social systems, such as supplementary taxation and accepting volunteers from areas with an adequate management population, in order to compensate for the economic externalities.

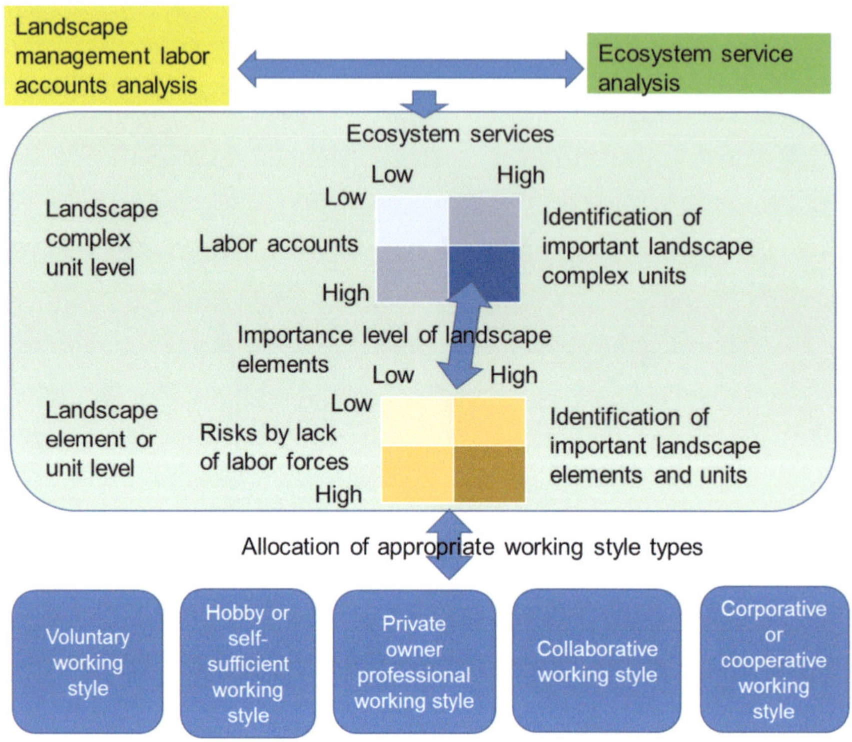

Fig. 17.19 Research development by the combination of landscape labor accounts analysis and ecosystem service analysis

17.5.3 Research Development by Combining Analysis of Landscape Management Labor Accounts and Analysis of Ecosystem Services

This book does not contain extensive descriptions of ecosystem services, as it was not the main purpose of the work to present such descriptions. However, the concept of landscape management labor accounts could be developed in combination with research on ecosystem services to facilitate sustainable and ecological landscape management, as shown in Fig. 17.19.

The landscape management labor accounts approach could indicate how much labor is needed for managing the targeted landscape area, and the ecosystem service approach could indicate the ecological service value of the area.

If the spatial distribution of ecosystem service amounts at a regional scale were known, a comparison could be made, by using cross tables, of the values of the landscape management labor accounts and the ecosystem services. If an area showed high values for both labor accounts and ecosystem services, it could indicate that such area should be managed intensively to maintain the ecosystem services. If an area showed a high value for ecosystem services and a low value for labor accounts, it could indicate that there had been no human interference to maintain the ecosystem services. If the area showed a low value for ecosystem services and a high value for labor accounts, the labor density could be reduced to some extent. An area showing low values for both labor density and low ecosystem services could be utilized for artificial land development. Alternatively, such an area could be improved to create more value in the ecosystem services by adding more labor accounts. Such an assessment is valuable, especially at the level of landscape complex units of towns and villages.

Such evaluation could be conducted in more detail at the element or unit level, taking into account the significance of the landscape elements or units and the risks posed by a lack of labor force. For instance, an urban forest has high cultural ecosystem value, as it is an important resource for outdoor recreation; therefore, it should be managed with adequate and appropriate care.

On the other hand, as shown in Chap. 13, the Jori-sei paddy fields that have high ecological value because of the rich biodiversity they sustain could be preserved or maintained by the addition of extra labor accounts.

The participation of residents in such an evaluation process is important, and the labor accounts data could facilitate consensus among the role players in such a participation process.

Such evaluation could be combined with a consideration of the appropriate allocation of working style types. Consequently, the allocation could play an important role to determine the sustainability and quality level of the maintenance.

The research on the landscape management labor accounts has just started and various challenges remain; nevertheless, the results of the current research could contribute to preserving the sustainability of valuable landscapes.

References

Kawaguchi N, Takatori C, Murayama A, Shimizu H (2016) Proposal and estimation of landscape management labor accounts in urban areas. J City Plann Inst Jpn (under review)

Kieninger PR, Yamaji E, Penker M (2011) Urban people as paddy farmers: the Japanese Tanada ownership system discussed from a European perspective. Renew Agr Food Syst 26(4):328–341

Mie Prefecture (2013) Forest register. Mie Prefecture

Ministry of Agriculture, Forestry and Fisheries of Japan (2005) Agriculture business management statistics investigation farm products life costs statistics. http://www.maff.go.jp/j/tokei/kouhyou/noukei/. Accessed 1 July 2015

Ministry of the Environment of Japan (2005) Vegetation survey, natural environmental information GIS, 1:25,000 and 1:50,000. http://www.biodic.go.jp/trialSystem/EN/shpddl.html. Accessed 1 June 2014

Nagoya city (2010) Green status survey. Licensed by Nagoya city

Nagoya city (2012) Basic survey of city planning. Licensed by Nagoya city

National land Information Division, National Spatial Planning and Regional Policy Bureau, Ministry of Land, Infrastructure, Transportation and Tourism (MLIT) of Japan (2009) National land numerical information, land utilization tertiary mesh data. http://nlftp.mlit.go.jp/ksj-e/gml/datalist/KsjTmplt-L03-a.html. Accessed 2015

National land Information Division, National Spatial Planning and Regional Policy Bureau, Ministry of Land, Infrastructure, Transportation and Tourism (MLIT) of Japan (2014) National land numerical information, land utilization tertiary mesh data. http://nlftp.mlit.go.jp/ksj-e/gml/datalist/KsjTmplt-L03-a.html. Accessed 2015

Part IV
Perspectives

Chapter 18
Urban Planning: Is a Networked Compact City Vision Realistic?

Nobuko Kawaguchi and Akito Murayama

Abstract The largest impact on Japanese cities is the arrival of the serious depopulating, hyper-aged society. Only scarce public money is available for landscape management, including the new development of parks and maintenance of existing parks as urban facilities. Urban greening policies must shift toward increasing greenspace along public-owned roads and on private property. Many cities pursue a vision of the networked compact urban society where an increase of population and building density center around public transit stations and decreases in areas under-served by public transit. The increase of greenspace will have significant influence on landscape management labor accounts, especially in de-intensifying urban areas. This leads to the question whether or not a vision of the networked compact city is realistic. The authors conclude that in order to answer this question, further consideration of urban block scale is necessary.

Keywords Depopulation · Hyper-aged society · Urban form · Intensification · Deintensification

18.1 Introduction

Japanese cities are facing both progressive and sudden risks (Murayama 2016). Some progressive risks include a decline of the labor force, hyper-aging, economic stagnation, widening disparity, governmental financial difficulties, and intensification of

N. Kawaguchi (✉)
Graduate School of Environmental Studies, Nagoya University,
D2-1(510), Furo-cho, Chikusa-ku, Nagoya, Aichi 464-8601, Japan
e-mail: nobqou@gmail.com

A. Murayama
Urban Land Use Planning Unit, Department of Urban Engineering,
School of Engineering, The University of Tokyo,
7-3-1 Hongo, Bunkyo-ku, Tokyo 113-8656, Japan
e-mail: murayama@up.t.u-tokyo.ac.jp

H. Shimizu et al. (eds.), *Labor Forces and Landscape Management*,
DOI 10.1007/978-981-10-2278-4_18

environmental problems such as climate change, energy shortages, food and water security, etc. Sudden risks include disasters such as major earthquakes, tsunamis, typhoons, torrential downpours, and volcanic eruption. The major challenge in urban landscape planning is to respond to these risks and to restructure urban areas to be resilient to both chronic stress and acute shocks (100 Resilient Cities 2016). The largest impact on Japanese cities is the arrival of intense depopulation patterns coupled with hyper-aging, a combination of which no other city in the world has previously experienced. More and more public money is consumed by medical and welfare-related projects, leaving scarce resources for landscape management, including new development of parks and maintenance of existing parks as urban facilities. Since there is a clear limitation, urban greening policies must shift toward increasing greenery along public-owned roads and on private property. This issue was discussed in recent working group meetings of the Nagoya City Council on Greenery.

Many cities, including Nagoya City, have a vision for a networked compact urban form that results in urban building and population densities around public transit stations with densities decreasing in areas under-served by public transit. This vision calls for a realistic way to pay for planning, developing, and managing urban greenspaces in areas both intensifying and de-intensifying. Thus, the concept of a landscape labor account plays an important role as a key indicator for considering the amount, distribution, and quality of greenery in transforming urban areas. This will contribute to the discussion "Is a networked compact city vision realistic?" from landscape management point of view.

18.2 Considering the Possibility of Urban Landscape Management in the Promotion of a Networked Compact Cities

This section's discussion will be based on case studies examining landscape management labor accounts in urban areas that were presented in Chaps. 10 and 17. These chapters assessed landscape management labor accounts for representative greenery in urban areas. The Social Labor Intensity (SLI) defined in these chapters will be used to discuss the reality of promoting a networked compact city in the age of a declining labor force.

Figure 18.1 shows the results of calculating population changes (a), the change in rate of the aging population (b), GLD (c) and SLI (d) for the landscape complex units in Nagoya City using the 2005 and 2010 National Census. GLD and SLI were calculated based on citywide landscape management labor accounts calculated in Chap. 17. First, the graph addressing population change (a) shows that people participating in urban green management were typically 20 years old or older. Therefore, the graph shows the total population divided into groups of people 20 years and older, 20 to 64 years old (labor force age), and 65 years and older.

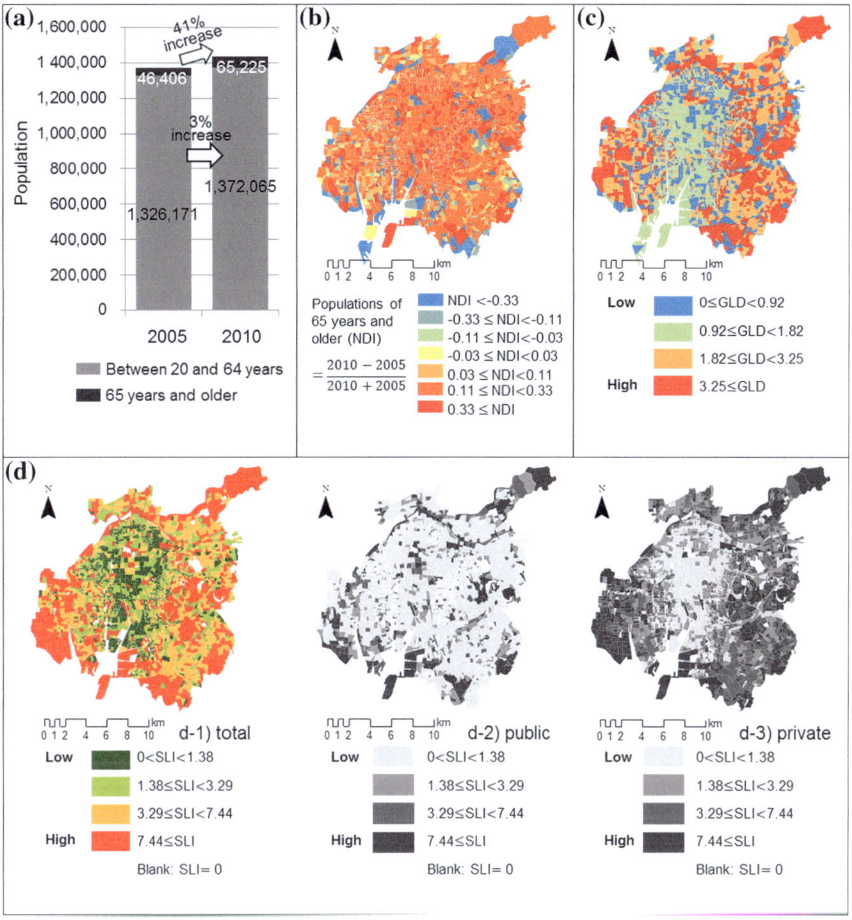

Fig. 18.1 Population change, GLD and SLI of the landscape complex units in Nagoya city. **a** Population change (2005–2010). **b** Population change ratio (2005–2010). **c** Spatial distribution of GLD. **d** Spatial distribution of SLI

According to this graph, the number of people capable of managing greenery, i.e., 20 years and older, slightly increases between 2005 and 2010. Next, the figures of spatial distribution (b, c and d) were compared. In the landscape complex units with increasing populations of 65 years and older, an increase of burden on greenspace management is expected due to aging. In the urban center with smaller GLD and SLI, the burden will be less significant. However, along city peripheries, landscape complex units have higher GLD and SLI. Aging in these units will result in an increasing burden on urban green management. It was also confirmed that in aging urban areas, SLI among private properties is significantly high.

Based on the above analysis, the possibility of future landscape management in urban areas will be discussed.

18.2.1 Landscape Management Labor Accounts for Private Properties

There are challenges ahead regarding managing urban greenspace of single-family home properties and agricultural land that is owned and managed by individual property owners. The maintenance of single-family home properties and agricultural land is a concern given the increase in aging, resulting in an increase of future burden for urban green management. On the other hand, in locations with multi-family housing, many people are able to take part in landscape management. If the area of green space increases in urban areas under-served by public transit, there will be a significant increase in landscape management. Landscape complex units requiring high landscape maintenance means that labor must be shared by many people.

It is important to conserve abundant greenery in the yards of single-family residential areas. The system to support the management burden of property owners would help, but governmental public funds are limited.

In agricultural lands, the government has some projects to promote agriculture as industry. However, agricultural lands within urban area are decreasing. In the future, agricultural lands for recreation or environmental contribution may increase, however agricultural land for production may decrease. The system of community gardens managed by government might play an important role in managing greenspace by people other than private property owners. However, it should be noted that people are aging and the individual contribution to landscape management will be limited.

18.2.2 Landscape Management Labor Accounts in Public Properties

Parks and street trees in public properties have high landscape management labor accounts. Parks and street trees are important in the conservation and creation of greenspace in urban areas and there are already public projects related to parks and street trees. However, the increase of public properties with greenspace will increase landscape management labor accounts. To lower SLI, landscape management labor accounts per capita, the role of landscape management activities by civic organizations would play an important role. However, again, with the aging and decreasing labor force, the sustainability of activities conducted by civic organizations is also an issue.

18.3 Conclusion

18.3.1 Future Imaginings of Green Space Management Responding to Population Change

As Introduced in Chap. 3, Nagoya City has a vision to concentrate populations around transit station areas. Figure 18.2 explains separately population and landscape management labor accounts surrounding transit station areas with increasing population concentrations and other areas with decreasing populations.

Population Concentration Areas

In areas with an increasing population, it is important to set aside areas in order to effectively plan greenspace. Open high-use urban centers, for example, can receive high-maintenance plantings. These plants will be managed with low landscape management labor accounts, similar to those seen in commercial facilities or multi-family housing. However, the amount of greenspace in these areas tends to decrease. Other than allocating open space in building properties, the quality of the urban center environment could be enhanced by planting many street trees. Street

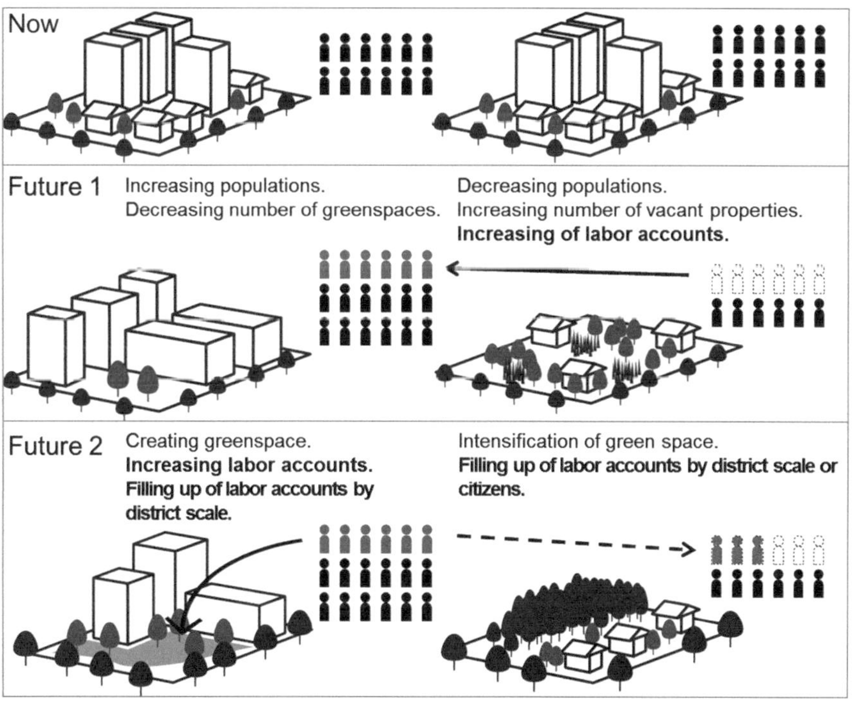

Fig. 18.2 Population and landscape management labor accounts of surrounding transit station areas and other areas

trees have the second largest landscape management labor account among public properties (see Chap. 17). The public fund for street trees decreases from year to year. In order to supplement this public fund shortage, it is necessary for private corporations and area residents to contribute to the system.

Population Decreasing Areas
In areas with a decreasing population, hollowing-out of properties is anticipated. There is a high possibility that illegal waste dumping and crime will increase with the increase of abandoned forests and grasslands. The situation where unmanaged properties are dispersed in an area is not efficient. Furthermore, the landscape management labor accounts increase if vacant properties are transformed to green space. In order to effectively manage greenery among a small population, it is necessary to zone green management. For example, one can imagine properly managing these properties by dividing a zone with concentrated grasslands and forests with low frequency of management and use from planted zones with high frequency of management and use. However, the concern is the shortage of labor to manage the large grasslands and forests in these less inhabited areas. A system is needed to allow not only property owners but also groups of residents or volunteers to take part in management. It is also necessary to consider the reduction and the exemption of property tax when open space is managed as greenspace or it is opened to the public as a quasi-park.

18.3.2 Is a Vision of Networked Compact City Realistic?

Contradictory to an optimistic vision of a networked compact city proposed by urban planners, the consideration of land management labor accounts suggests that an increase of greenspace in urban areas under-served by public transit would be a large burden for the area's aging and decreasing population. This is one of the major side effects of promoting a networked compact city. Should planners give up the vision? Authors think it is too early to answer this question and further consideration on urban form, including the spatial distribution of greenspace and quality of urban green, is necessary.

In the transformation process of urban form especially in de-intensifying urban areas, it is important to look at urban block scale. First, a vision for the ideal physical environment of urban blocks consistent with numbers and trends of population and households should be developed. Second, the transformation process of urban blocks and guide designs of buildings and open spaces should be designed and managed. Lastly, the whole and the parts of urban form that emerges through the transformation of urban blocks should be coordinated. Though this process can be conducted under the vague conceptual vision of a networked compact city, it provides reality to the various physical and social environments of urban blocks.

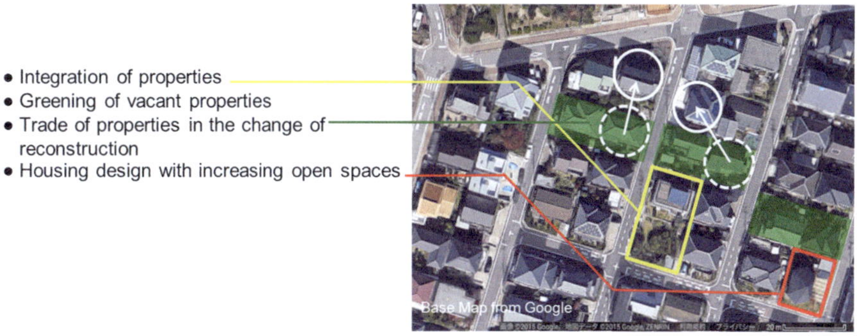

Fig. 18.3 Model of de-intensifying Urban Blocks with Increasing Greenery

Figure 18.3 shows the model of de-intensifying urban blocks with increasing greenspaces. Various property transformations are possible including property integration, vacant property greening, and property trading, with the aim of building reconstruction and housing designs with increasing open spaces. Landscape management labor accounts and their burden for residents might differ depending on the pattern and quality of increasing greenspaces. Resident age and lifestyle also influences the method of managing green and the quality of that management.

While understanding the overall relationship between a vision of the networked compact city and the increase of greenspaces and landscape management labor accounts, we must discuss urban block scale before giving up the vision. This is dependent upon landscape management becoming sustainable and keeping or even increasing property values in de-intensifying urban areas.

With the stagnation of economic growth and the increase of an aging population, investable government funds are decreasing. In the future, the important issue is to keep greenspace among private properties. In addition, we need to consider the extent of available citizen volunteerism for supplementing labor shortages for managing public greenspaces. The landscape labor account is an indicator to measure the population's labor amount. It is well appropriate to explain that there is a major labor shortage of individual persons due to aging populations to manage private gardens, agricultural lands and public greenspaces conserved by civic organizations.

In order to properly consider urban green management requirements, the legislative system and effective public investment strategies must be explored. For example, Funabiki (2009) investigated measures to maintain greenspaces from a cost point of view. Funabiki compared the cost under different systems and estimated the cost benefit of greenspace conservation. In an aging city, the maintenance of greenspace through stricter land use regulation will increase the burden to citizens. In the future, it is necessary, therefore, to consider the appropriate method for implementing public support in response to possible citizen labor requirements using the concept of landscape management labor cost.

References

100 Resilient cities. https://www.100resilientcities.org. Accessed 1 April 2016

Funabiki T (2009) A study on the acquisition and enhancement policies of green space in japan from the view of financial strategies. J Jpn Inst Landscape Archit 72(5), 793–798 (Landscape Research Japan)

Murayama A (2016) Land-use planning for depopulating and aging society in Japan. In: Yamagata Y, Maruyama H (eds) Urban resilience—a transformative approach. Springer (in press)

Chapter 19
Urban Periphery Planning: Concept to Link Urban and Rural Communities in the 21st Century

Toru Terada

Abstract Urban-rural mixture, a typical scene of Japanese urban periphery, is often recognized as a failure of modern urban planning, since urban–rural division by zoning is often a goal of city planning during expansion periods. In this chapter, a new planning concept for use during city shrinkage periods, which uses urban–rural mixture as a potential for enhancing landscape sustainability, is discussed. The discussion begins by investigating what would happen if the current trend is followed, without proactive policy intervention (i.e., worst-case scenario). Then, preferable future goals and required measures required to achieve them are discussed as alternatives. The chapter concludes that orchestrating an integration of urban and rural, rather than dichotomizing them, is key for overcoming conventional modern urban planning dilemmas of the 20th century.

Keywords City shrinkage · Land use policy · Farmland and forest management · Urban-rural orchestration · Living in the garden

19.1 Introduction

Japanese urban periphery has "land use confusion" with its urban–rural mixture, resulting from loose land use regulations and high priority of individual property rights (see Chap. 4 of this book). The urban–rural mixture is sometimes recognized as a failure of modern urban planning, since urban–rural division (aiming for maximization of each function) is often a goal of city planning during expansion periods. In the near future, many cities in developed countries will be faced with city shrinkage (Oswalt and Rieniets 2006). Although models for challenging city shrinkage have been proposed—the compact city (Dantzig and Saaty 1973; Burton et al. 2003), a proactive approach for reconstructing city form; perforated city

---

T. Terada (✉)
Department of Urban Engineering, School of Engineering,
The University of Tokyo, 7-3-1 Hongo, Bunkyo, Tokyo 113-8656, Japan
e-mail: terada@k.u-tokyo.ac.jp

H. Shimizu et al. (eds.), *Labor Forces and Landscape Management*,
DOI 10.1007/978-981-10-2278-4_19

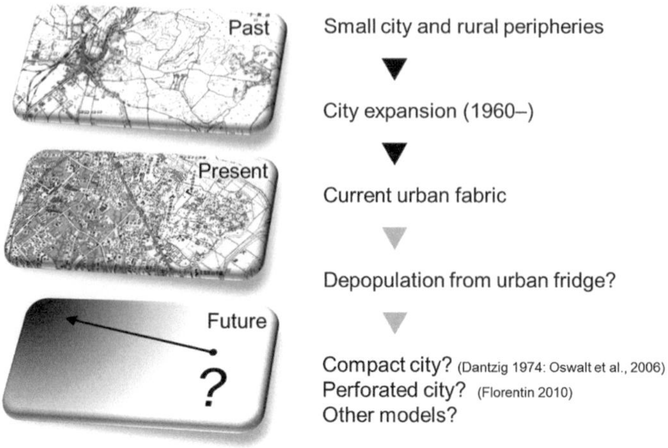

Fig. 19.1 What city form should be assumed for the future planning in the city shrinkage period? (Map: Geospatial Information Authority of Japan)

(Florentin 2010), a passive and incremental approach to manage city shrinkage—no single successful model has emerged; rather, individualistic approaches considering each city's characteristics should be used in Fig. 19.1.

What are some actions that Japanese urban and landscape planners should do during periods of city shrinkage? First, it can be expected that measures aimed at controlling city expansion will gradually lose effectiveness; as such, new goals and measures should aim to proactively head off shrinkage, or support comfortable living in areas that have already experienced shrinkage. In general, city shrinkage should be mitigated in existing, highly urbanized areas that already contain high-performance urban infrastructure, while comfortable living should be promoted in urban periphery, where the shrinkage is likely to occur; in addition, farmland and forest there should be used as potential ways to enhance environmental sustainability.

The following sections of this chapter discuss how urban periphery can best achieve comfortable living during periods of city shrinkage. Emphasis is placed on a key concept for urban periphery planning: "living in the garden," which envisions a community that is functionally, physically, and socially combined with farmland and forest ecosystems, rather than a simple and confused urban–rural mixture.

Discussions are based on Area Divisions by the City Planning Act, which divides a city planning area into two parts: Urbanization Promotion Area (UPA) and Urbanization Control Area (UCA) (Nakai 1988; MLIT 2003). This area division has heavily impacted urban periphery landscape; therefore, even this scheme should be re-examined, as it is a typical, conventional measure enacted during a city expansion period, the area division should be considered as a basic frame when prospecting future situations. The discussion begins by investigating what would happen if the current trend is followed, without proactive policy intervention (i.e., worst-case scenario). Then, preferable future goals and required measures to achieve them are discussed as alternatives.

19.2 Landscape Management of Inner Periphery (UPA)

19.2.1 What Happens if No New Measures Are Taken?

Once land is designated as UPA, it should be developed for urban areas within 10 years; however, especially within the inner edge of UPA, or inner periphery, development is often still ongoing after this period and not likely to be completed successfully, resulting in a confusing and partially completed land use mix of building, farmland and forest. This section describes an anticipation of the future of inner periphery, assuming if the present trend is followed.

A typical characteristic of inner periphery is its affordable housing thanks to low land prices resulting from the area's inefficient access to the center of the city. Japanese young families often hope to live in a detached house, even if it is located in an inconvenient area. Responding to this, a "fast and cheap" housing supply system (e.g., prefabrication) has been developed to meet this demand (Noguchi 2003). Such housing supply is sometimes called *minikaihatsu* since it tends to be carried out on a small scale (i.e., less than 1000 m^2) in order to abide with the City Planning Law (Sorensen 2001). Since such demand still exists, the housing supply in inner periphery may continue in the future. Meanwhile, many protected farmlands designated as Productive Green (PG) in UPA are expected to be sold around 2021 (as discussed in Chap. 4) and developed with buildings. Unless countermeasures are taken, further sprawl of such development will continue into farmland in UPA.

It is also predictable, however, that the population in Japan will continue to decline, especially in younger-generation demographic groups (IPSS 2012). From this prediction, it is foreseeable that the affordable housing supply in inner periphery for young families will gradually decrease and vacant or abandoned land will increase correspondingly. In addition to this, quality of existing infrastructure and housing in inner periphery are often poor, since the development is fast and not well-controlled by urban planning but rather simply follows the housing market, accepting sprawl development. It is doubtful that such poor housing will succeed to the next generation. Also, encroached farmlands by sprawl development are not suitable for professional farming; therefore, they might be simply abandoned. Existing forest is also in the same situation, as there are no incentives for land owners to maintain it, causing abonnement without any linkage to the urban environment (e.g., urban forest for scenic beauty or recreational purposes).

19.2.2 A Preferable Future and Relevant Measures

In order to sustain the environment of inner periphery during city shrinkage periods, the following two strategies are essential: (1) preventing additional urban sprawl, and (2) conserving farmland and forest as urban greenspaces to improve dwelling environments. The inner periphery should be re-designed as a place for "living in the garden," a concept that advocates that farm and wooded landscapes should be functionally, physically, and socially integrated into urban communities so that the residents there are able to enjoy green lifestyles. Concrete tactics to achieve this concept are discussed below.

First, to prevent additional sprawl, the District Plan by the City Planning Act should be applied. This plan is not a mandate, but an optional planning scheme for neighborhood communities that invites ordered development by determining rules for new developments such as minimum requirements for the site area, road width, amount of greenspace, etc. The District Plan is an integrated plan that can determine a series of measures for achieving the preferable city form based on neighborhood consensus. The plan is essentially for controlling city expansion rather than for managing city shrinkage, which means that the plan is not yet linked to the measures for shrinkage management. For example, the adjacent site-purchase scheme, which involves financial support for obtaining an adjacent vacant site to make the housing property wider, should be developed and integrated into the new district plan (Katsumata 2013). The current study found that there were inconvenient residential districts more likely to start shrinkage and also that residents living in such areas do not have positive willingness to continue living there (Yamada et al. 2016). An alternative model for urban dwelling, such as the affluent urban–nature life under large property, is vital to sustainably manage such areas (Fig. 19.2).

To conserve farmland and forest as urban greenspaces, aggregated farmlands whose landowners have a desire to continue their farming practice should first be conserved as PG, taking advantage of consumer proximity. Running allotment gardens for urban residents is currently in demand and could be an alternative to conventional farming (Henderson and Hartsfield 2009; Shimpo et al. 2014). Such neighborhood-scale farming can create micro-level "food sheds," which have high potential for provisioning food that can enhance local resilience (Kurita et al. 2009). The issue, however, is how the other abandoned farmland and forest can be maintained that are not suited for conventional agriculture and forestry, rather than being recognized as potential building sites. Public intervention through purchasing the land for creating public parks or greenspaces is not an option due to declining government budgets affected by depopulation. A possible measure might be applying a land rent scheme, which is a system to separate land utilization rights from land ownership and that promotes active participation of urban residents to manage neighborhood farmland and forest. The *Kashi-niwa* (rental garden) program, operated by Kashiwa City, a suburban local municipality in the Tokyo Metropolitan Region, is a leading example. The *Kashi-niwa* program promotes the renting of abandoned greenspaces such as farmland, forest, or urban open spaces

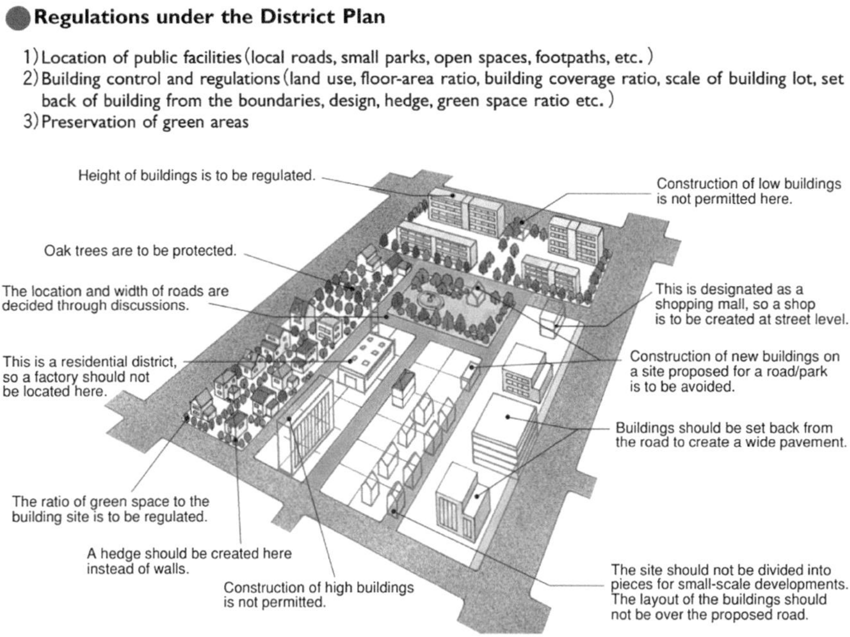

Fig. 19.2 Current district plan; a holistic scheme (to be further improved) to reduce shrinkage in the neighborhood scale (Illustration: MLIT (Ministry of land, infrastructure, transport and tourism) 2003)

(e.g., vacant land in residential district) to urban residents who want to manage such lands as "neighborhood commons" (Terada et al. 2012). The city coordinates between landowners and urban residents and supports finding optimized land as well as executing the contracts between the parties. Based on these contracts, land owners are able to get subsidies for reducing their burden of land property tax, and land users can also get financial support for their green maintenance activities. Thanks to this economic benefit, most land owners rend out their land to urban residents without any charge. The duration of such contracts is typically one year with automatic revision, so if land owners do not agree with revising the contract they can shortly return the land to being used for their own purpose. As of March 2016, 47 sites were registered as neighborhood commons and managed voluntary by urban residents as communal greenspaces (Kashiwa City 2016). Such temporal, incremental, and experimental responses were often ignored during past periods of city expansion, but during city shrinkage, such responses are highlighted as alternatives to more permanent solutions (Németh and Langhorst 2014; Terada et al. In print) (Fig. 19.3).

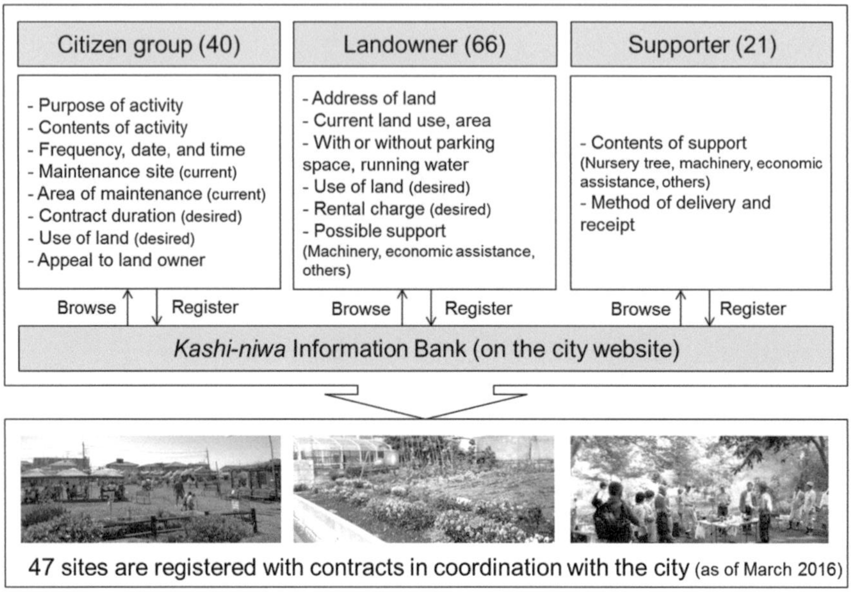

Fig. 19.3 Schematic overview of Kashi-niwa program (current registered number of each is shown in the parentheses)

19.3 Landscape Management of Outer Periphery (UCA)

19.3.1 What Happens if No New Measures are Taken?

Maintaining agriculture is the most important factor for maintaining sound landscapes in peri-urban UCA designated areas, or outer periphery. However, it is not easy to maintain these landscapes due to complex negative factors such as aging of farmers, no successors, an increasing number of part-time farmers, and a decrease in farmers' motivation due to gain of non-agricultural incomes (e.g., real estate investment). Investment in agricultural infrastructure is not prioritized in outer periphery because the farmlands there are threatened by urbanization; as such, they are often not designated as Farmland for Agricultural Promotion (i.e., FAP, or prime farmlands, based on the Act on Establishment of Agricultural Promotion Regions). This disadvantage makes farmland improvement difficult.

If the current trend is followed, landscapes in outer periphery are likely to degrade in the future. Existing research has identified that small paddies in the narrow valley, which are a component of *yatsu* landscape in Tokyo region (e.g., continuous upland and lowland landscapes formed by denudation by small rivers), tend to be abandoned due to disadvantages from machinery-based modern farming points of view (Takahashi and Hara 2013). Without improvement, even crop fields on upland would also be abandoned and negatively impact the entire area. Current urban sprawl does not positively affect surrounding farmland and forest, and does not encourage new farmers to settle there, contributing to a lack of maintaining these landscapes. Worse, housing in outer periphery, or the frontier of urbanization, are most likely to be abandoned in the near future due to city shrinkage. This means that the emergence of vacant land and housing will become another serious problem for maintaining outer periphery landscapes.

In short, the responses to degraded outer periphery landscapes are not focused on either agricultural or urban development side; rather, these half-effort responses make the situation worse, resulting in overall decline of social vitality.

19.3.2 A Preferable Future and Relevant Measures

The following proposals aim to restore outer periphery landscapes and are discussed from three aspects: (1) revitalization of peri-urban agriculture, (2) utilization of abandoned forest, and (3) an alternative vision for peri-urban planning.

First, in terms of restoring agriculture, the multifunctional peri-urban agriculture should take advantage of its urban proximity by responding to the requirements of urban society. Specific goods and services that peri-urban agriculture can service can count as adding environmental quality, cultural landscape, leisure and recreation, and regional food supply (Zasada 2011; Lovell 2010). Regarding the regional food supply, growing high-value added products such as fresh, fragile vegetables and fruits have advantages in terms of proximity to consuming areas. Regarding leisure and recreation, farmers can invite urban residents to pick-your-own farm events or provide a "farming experience." To strengthen business foundations, organizing a farmer's cooperative could be an option in order to aggregate farmlands, undertake product marketing and distribution, or carry out corrective works. While the Agricultural Land Act gives great weight to protecting individual farmers and conventional agriculture, the emergence of private enterprises should be more encouraged from a landscape management point of view. For example, in recent years, some enterprises began running "rental farms" by borrowing farmland from local farmers. In this system, urban residents can enjoy growing their own vegetables in an allotment on the rental farm, with technical support from the company's staff. In some cases, farmers actively joint-operate the farm; such corroboration among multiple stakeholders is expected to be a new style for maintaining peri-urban farmland (Mitsuhashi 2016).

Second, to re-utilize abandoned forest, multi-functional forest use should be considered (similar to the case of farmland). In addition to recreational uses such as creating public footpaths (e.g. originated in the UK), biodiversity conservation, mitigating urban heat islands, carbon management, and locally provisioning timber and wood fuel can also play important roles (Vejre et al. 2010; Terada et al. 2010; Bruckman et al. 2011). Even as peri-urban forests can have multiple functions related to urban society, applying only the measures of urban green conservation is not enough to restore forest management. Specific tactics can be, for example, the

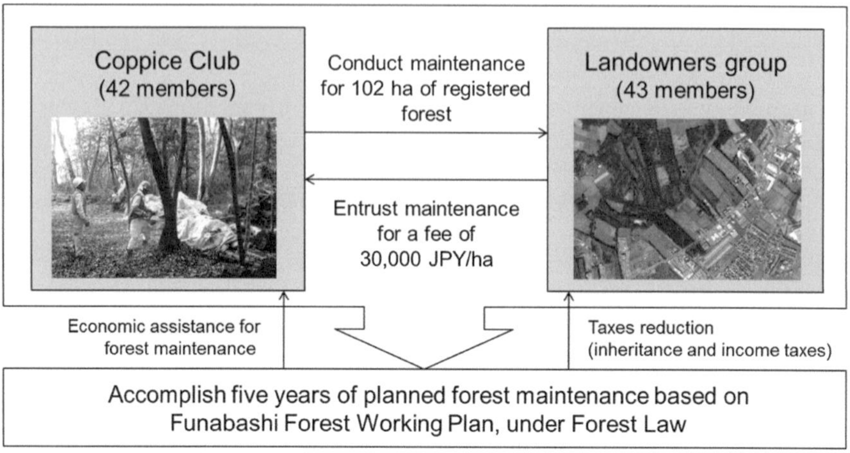

Fig. 19.4 Schematic overview of funabashi forest working plan

aggregation of small scale land owners' rights by formulating a Forest Working Plan (i.e., on-site forest management plan based on the Forest Law), and entrust actual forest management to forestry cooperatives or other expert groups. A leading example can be seen in Funabashi City, a suburban community in the Tokyo Metropolitan Region, which formulated a forest working plan for 102 ha of peri-urban forest, or one-fifth of whole forested area of the city. This area is managed by a semi-professional citizen group named "Coppice Club" under entrustment from 43 small forest land holders (Terada et al. 2010; Bolthouse 2013). Land owners pay expenses toward the entrustment, but can get reduction of inheritance tax, which is relatively higher in peri-urban areas (Fig. 19.4).

Third, urban planning should be modified to improve the entire environment of peri-urban areas. Under the current urban planning system, peri-urban areas are recognized as land reserved for development and disordered developments such as urban sprawl, construction-material yards, and manufactural buildings have been accepted. To prevent these negative impacts, some local municipalities have established local ordinances that aim to guide appropriate land use by formulating community district improvement plans based on the Community Areas Development Act (Mashima et al. 2012). Based on the plan, for example, minimum housing site areas can be determined to encourage larger sized house with accompanying gardens, and also have aims to prevent further urban sprawl.

The above three tactics should be linked with one another and applied effectively in combination. In Japan, there is no comprehensive planning in place for improving peri-urban landscapes (i.e., like the town and country planning in the UK). Planning that combines urban and rural dimensions is essential to resolving discrepancies between the two areas, and implementing the "living in the garden" concept on the ground.

19.4 Conclusion

To improve the environment of urban periphery, urban and rural must be orchestrated in integrated manner rather than dichotomized. Farmland and forest management needs an urban dimension; similarly, urban development needs a rural dimension. Urban periphery areas, which are considered the frontier of urban expansion, would be the most affected areas of city shrinkage in the near future. City shrinkage, however, should be recognized as catalyst to improve urban periphery environments, by enhancing ecological functions of farm and wooded landscapes.

Landscape management labor accounts, one of the key ideas of this book, could play an important role when discussing future management plans of various landscapes. It would be very difficult to gain consensus among various stakeholders such as farmers (i.e., landholders), urban residents, local government, and planners; however, if the future required labor force could be estimated using the landscape management labor accounts, this would be a great support to making discussions smoother and future plans more concrete.

When considering fauna and flora, it is well-known that species diversity becomes higher in the ecotone, a transitional zone between the two different ecosystems. Urban peripheries also should be designed such transitional zones, not as the areas that illustrate the contradictions between two different social systems, but for the areas' emerging "gardens," where urban and rural environments are organically combined. The garden city that Ebenezer Howard proposed in the late 19th century was a conceptualized utopia aimed to be implemented as new towns in suburban London, triggered by environmental deterioration in industrialized London during that time. The garden city of the 21st century in Japan may emerge as the transitional ecotone for the mutualism of urban and rural landscapes, which is realized by the "gardenization" of existing communities.

References

Bolthouse J (2013) End of tradition, reworking of custom: re-assembling satoyama woodlands on Tokyo's urban fringe. In: Rotherham ID (ed) Cultural severance and the environment. Springer, Netherlands, pp 387–399

Bruckman VJ, Yan S, Hochbichler E, Glatzel G (2011) Carbon pools and temporal dynamics along a rotation period in quercus dominated high forest and coppice with standards stands. For Ecol Manage 262(9):1853–1862

Burton E, Jenks M, Williams K (eds) (2003) The compact city: a sustainable urban form?. Routledge, London

Dantzig GB, Saaty TL (1973) Compact city: a plan for a liveable urban environment. W.H. Freeman and Company, New York

Florentin D (2010) The "Perforated City:" leipzig's model of urban shrinkage management. Berkeley Plan J 23(1):83–100

Henderson BR, Hartsfield K (2009) Is getting into the community garden business a good way to engage citizens in local government? Natl Civic Rev 98(4):12–17

IPSS (National institute of population and social security research) (2012) Population projections for Japan 2011–2060. http://www.ipss.go.jp/site-ad/index_english/esuikei/gh2401e.asp. Accessed 1 Mar 2016

Kashiwa city (2016) The list of registered sites of Kashi-niwa program. http://www.city.kashiwa.lg.jp/soshiki/110600/p006778.html. Accessed 5 Mar 2016 (in Japanese)

Katsumata W (2013) Jutakuchi no shukutai kanri no kanten kara mita daitoshiken kougai no machidukuri no houkousei (Direction of community development for suburbs of large metropolitan regions from the perspective of managing shrinking residential areas). J Land Inst 21(4):37–44 (in Japanese)

Kurita H, Yokohari M, Bolthouse J (2009) The potential of intra-regional supply and demand of agricultural products in an urban fringe area: a case study of the kanto plain. Jpn Geografisk Tidsskrift-Danish J Geogr 109(2):147–159

Mashima T, Kawakami M, Rachi M, Katagishi M (2012) Study on actual condition and problems of development permission ordinance for land use control in urbanization control area: mainly focused on municipalities' ordinances about article 34-12 of city planning law in kanto region. J City Plan Inst Jpn 47(3):439–444

Mitsuhashi T (2016) A study of the involvement of farmers and a business enterprise in running the gardens for farming experience. Master thesis of the Department of Natural Environmental Studies, the University of Tokyo

MLIT (Ministry of Land, Infrastructure, Transport and Tourism) (2003) Introduction of urban land use planning system in Japan. http://www.mlit.go.jp/common/000234477.pdf. Accessed 1 Feb 2016

Nakai N (1988) Urbanization promotion and control in metropolitan Japan. Plan Perspect 3(2):197–216

Németh J, Langhorst J (2014) Rethinking urban transformation: temporary uses for vacant land. Cities 40:143–150

Oswalt P, Rieniets T (2006) Atlas of shrinking cities. Hatje Cantz Verlag, Berlin

Shimpo N, Stiles R, Yokohari M, Miyamoto M (2014) The influence of the Viennese allotment gardens act on the housing renovations in Kleingärten in Vienna. Austria Urban Reg Plan Rev 1:99–114

Sorensen A (2001) Building suburbs in Japan: continuous unplanned change on the urban fringe. Town Plan Rev 72(3):247–273

Takahashi K, Hara, K (2013) Changes in Yatsu landscape structure according to topography and land ownership: an example in an urban fringe from central Honshu, Japan. Landscape Ecol Manag 18(1):57–72 (in Japanese with English abstract)

Terada T, Amemiya M, Hosoe M, Yokohari M, Asami Y (2012) A study of the maintenance and management scheme for provisional open spaces. J Jpn Inst Landscape Archit 75(5):651–654 (in Japanese with English abstract)

Terada T, Yokohari M, Bolthouse J, Matsumoto R (2010) A citizen-based maintenance scheme for conservation of suburban satoyamas based on the forest working plan. J Rural Plann 29: 179–184 (in Japanese with English abstract)

Terada T, Yokohari M, Bolthouse J, Tanaka N (2010b) "Refueling" satoyama woodland restoration in Japan: enhancing restoration practice and experiences through woodfuel utilization. Nat Cult 5(3):251–276

Terada T, Yokohari M, Amemiya M (In print) Urban farming in Tokyo: towards an urban-rural hybrid city. In: Lewis T Chandola T (eds) Green Asia: ecocultures, sustainable lifestyles and ethical consumption. Routledge, London

Vejre H, Jensen FS, Thorsen BJ (2010) Demonstrating the importance of intangible ecosystem services from peri-urban landscapes. Ecol Complex 7(3):338–348

Yamada C, Terada T, Tanaka T, Yokohari M (2016) Directions for vacant lot management in the outer suburbs of the Tokyo metropolitan region. Urban Region Plan Rev 3:66–84

Chapter 20
Rural Planning: Sustainable Management in Collaborative Activities

Kazu Hagihara, Nobuko Kawaguchi and Noriyuki Kawamura

Abstract This chapter outlines several illustrative cases for proposed future rural planning based on the landscape labor accounts. At first, the three case-studies for Asami, Nyu, and Kayumi in Part 3 (see Chaps. 13–15) are compared, and characteristics or issues of labor accounts in agricultural areas are described. Second, the general characteristics of community management based on the other cases will be shown; finally, appropriate policies, social systems, and initiatives for sustainable management of agriculture and rural areas are discussed.

Keywords Landscape labor accounts · Cooperative management · Ecological system · Land improvement districts · Community farming · Corporate farming

20.1 Introduction

In this chapter, several illustrative cases for proposing future rural planning based on landscape labor accounts will be shown. First, the three case-studies for Asami, Nyu, and Kayumi that were discussed in Part 3 (see Chaps. 13–15) are compared to one another, and the characteristics and issues of labor accounts in agricultural areas will be described. Second, the general characteristics of the community

K. Hagihara (✉)
Center for Community Co-design, The University of Shiga Prefecture,
2500, Hassaka-cho, Hikone, Shiga Prefecture 522-0057, Japan
e-mail: hagihara.k@office.usp.ac.jp

N. Kawaguchi · N. Kawamura
Graduate School of Environmental Studies, Nagoya University,
Furo-cho, Chikusa-ku, Nagoya, Aichi Prefecture 464-8601, Japan
e-mail: nobqou@gmail.com

N. Kawamura
e-mail: kawamura.noriyuki@e.mbox.nagoya-u.ac.jp

H. Shimizu et al. (eds.), *Labor Forces and Landscape Management*,
DOI 10.1007/978-981-10-2278-4_20

management will be shown, based on other cases. Finally, the appropriate policy, social system, and initiative for the sustainable management of agriculture and rural areas will be discussed.

20.2 Comparison of Three Cases of Landscape Labor Accounts in Agricultural Areas

20.2.1 Viewpoint of Organization Types and Geographical Conditions

Figure 20.1 shows a scatter diagram of the labor densities and labor accounts per person for three types of paddy fields. In the two corporate organizations in the Asami and Kayumi districts, the labor accounts per person are in the range of 2,000–2,500 h. This corresponds to working 6 days per week and seven and a half hours per day for 1 year. These organizations can contract with individual farmers for either all kinds of farm work or for a particular task, like mowing and harvesting, and they operate larger-scale farmlands than do individual farmers. Regarding the geographical condition, the labor density of Kayumi, which is located on slope land, is larger than that of Asami, located on flatland.

For the five individual farmers in the Asami, Nyu, and Kayumi districts, the labor accounts per person are in the range of 50–1,000 h. Typically, labor accounts for the modern type of farm work in the Asami district and the early 20th Century type in the Nyu and Kayumi districts comprise 3 days per week and 3 h per day,

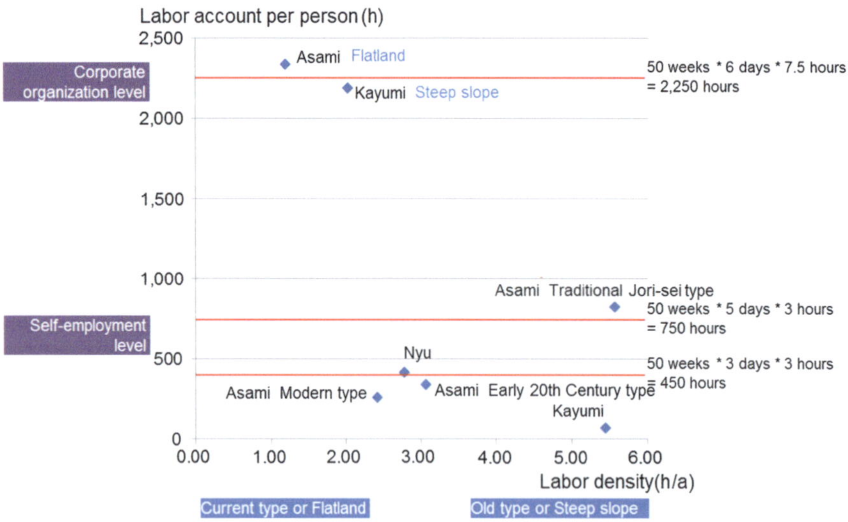

Fig. 20.1 Labor density and labor account per person in paddy fields

totaling 450 h a year. In addition, unimproved land (the traditional Jori-sei type paddy field) in the Asami district is typically worked 5 days per week and 3 h a day, totaling 750 h a year. Those examples are shown in Fig. 20.1. Many individual farmers work with the purpose of maintaining their health during their elder years. Most of them were farmers with a side job from which they received income. After retirement, they became full-time farmers.

When we compare the five individual farmers from the viewpoint of geographical condition and land improvement, the most efficient one (with the lowest labor density) is that of the modern type paddy field located on flatland, and the second most efficient are both the farmers in Nyu and Kayumi districts located on sloping land and the one with the early 20th Century type paddy field in Asami on flatland. Ranked last is the traditional Jori-sei type paddy field in Asami that was developed by improvement of the old land. This may be related to the tendency of the farmland to be located in a hilly and mountainous area that is smaller than that in flatlands. The value of labor density appears to relate to the farmland area or freshness of the facilities, which reflects the time of land improvement (from the Asami cases).

In the Asami and Kayumi districts, the individual small-scale farmers have a serious aging problem. Today, many individual farmers consign farming to corporations (companies) consisting of large-scale farmers. However, in the future, with the aging of the population leading to an increase in non-cultivated farmland area, the present number of corporations and the capable scale of cultivation will not be sufficient to meet their needs. On the other hand, in the Nyu district, each community has started a farming organization to actualize cultivation and the marketing of harvests. From now on, community management is needed as well as that of independent companies.

20.2.2 Viewpoint of Cooperative Management on the District/Settlement Scale

The second issue is the system for cooperative management on a district/neighborhood scale. Table 20.1 shows such systems in the three districts. The agricultural water channels, ponds, etc. are managed by each agricultural community, which is limited to the farmers. In the Nyu district, a management organization called "Midori-Net Nyu," conducts the management of agricultural water in each land improvement district, which multiple agricultural communities join. In the Asami district, main water channels are managed by the *Asami District-Scale Neighborhood Association*, and other small water channels are managed by each agricultural community. On the other hand, the Kayumi district has not had a joint management system for a very long time, and, recently, the extent of the management of many ponds is limited to several leading members of the irrigation association mowing around the pond, which is not in use now. In general, the tea

Table 20.1 Cooperative management system

	Agricultural water channel		Agricultural pond	Agricultural road	Domestic wastewater channel
	Main channel	Channel			
	Mowing of grasslands and cleaning	Mowing of grasslands, cleaning, repair of channel	Mowing of grasslands	Mowing of grasslands and repair of gravel pavement	Mowing of grasslands and cleaning
Agricultural communities		A, N	K	A, N	
Neighborhood association					A, N
District-scale neighborhood association or irrigation association	A, N				

A Asami district, *N* Nyu district, *K* Kayumi district

plantations do not use irrigation facilities such as water channels. Consequently, in the case of the Kayumi district, there is no active community management system for the water channels. The water channels located in front of the paddy fields have been managed by individual farmers in the style of tea plantation management. Many farmlands have promoted consigned cultivation with a few corporations and have changed to new farmland types that are easier to manage because of the aging population. People are worried that traditional and local community management is going to be lost. Under such a trend, irrigation channel maintenance and its habitat conservation for fish in the Asami district are promising new district-scale management systems.

20.2.3 Viewpoint of Ecological System and Cultural Landscape Management

The third issue is preventing the deterioration of biodiversity when undertaking land improvement development for effective production and the deterioration of the cultural landscape of farmland with the increase of abandoned fields. In addition, it is necessary to promote new industrial development (support economic development of the district) together with the richness of the ecosystem and conservation of the cultural landscape.

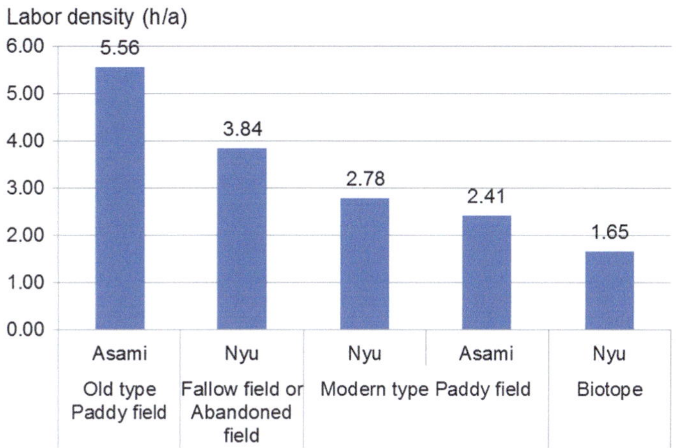

Fig. 20.2 Labor density in paddy fields

Figure 20.2 shows the labor density for paddy fields under each method of administration. In the Asami district, movement of fish between old Jori-sei type canal and paddy fields is possible, water is left in the canal during winter, and the area for vegetation is large. It is confirmed that these situations realize a rich ecosystem (see Chap. 13), which leads to a large labor density value. On the other hand, paddy fields of the modern type have fewer organisms, which results in a low value of labor density. In the activities necessary to maintain the rich ecosystem of the Jori-sei paddy fields, the large labor density value is a serious issue.

In the Nyu district, Midori-Net Nyu has continued activities to maintain the cultural landscape and preserve the ecosystem. It is shown that the lowest labor density is present in cases where water surfaces are kept as biotopes. Fallow fields utilized as grass fields get more labor density points than when used as cultivated paddy fields and biotopes filled with water. The Nyu district is prospective in that it perceives preservation of the ecosystem and cultural landscape as tourist attractions, which results in the realization of the sixth industrialization of agriculture.

20.2.4 Viewpoint of Kitchen Garden Management

The fourth issue is the high labor accounts of kitchen garden management. In the three districts, not only cultivated land for sale but also many kitchen gardens for household consumption product are widely distributed. Figure 20.3 shows the labor accounts per person and the labor densities in kitchen and residential gardens of the three districts. Labor accounts and densities of kitchen gardens without commercial purposes largely differ from person to person, and such residential property is similar to the case of management of residential gardens. Because kitchen gardens

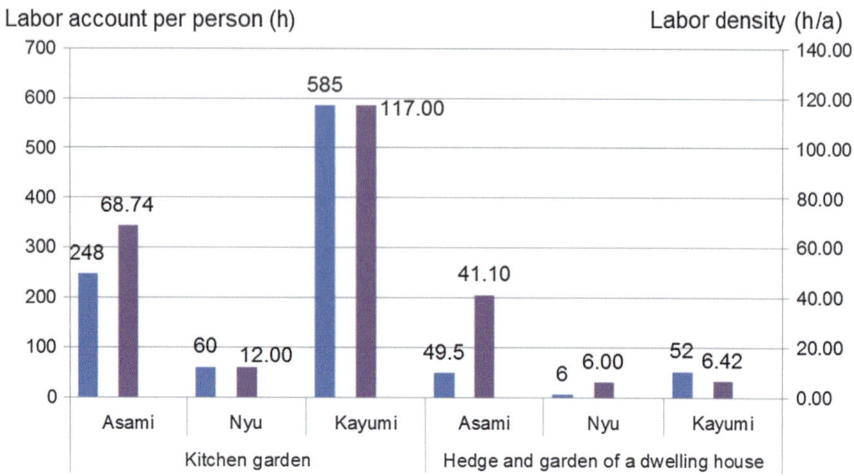

Fig. 20.3 Labor density and labor account per person of kitchen gardens and dwelling house gardens

have larger values of labor densities than do paddy fields and farmlands, they require large work times, even in small-scale farmland areas. Aging sometimes precedes the change of paddy fields and farms used as managed cultivated lands into non-managed farmland/kitchen gardens. The Nyu district conducts the selling of harvests from such kitchen gardens at retail stores.

20.3 Foresight in Agricultural Field Improvements in Japan from the Viewpoint of Research Review

20.3.1 Establishment of Land Improvement Districts Based on Agricultural Field Improvement Projects

The national government and local governments have been implementing agricultural field improvement projects since the end of World War II. Japan's land improvement districts are, in large part, responsible for promoting and maintaining the country's infrastructure. These districts can be classified as follows (Fig. 20.4; MAFF 2015a).

Organization of the creation of land was classified into three types: those managed by the central government, the local government, and land improvement districts. Most organizations have been managed by local governments and land improvement districts. In addition, in many cases, local governments play the role of consigning business activities; so, in these cases, the land improvement district is an outsourcing company.

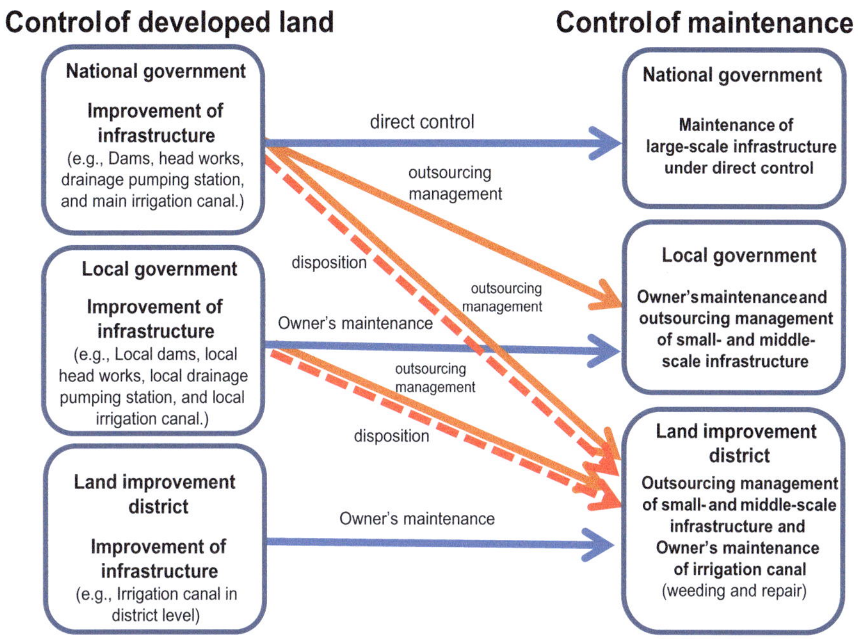

Fig. 20.4 Management system of developed land in Japan (MAFF 2015a) (Some parts of this figure were altered.)

As mentioned above, land improvement districts not only manage the civil infrastructure but also join in the cooperative effort of sharing traditional collaborative activity in local communities. All such activities are supported by volunteers. (Engineers CPD Organization for Rural Development 2008). In particular, collaborative activity in Japan is deeply involved in landscape management. As pointed out in Chap. 5, sometimes a land improvement district receives a direct payment.

20.3.2 Actual Status of Japanese Community Farming and Corporate Farming

Japanese community farming is defined as follows: "An agricultural production corporation based on a village, and this corporation unifies the production process in whole or in part" (MAFF 2015b). However, in reality, the organization and function of community farming takes a variety of forms from region to region and village to village. The effort of the Fujinoki district in the Matsusaka City of Mie Prefecture is a typical example of Japanese community farming (Kawamura 2012). The report by COSMOS (a voluntary organization for agricultural producers) in 2012 shows that the Fujinoki district consists of 230 people in 49 households. There are 46 households engaged in agriculture (6 full-time farmers and 40 part-time farmers), and the total farming area is 63 ha (55 ha of paddy fields, 5 ha of cultivation, and 3 ha of orchards). COSMOS was established by a prefecture-run project for agricultural field improvement in 1988. Later, in 2001, the organization was re-established as "the general appearance of Agricultural Producers' Cooperative Corporation."

This organization has four dimensions.

(1) It has jointly-owned agricultural tools and machinery that it maintains.
(2) It has the power of disposition regarding agricultural crops.
(3) It accepts the obligations pertaining to agricultural work and farming.
(4) The operators (6–7 people) in this association work in the fields in the district.

For this reason, the certified farmers engaged in farming are residents of their fields. The Koishiro district, as explained in Sect. 20.3.3, is engaged in community farming that is organized by the joint efforts of several rural communities.

On the other hand, by farm mechanization and the incorporation of farm management, large-cable agricultural production could solve the problem of labor force in agriculture. Fewer people would be needed to manage farmland in Japan.

However large-scale agricultural production has limitations in the following respects:

(1) Whether large-scale agricultural production is promoted or not depends on matching farmers who want to increase their farmland with those who want to abandon it. The bad conditions of farmland located in slope lands make them likely to be abandoned.
(2) After large-scale agricultural production is promoted, it is important to determine who or which group will conduct maintenance of common facilities, such as the channels. Responsible entities, such as corporate farmers and companies, should conduct maintenance work.

For these reasons, the Japanese-style direct payment system is used to pay for community activities, rather than supporting farmers by sustaining agricultural product prices.

20.3.3 The Actual Status of Cooperation Among Various Actors in Collaborative Activities

In this section, we focus on the Koishiro district in Matsusaka City of Mie Prefecture, which established a new organization for local communities promoting the involvement of non-farm households (Hagihara et al. 2015).

This district is located in flatlands surrounded by the Kushida and Harai Rivers. The district comprises the seven upstream villages of Houda, Iseba, Yokoji, Meda, Hayamase, Inagi, and Takagi. The scale of farming is comparatively large, with the

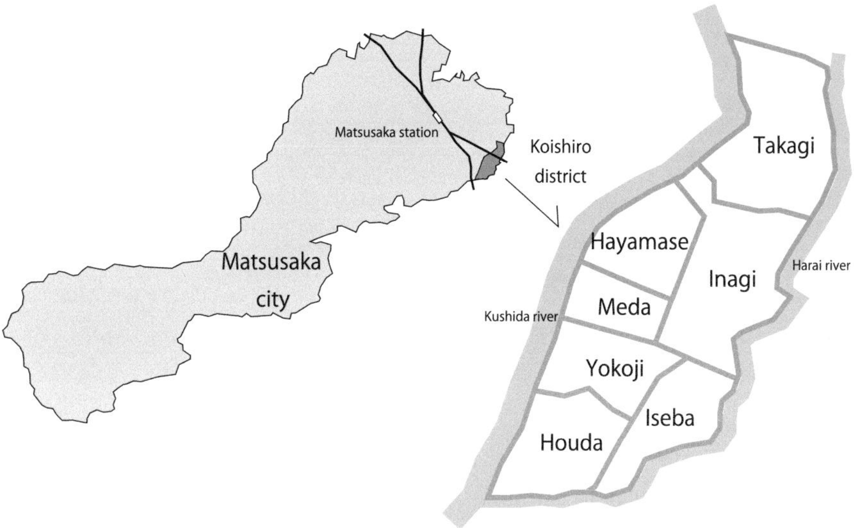

Fig. 20.5 Field areas in Koishiro district

Table 20.2 Summary of Koishiro district

Theme of Machizukuri council in Koishiro	Area (ha)				Farming scale (ha)	Population	The number of households
	Total	Paddy field	Dry field	Orchard			
The development of general coordination and policy planning	282	260	19	3	1.03	2,148	796
The development of local contribution activities							
The development of welfare							
The development of health and sports							
The development of greenspaces							

average farming household having cultivated acreage of 1.03 ha (Fig. 20.5 and Table 20.2; Hagihara et al. 2015).

Koishiro Clean Net 21, which is the equivalent of a land improvement district, has close ties to the "General Coordination and Policy Planning" and "Greenspaces" task forces in this Regional Residents Committee.

Table 20.3 Actual status of collaborative activities

Type	Frequency	Houda Iseba Yokoji	Meda Hayamase	Inagi	Takagi
Farmers only	Frequency	1 time	1 time (Hayamase: non)	1 time	3 times
	Participants for each cycle	20–60 persons (Iseba: non)	20 persons (Hayamase: non)	20 persons	33 persons
	Pay a fine	Non (Yokoji: 3,000 yen)	Non	Non	Non
Farmer and non-farmer	Frequency	4–6 times	4–5 times	3 times	5 times
	Participants for each cycle	20–90 persons	20–30 persons (Hayamase: 70 persons in MARCH only)	40 persons	75 persons
	Pay a fine	About 3,000 yen	Meda: 10,000 yen/year Hayamase: 2,500 yen	Non	Non
A village community (add to PTA)	Annual cycle	2–3 times (Iseba: 12 times)	Meda: 12 times Hayamase: 1 time	Non	6 times
	Participants for each cycle	20–50 persons	5–20 persons	Non	45 persons

Table 20.3 and Fig. 20.6 show the actual status of the collaborative activities by each of the seven villages. As observed above, collaborative activities are distinguished by three characteristics.

(1) There is a great divergence in the frequency and scale of collaborative activity performed by the seven villages.
(2) There are three types of collaborative activity management styles: management by farmers only, management by farmers and non-farmers, and management by a village community.
(3) There is a great difference with respect to the imposition of a penalty for the nonperformance of a collaborative activity among the seven villages.

All indications are that the seven villages have a strong tendency to maintain traditional customs. To involve a more diverse range of agencies in local communities, each resident needs to rethink long-standing practices.

Currently, the network of collaborative activities in the Haraigawa River basin is based on a diverse range of agencies in the local community. Not only local residents themselves, but residents from other areas, were engaged in dealing with the conservation of rare creatures in 2003 (see Chap. 22).

Fig. 20.6 Field of collaborative activity in Koishiro district

20.3.4 Landscape Labor Accounts Have Contributed to Actual Working Conditions

This section correlates and presents the requirements for the reallocation of landscape labor accounts based on an oral survey about the actual conditions in rural areas.

Table 20.4 describes the cooperative management system in Koishiro by touching on some details shown in Table 20.1.

The management system in Koishiro follows a similar pattern to that of Asami. However, unlike the case of Asami, because the Agricultural Field Improvement Project in Koishiro was already completed, the land improvement district occupies an important place there.

Table 20.5 shows the landscape units in four districts based on this cooperative management system. Basically, a farming season is divided into a busy period and a fallow period, and different organizations handle the farming tasks during these different periods. Take, for example, paddy fields and cultivated fields in rural areas. During the busy period, individuals and organizations (individual farmers, community farming groups, agricultural cooperatives, land improvement districts, and so on) are mainly responsible for farm work. On the other hand, during the fallow period, the local community—including non-farmers—are mainly responsible for

Table 20.4 Cooperative management system (added data of Koishiro district)

	Agricultural water channel		Agricultural pond	Agricultural road	Domestic wastewater channel
	Main channel	Channel			
	Mowing of grasslands and cleaning	Mowing of grasslands, cleaning, repair of channel	Mowing of grasslands	Mowing of grasslands and repair of gravel pavement	Mowing of grasslands and cleaning
Agricultural communities		A, N, Ko	Ka	A, N, Ko	
Neighborhood association					A, N, Ko
District-scale neighborhood association or irrigation association	A, N, Ko				

Ko Koishiro district, *A* Asami district, *N* Nyu district, *Ka* Kayumi district

cooperative activity. In particular, the Machizukuri Council, the elementary school (including the PTA), and the local government are all members of the local community.

As mentioned above, in Japan, cooperative activity during the fallow period is standard practice. In Kayumi, only full-time farmers play a major role in landscape labor, but this is a particular case. Another topic is that individual farmers cannot take on the task of solving the problem of abandoned farmland. At the same time, the problem cannot be addressed without cooperative activity by the local community. The local community in rural areas needs to focus on establishing a broad network of cooperative activity.

In light of the above, from what perspective should sustainable agriculture and rural areas be kept? As a first step to answering this question, we should focus on Fig. 20.7 (system chart based on social background). In concrete terms, it includes the following explanation.

First, the demands of the marketplace promoted the mechanization of farming, the use of agricultural chemicals and artificial fertilizers, land improvement projects, community farming, and companies entering the agribusiness. As a result of this, the labor account was gradually lost, and the number of part-time farmers increased.

On the other hand, practical farmers have contributed to covering the shortfalls caused by the labor force reduction in rural areas. However, for an expensive business investment, the rationalization of farming invited the result of farm retirement.

Table 20.5 Work item and operating body of rural area landscape elements

Landscape element		Work item	Operating body	Item trait	District
Paddy field cultivation	Busy period	Prepare soil for next season, maintenance of pond, irrigation, rice planting, settled planting, pull weeds and cut grass, additional fertilization, pesticide application, harvesting operation	Individual farmers, community farming, agricultural cooperative, land improvement district	Farmer actively works (including part-time farmers)	Asamai Nyu Koishiro
	Fallow period	Cleaning agricultural water, cut grass, building fence to prevent bird and animal damage	Individual farmers, community farming, agricultural cooperative, Machizukuri council, land improvement district, primary school (including PTA), local government	The cooperative activity of the local community, included a non-farmer	
	Other items	Winter-flooded rice paddies and biotope space	Machizukuri council, land improvement district, primary school (including PTA), local government		
Orchard	Busy period	Pruning, additional fertilization, pull weeds and cut grass, pesticide application, covering harvest operation	Individual farmers community farming agricultural cooperative land improvement district	Farmer actively work (included part-time farmer)	Asamai Nyu Koishiro
	Fallow period	Cleaning agricultural water, cut grass, and build a fence to prevent bird and animal damage	Farmers, farmers and non-farmers, Machizukuri council, land improvement district, primary school (including PTA), local government	The cooperative activity of local community, included a non-farmer	

(continued)

Table 20.5 (continued)

Landscape element		Work item	Operating body	Item trait	District
Tea plantation	Busy period	Additional fertilization, first picked tea leaves, second picked tea leaves, pruning	Individual farmers, community farming	Farmer actively worked (included part-time farmer)	Nyu Kayumi
	Fallow period	Prepare soil for next season, winter pruning	Individual farmers, community farming	Farmer actively worked (included part-time farmer)	
Abandoned farmland		Weeding and tree trimming, cultivated soil restoration	Farmers, farmers and non-farmers, Machizukuri council, land improvement district, primary school (including PTA), local government	The cooperative activity of the local community, included a non-farmer	The general characteristic
Pond irrigation	Busy period	The maintenance of pond irrigation	Individual farmers, community farming, agricultural cooperative, land improvement district	Farmer actively worked (included part-time farmer)	The general characteristic
	Fallow period	Draining and dredging up	Farmers, farmers and non-farmers, Machizukuri council, land improvement district, primary school (including PTA), local government	The cooperative activity of the local community, included a non-farmer	The general characteristic

Currently, for the conservation of biodiversity, Japanese rural areas are accelerating the introduction of environmental agriculture. This trend has led to the production of high value-added agriculture. The branding and differentiation of local resources, essential to the raising of agricultural product prices, requires collaboration with actors in the community, such as tourism. In addition, rural areas play an important role in the conservation of biodiversity, but this value is not

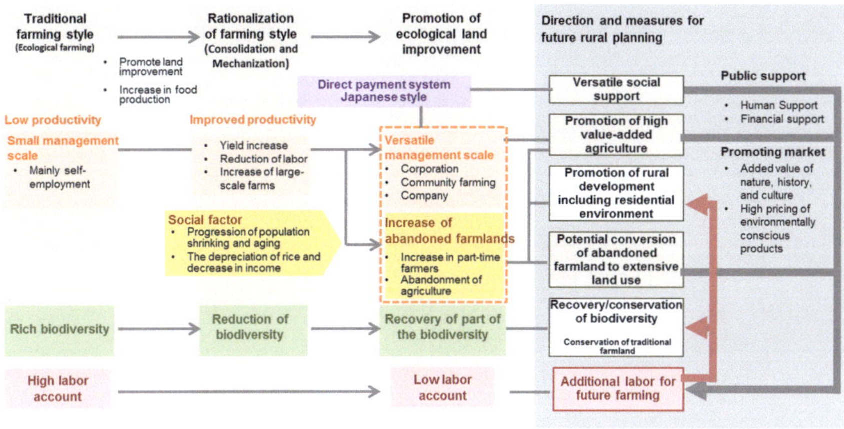

Fig. 20.7 System chart based on social background

internalized in the market mechanism. Therefore, individual farmers will receive a great deal of social support, such as the direct payment system. Chapter 5 discussed the direct payment system based on a research review conducted during recent years. Therefore, to conclude, we intend to suggest a vision that compensates for the lack of landscape labor accounts.

Figure 20.8 shows the difference between the current and the required labor accounts. The burden charge for this absence should not be experimentally priced by each village. The costs of landscape labor should be priced, in principle, based on a unified scale value. In this case, a flat-rate cost in the range of an elementary school district is calculated by averaging the calculated costs based on each village. The reason for this is that the aging of the population prevents the village from managing landscape labor on its own. Recently, in Japan, the range of the elementary school district has been considered the most appropriate scale.

On a related point, the financial contributions of the central and local governments can only be accomplished by sharing the significance of fundamental landscape management labor. Take Koishiro as an example; this district might need to calculate all of the convenience costs in the district alone. However, increasing the scope to include Nyu and Kayumi—not only the Koishiro district but also the Kushida River basin—might require calculating all of the convenience costs throughout the entire area.

In conclusion, the calculating of landscape management labor accounts can become an interconnected cooperative activity in rural areas by using a unified method. Furthermore, this method is able to contribute to the formation of a national spatial strategy.

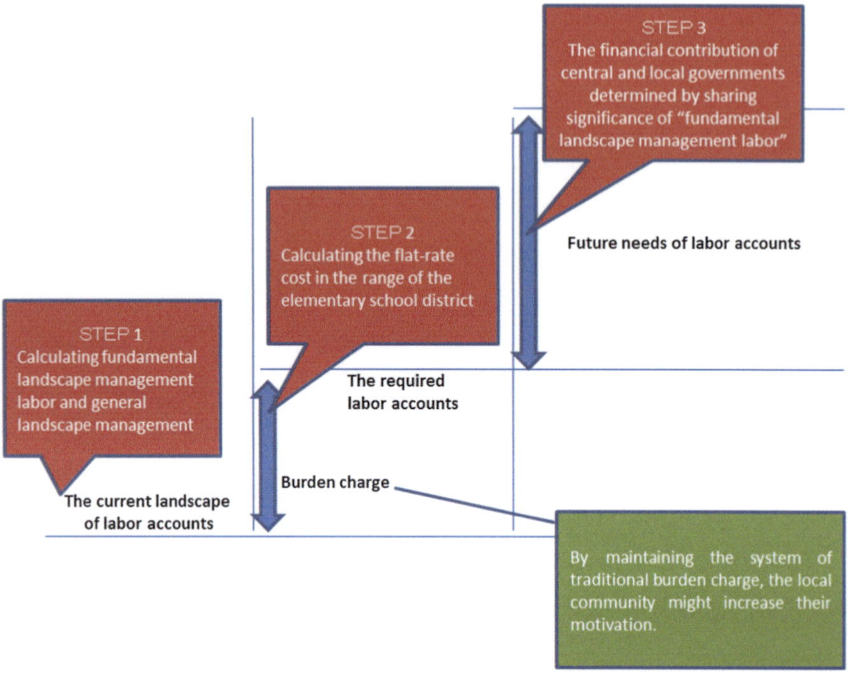

Fig. 20.8 Difference between the current landscape of labor accounts and the required labor accounts

References

Engineers CPD Organization for Rural Development (2008) The study group about farm land conservation in 2008. http://www.bio.mie-u.ac.jp/kankyo/chiiki/ryuiki/NouchiHozen_29.pdf

Hagihara K, Tomiyoshi M, Kawamura N, Nakatsuji C (2015) The determinants for the awareness of landscape change as seen from the characteristics of the landscape management in the field after farmland consolidation: a case study of district council town planning Koishiro district Matsusaka, Mie Prefecture. J Reg Policy Stud 7:11–18

Kawamura N (2012) Globalization and regional agriculture: a case study of collective farming management. Sociol Rev Nagoya Univ 33:67–84

MAFF (2015a) For land improvement districts. http://www.maff.go.jp/j/nousin/kikaku/dantai/. Accessed 31 Oct 2015

MAFF (2015b) For community farming. http://www.maff.go.jp/j/kobetu_ninaite/n_seido/seido_syuuraku.html. Accessed 31 Oct 2015

Chapter 21
Forest Planning: Human Resource Development for Sustainable Forest Planning

Kazukiyo Yamamoto and Yasuhiro Hirano

Abstract Due to lower timber prices and high costs for management with small-scale forest owners, the area classified unmanaged plantation forests has increased and could lead to ecosystem disservices such as soil degradation and landslides. Forest planning for the plantation forests depends on government subsidies, and the usage trends of these subsidies have been toward afforestation in the 1950s to weeding and thinning and to wood production to enhance self-sufficiency rates. Ecosystem disservices can be caused by deforestation for use of wood biomass materials and increased populations of Sika deer in the rural area. To conduct sustainable forest planning in the future, human resource development of labor such as "foresters," who have special knowledge and skills for forest management, will be important on both regional and global scales.

Keywords Ecosystem disservice · Forester · Forest management · Sika deer · Unmanaged forests

21.1 Current and Future Issues of Forest Planning

Forest management activities have stagnated under the long-term decline of domestic wood prices, which has caused an increase in unmanaged forests. It accelerated the depopulation of the mountain villages and increased the number of small-scale forest owners who live away from their forests (Watanabe 2015). In Japan, about 75 % of forest land owners have small-scale forest areas of less than

K. Yamamoto (✉)
Graduate School of Bioagricultural Sciences, Nagoya University,
Furo-cho, Chikusa, Nagoya 464-8601, Japan
e-mail: kazukiyo@agr.nagoya-u.ac.jp

Y. Hirano
Earth and Environmental Sciences, Graduate School of Environmental Studies,
Nagoya University, Furo-cho, Chikusa, Nagoya 464-8601, Japan
e-mail: yhirano@nagoya-u.jp

H. Shimizu et al. (eds.), *Labor Forces and Landscape Management*,
DOI 10.1007/978-981-10-2278-4_21

5 ha (Table 21.1) (Forestry Agency of Japan 2015). In the forests of those owners, the roads have not sufficiently been developed. In addition, there are many areas where the forest ownership boundaries are not clear or are widely-scattered (Kishi 2015). In Japan, weakened interest in the forest management of small-scale owners is one of the most critical problems today. Because of lowering timber prices and increasing management costs, many forest owners have abandoned managing their forestland. In many places throughout the country, these abandoned forests cause danger in the form of windfalls, soil degradation, and landslides, leading to ecosystem disservices (Ota 2013).

Planted forests account for about 40 % (approximately 10 million ha in 2014) of the total forested area (approximately 25 million ha in 2014) in Japan (Forestry Agency of Japan 2015). Expansive afforestation was concentrated in the 1950s and 1960s, and the age class distribution of the planted forests are quite clustered, as shown in Fig. 6.2 (Ota 2013). Expansive afforestation was mainly conducted by a large number of small-scale forest owners using silvicultural subsidies. In Japan, more than 60 % of artificially planted forests are private, and a large number of them are owned by small-scale forest owners (Forestry Agency of Japan 2014). Even under the stagnation of forest management activities, forests planted during the period of expansive afforestation have continued to grow and require management. Therefore, weeding, improvement cutting, and thinning were added to the list of subsidy-eligible afforestation project items (Forestry Agency of Japan 2014).

The Forest Owners' Cooperatives have been major contractors, accounting for more than half of forestry activities, including planting, weeding, and thinning (Forestry Agency of Japan 2014). The Forest Owners' Cooperatives were expected to become major forest management bodies on behalf of individual forest owners, not only by contracting with them but by integrating their forest land and then contracting long-term forest management (Hirata 2013). However, the actual situations of small-scale forest owners and unnecessary governmental subsidies have

Table 21.1 Area of forest households in Japan (2010)

Owned forest area (ha)	Forest households	
	Area (ha)	Cumulative (%)
1–3	851,646	16
3–5	571,675	27
5–10	775,994	42
10–20	825,595	58
20–30	442,379	67
30–50	462,189	75
50–100	434,883	84
100–500	555,512	94
500–1000	131,054	97
Over 1000	161,632	100
Total	5,212,559	

Source Ministry of Agriculture, Forestry and Fishery (2010) Census of Agriculture and Forestry been making it difficult (Hirata 2013; Ota 2013). The expansion of subsidy-eligible items in afforestation projects made forest sectors strongly dependent on subsidies because subsidy-based forest management was easier than that based on wood production and advanced business management (Hirata 2013).

Under such circumstances, the Japanese government launched a relatively strong initiative to increase domestic forest production, named the "Forest and Forestry Revitalization Plan," in 2010 (Ota 2013). This plan was intended to enhance the wood self-sufficiency rate by 50 % or more by 2020, especially focusing on the development of a mechanism through which an ambitious entity could advance the intensive forestry and human resources needed to support sustainable forest management and forestry vitalization (Kakizawa 2013). For intensive forestry, woodland integration is one of the key elements of this plan as well as mechanization and personnel training; however, standardization of forest practices under the name of woodland integration may influence small-scale owners (Ota 2013). The regulation was amended in 2012, and individual forest owners were required to have a forest area of at least 100 ha to be granted a government subsidy to make a forest management plan. However, more than 95 % of forest land owners and 80 % of total planted forest areas in Japan are of less than 100 ha of forest area (Table 21.1).

21.2 Ecosystem Disservice and Future Improvement of Plantation Forests

Utilization of woody biomass as a part of alternative fossil-fuel energy resources have recently been focused upon in several rural areas of Japan, including from the viewpoint of preventing global warming. Forest resources are renewable energy and can contribute to the promotion of a sound material-cycle society. It would be effective for the sustainable management of Japanese plantation forests to solve the present problems of the accumulated biomass resources, but it is too late for thinning. Planning for the construction of power generation system plants in order to use woody materials has recently been increased in several rural areas because of the Feed in Tariff (FIT) program for renewable energy that has been introduced by the government (Kamimura et al. 2012). One of the primary aims of this program is the acceleration of use of the presently unutilized wood materials that were left after the thinning of the plantation forests. However, in some rural areas, the disorderly planning, such as more than one new plant in the same area, can lead to a shortage in the supply of wood materials (Kamimura et al. 2012), indicating the need for further use of imported wood materials to maintain the new wood biomass plant. This contradiction should be considered, and information on future schemes should be shared between adjacent rural communities beyond the municipalities.

For the forest soils, in the short term, deforestation for use of wood biomass materials can enhance dry soil conditions and higher soil temperatures, which can induce the outflow of nitrate-nitrogen into streams (Fukuzawa et al. 2006).

The enhancement of nitrate-nitrogen could decrease the water quality of streams. During the harvest, the forestry machinery can also damage the physical properties of forest soils that play important roles in ecosystem services, such as water storage, nutrient cycling, and biodiversity. In the long term, the direction of the future soil degradation can be determined by which tree species are replanted after the harvests, in particular, the western Honshu and Shikoku area have forest soils with low acid buffering capacities. Replanting of Japanese cypress can accelerate further soil acidification within decades (Tanikawa et al. 2014), whereas other tree species that can induce Ca accumulation in the soil can be an alternative selection for planting to ameliorate further soil degradation. The bamboo was planted in the rural area but was left unchecked and has now invaded forest areas dominated with native or plantation tree species because they grow too fast to reach the tree height levels. One positive aspect is that bamboo can accumulate Ca in the soil, which can improve soil acidification (Umemura and Takenaka 2015). However, when the labor accounts are too short to manage bamboo production, the introduction of further bamboo planting should be careful considered in light of the present invasive situation.

The amount of carbon storage in the forest soils is the result of the accumulation over long-term periods of more than one-hundred years, and is two times higher than that of vegetation, including trees. The carbon storage in forests is considered one of the supporting services in an ecosystem. The monitoring of forest soil carbon storage has continued in order to report the changes in forest carbon storage for an international agreement intended to reduce greenhouse gases (Takahashi et al. 2010). Globally, it is obvious that land use and land use change by humans affects forest soil carbon storage. We need to continue to monitor the local area so that the changes in land use and over-yield harvests do not decrease forest carbon storage and enhance greenhouse gases.

The damage to forest ecosystems by the grazing of sika deer has been serious in several Satoyama and forest areas in Japan because of the increase of the deer population (Forest Agency of Japan 2015). This can induce not only the loss of vulnerable alpine plants and native tree seedlings but also the erosion of the surface soil, which has higher nutrient contents for plants and water holding capacity (Koizumi 2012). The practical treatment to protect the loss of vegetation is to build deer preventive fences or nets along the boundary areas. This is highly effective but expensive. In addition, the labor accounts need to build it and then take care of it annually. In particular, the workers who manage both forests and farmlands in the Satoyama area have had to treat this problem only recently, meaning that they need more labor accounts than they did a decade ago. The ecological treatment to prevent surface soil erosion has focused on plants that are unpalatable for deer. After the deer graze in forests, only unpalatable plant species such as *Hypolepis punctate*, *Macleaya cordata*, and *Pieris japonica* will remained. Although the biodiversity decreases under this condition, the ability to protect the surface soil structure can be retained (Ishida et al. 2008; Yamase et al. 2015). To prevent the indirect degradation of ecosystem services, management to control the population of wild animals at the regional and river-basin levels is urgently needed because it is presently only managed at the prefecture level; in particular, there could be a problem with the in and out flow of sika deer at the boundary of the prefectures. Moreover, the number of hunters has decreased because of the shrinking and aging of the population in rural areas as well as the changes in food ways brought about by globalization. Education and training to increase the number of hunters is challenging to activate in the rural area community. Not only would more hunters decrease and control the sika deer population but efficient use of hunted animals should be considered, such serving game meat at restaurants and producing leather products from their hides (Hirano 2014).

21.3 Human Resource Development for Future Forest Planning

For the promotion of current policy, the Forestry Agency of Japan created human resource development programs for "Forest Management Planners," who should lead proposal-based coordination with forest owners and consolidation of forestry practices, taking major responsibility for the operation of the forest management planning system under the "Forest Act" (Forestry Agency of Japan 2015) and for "Foresters," which was created to support local governments and local forest management as leading experts based on the "Forest and Forestry Revitalization Plan" (Forestry Agency of Japan 2013; Oishi et al. 2014). Because the qualification systems in the program have just been built, it is suggested that there is a lack of professional higher education institutions or professional programs in educational institutions; therefore, they have not yet achieved mainstream recognition in Japan (Aikawa and Kakizawa 2015). As the commissioned ratio of forestry operation is quite different by holding scales, type of operation, and also regions (Sato 2010), the forestry activities differ between regions and also composition of holding scales. Therefore, further capacity building and the building of networks through their work is desirable (Oishi 2014) because the process of woodland integration should differ by region according to the forestry activities of local communities.

Currently, most of forest resources that were planted post World War II have shifted to harvesting from tending, and pressure to log them has been growing. The self-sufficiency rate and the volume of domestic timber production shows a steady upward trend since the early 2000s (Ota 2015). By a drop in the value of the yen and a strong wood demand in China, the price of domestic timber is also rising along with the imported timber price (Ota 2015). Since the forest management activities in domestic forestry are largely stagnant under the long-term declining of domestic wood prices, the rejuvenation and development of human resources involved with domestic forestry are required along with the cost reduction of timber production and the development of new demand for timber promoted by government policy. In addition to these domestic problems, the forestry of Japan should also respond to global needs, such as global warming, deforestation, and biodiversity. Thus, the role of "Foresters" as leading exporters that support local governments and local forest management will become increasingly important; therefore, the continuous development of their skills will be required. From the global viewpoint regarding the long-term sustainable management of forest ecosystems, monitoring soil carbon, nutrients, and biodiversity, including wild-life management in the local area, are needed to deal with ecosystem management. Namely, the development of human resources who can treat with both global and local matters is urgently needed in each rural and river basin area. The Forestry Agency of Japan has tried to increase the number of "Foresters" who can treat and build consensus regarding local and global matters. However, many foresters came from each prefectural level, and the present system for the development of human resources is still using the "top-down" system, from the global, national, or prefectural levels to local regional areas. The number of foresters who come from rural or regional areas shall be increased in the near future. The newer "bottom-up" stream can be utilized to share clinical environmental problems regarding the deficiency of labor accounts at the boundary between national and local field area levels efficiently. There are a number of specialists who have forest management skills and knowledge of forest environments, such as topology, climate, and soil productivity. However, these specialists have been aging, and, therefore, the handing down of their traditional skills is urgent. Otherwise, we will lose such knowledge. Foresters who are knowledgeable about both the specific rural region and global-national levels should be educated and developed to manage sustainable forests in the near future.

In the near future, foresters will be able to evaluate the current and future status of ecosystem services and disservices in planted and secondary forests in each region and then calculate how landscape management labor accounts need to manage sustainable forests. In other words, foresters who are trained in special skills will be able to consider not only conventional management but also landscape management labor accounts, as mentioned in this book, to build regional schemes for forest planning and management. To achieve future sustainable forest planning, we need to accumulate various regional case studies, as shown in Part III, in rural regional areas to clarify the current amounts and potentials of landscape management labor accounts in forest regions, including the rural Satoyama area.

References

Aikawa T, Kakizawa H (2015) Forestry education and qualification systems in developed countries: states and direction of recent change. J Fo Econ 61:96–107

Forestry Agency of Japan (2013) Annual report on trends in forest and forestry in Japan FY2012, Tokyo

Forestry Agency of Japan (2014) Annual report on trends in forest and forestry in Japan FY2013, Tokyo

Forestry Agency of Japan (2015) Annual report on trends in forests and forestry FY2014, Tokyo

Fukuzawa K, Shibata H, Takagi K, Nomura M, Kurima N, Fukuzawa T et al (2006) Effects of clear-cutting on nitrogen leaching and fine root dynamics in a cool-temperate forested watershed in northern Japan. For Ecol Manage 225:257–261

Hirano Y (2014) Clinical environmental studies on Kushida-river basin. In: Watanabe S, Nakatsuka A, Ow T (eds) Clinical environmental studies. The University of Nagoya Press, Nagoya, pp 168–179

Hirata K (2013) The human resource development in the forest and forestry revitalization plan and the mean of Japanese type forester. J For Econ 59:27–35

Ishida H, Hattori T, Kodate S, Kuroda A, Sawada Y, Matsumura T et al (2008) The Hypolepis punctata community established under severe feeding pressure by sika deer and its application to revegetation. Jpn J Conserv Eco 13:137–150

Kakizawa H (2013) Trajectory and challenges of forest policy of Japan. Nougyou-to-keizai 79:5–14. (In Japanese, tentative translation of the original Japanese title)

Kamimura K, Kuboyama H, Yamamoto K (2012) Woody biomass supply costs and potential for biomass energy plants in Japan. Biomass Bioenergy 36:107–115

Kishi S (2015) Forestry regeneration strategy of Japan to learn from Germany forestry. Nouson-to-toshi-wo-musubu 4:15–24

Koizumi T (2012) Effects of increasing the population of sika deer. Shinrinkagaku 61:2–63

Oishi T, Tamura N, Hirata K, Okuyama Y (2014) Actual activity of Japanese type forester candidate: as for apprentice forester of prefectural officials. J For Econ 60:33–42

Ota I (2013) Present status around small-scale forestry in Japan. In: IUFRO 3.08 and 6.08 joint conference, future direction of small-scale and community-based forestry proceedings, pp. 11–21

Ota I (2015) Opportunity incoming to the wood production and export industrialization. Nougyou-to-keizai 6:67–71. (In Japanese, tentative translation of the original Japanese title)

Sato N (2010) Plan that there is no "life of people" in sight. For Econ 63:7–9

Takahashi M, Ishizuka S, Ugawa S, Sakai Y, Sakai H, Ono K et al (2010) Carbon stock in litter, deadwood and soil in Japan's forest sector and its comparison with carbon stock in agricultural soils. Soil Sci Plant Nutr 56:19–30

Tanikawa T, Sobue A, Hirano Y (2014) Acidification processes in soils with different acid buffering capacity in *Cryptomeria japonica* and *Chamaecyparis obtusa* forests over two decades. For Ecol Manage 334:284–292

Umemura M, Takenaka C (2015) Changes in chemical characteristics of surface soils in hinoki cypress (*Chamaecyparis obtusa*) forests induced by the invasion of exotic Moso bamboo (*Phyllostachys pubescens*) in central Japan. Plant Species Biol 30:72–79. doi:10.1111/1442-1984.12038

Watanabe S (2015) A paradox between the forestry basic act and the forest policies of national and local government. For Econ 37:1–8

Yamase K, Todo C, Hirano Y (2015) Investigation of soil reinforcement by root system of 2 shrub species. J Jpn Soc Reveg Technol 41:15–20

Chapter 22
Inland Water System Planning: Management for Ecosystem Rehabilitation

Jyun-ichi Kitamura, Takashi Tashiro and Shigeya Nagayama

Abstract This chapter introduces the management of inland water systems where the ecosystem has deteriorated due to serious anthropogenic alterations. Inland water systems were originally composed of rivers, streams, ponds, and floodplains, and now include paddies, canals, and reservoirs for agricultural irrigation, hydropower generation, flood mitigation, and other water uses. These paddies, canals, and reservoirs have functioned to compensate the floodplain-dependent habitats. Related to the natural and social backgrounds there (see Chap. 7), the deterioration processes and measures for sustainable ecosystem management are reviewed in this chapter along with conventional literature that discusses each of the landscapes. This chapter also provides an overview of the management situations in each of the inland water landscapes, with particular reference to ecosystem rehabilitation. It also discusses how to sustainably conserve the ecosystem in some of the landscapes.

Keywords Floodplain · Inland water system · Riparian forestation · Rehabilitation · Management · Literature review

J. Kitamura (✉)
Mie Prefectural Museum, 3060 Isshiden-Kouzubeta, Tsu, Mie 514-0061, Japan
e-mail: kitamura@bio.sci.toho-u.ac.jp

T. Tashiro
Disaster Mitigation Research Center, Nagoya University,
#401 Disaster Mitigation Research Building, 1 Furo-cho,
Nagoya, Aichi 464-8601, Japan
e-mail: ttashiro@nagoya-u.jp

S. Nagayama
Aqua Restoration Research Center, Public Works Research Institute,
Kawashimakasada, Kakamigahara, Gifu 501-6021, Japan
e-mail: s-naga77@pwri.go.jp

H. Shimizu et al. (eds.), *Labor Forces and Landscape Management*,
DOI 10.1007/978-981-10-2278-4_22

22.1 Introduction

Rivers swell due to seasonal rains and changing channels, freely flowing from highlands to lowlands, forming and destroying each of the riverscape units such as main channels, creeks, and ponds. Although most of Japan's rivers have been affected by flow regimes due to anthropogenic alterations (Yoshimura et al. 2005), these processes have consequently reproduced riverscape complexes and aquatic habitats, where various organisms have adapted to site-specific variations through evolution. The biodiversity of a river community depends on and has been maintained by these kinds of systems.

Human beings have developed river infrastructure, not only to prevent disastrous flooding, but also to employ water resources; this often results in controlling river flow fields and regimes with the development and installation of durable infrastructure, such as dams and levees. Hence, river flow has been severely suppressed in inter-levee channels with the floodplains in outer-levee areas being terrestrialized, and with those in inter-levee areas forming as floodplain-dependent microhabitats (Nagayama et al. 2015). Furthermore, the inter-levee floodplain has also been gradually terrestrialized with riparian forestation, which has degenerated floodplain-dependent microhabitats (Kitamura 2008) (see Chap. 7).

Paddies, canals, and reservoirs constructed for agricultural irrigation have begun to function as secondary natures and to compensate the floodplain-dependent microhabitats. Recently, the "Agricultural and Rural Development Project" has been promoted in order to achieve "sustainable development of agriculture" and "stable provision of food". This involves making the ridges of paddies and canals more durable, however it also makes variable aquatic conditions monotonic and discontinuous due to channelization of canals and cutting off the water intake and supply during non-irrigation periods, which then decreases the microhabitats of aquatic organisms. Consequently, many floodplain-dependent species are threatened with extinction (Kitamura 2008).

Although inland waters have been heavily utilized to promote the convenience of our daily life, some policies and measures exist not only for our convenience but also for nature conservation, such as the "River Law" and the "Food, Agriculture and Rural Areas Basic Act". In the following discussion we give overviews of current situations facing each of the landscapes of inland waters, such as river and stream, paddy and canal, and reservoir, with particular reference to primary freshwater fish.

22.2 Rivers and Streams

River improvements, such as channelization, dam and levee construction, and so on, designed to prevent damage due to floods and promote public welfare and safety, have continuously eroded riverine habitat and ecologies along both lateral and longitudinal directions. The continuous levees mostly with channelization have especially separated their floodplains from river channels and decreasing the number of organisms dependent on floodplain environments; whereas the installation of dams or weirs have prevented migration of aquatic organisms and have formed lentic environments with sediment depositions above and vertical gaps with riverbed degradation below. According to the River Law amended in 1997, the purpose of river improvement and management newly covers conservation of the fluvial environment (Infrastructure Development Institute, Japan 1999), hence nature restoration projects are conducted to recover deteriorated environments as previously mentioned in Chap. 7. There are many examples of river restorations in Japan, as it was reported that the number of the "nature-oriented river works" (*Tashizen* (*Gata*) *Kawa Zukuri*) reached about 20,000 between 1991 and 2000 (Nakamura et al. 2006). In this section, we introduce management measures preventing erosion of two types of riverscapes: alluvial rivers and mountainous streams.

Most alluvial rivers in Japan have been constrained by levees to protect agricultural and urban areas from floods. Large rivers generally have a wide floodplain, which is inundated during floods, between levees, although artificial flood channels are often constructed along the river channel. However, there are less inter-levee floodplains in smaller streams due to intensive constriction by levees and bank protection.

Regarding large-scale alluvial rivers (mostly in Class A), Nagayama et al. (2015) has summarized and clarified their historical changes illustrating differences in characteristics of pristine and artificial riverscapes with conventional literatures. Figure 22.1 shows the typical patterns of landscape distributions in both pristine floodplains with meandering channels and inter-levee floodplains with straightened channels. In recent alluvial riverscapes with constructed levees, riparian forestation has been developed in the narrow inter-levee floodplains as was also described in Chap. 7. These processes result in fine sediments being transported in flood events, which then end up stagnated and deposited between the levees, which may promote the development of bank with riparian forestation, as illustrated in Fig. 22.2. Because the riparian forestation mostly causes a decrease in the area of active zone and increases in the water level during flood events, many studies have been undertaken to reveal its mechanism and to control its development in the field of river engineering (Miyamoto et al. 2013). According to the review of these conventional works, it could be understood that there are three main causes of riparian forestation, such as eutrophication of water and sediment, changes in flow and sediment regimes, and invasions of exotic species (Miyamoto et al. 2013). As a practical example, Tashiro (2004) revealed that the riparian forestation had coincided with the changes in topographic features and sediment compositions, such as riverbed degradation with armoring, which had decreased the chance of "riverbed disturbances" in the alluvial reach of the Yahagi River (as shown in Fig. 22.3). These effects may be caused by regulating flow and sediment with the installations of upstream large dams (e.g. Yahagi Dam) and many small check dams preventing flood and debris flow disasters, and by mining riverbed gravels to be used for construction materials (Kitamura et al. 2001; Tashiro 2004). The riparian forestations subject to fewer disturbances are currently observed in many fields of large-scale alluvial rivers in Japan (Miyamoto et al. 2013). Consequently, the large alluvial rivers, especially those that are floodplain environments, have been degenerated with irreversible processes followed by artificial alterations.

To restore the original floodplain environment with various water landscapes connected during flood, nature-oriented works have been developed. These include reintroducing channel meandering (Nakamura 2003), reforming inter-levee floodplain by cutting riparian trees and/or "flood-channel" (see Fig. 22.1) excavation (Sanuki et al. 2010; Nagayama et al. 2015), and connecting temporary water areas

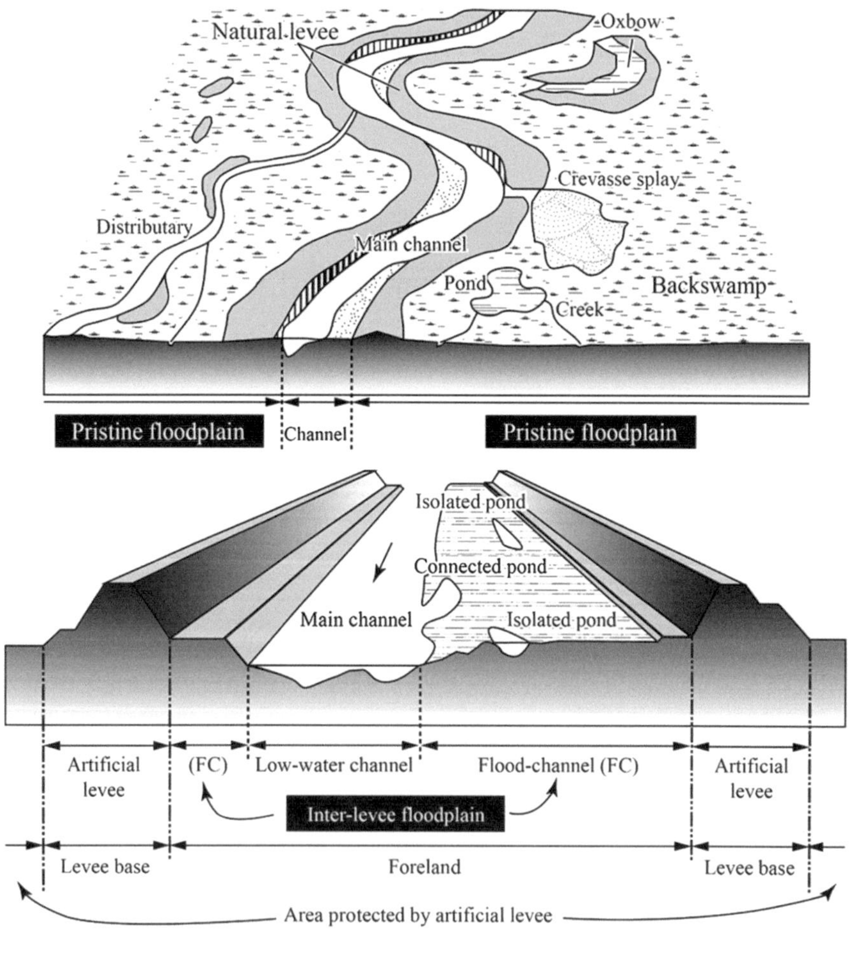

Fig. 22.1 Landscape distributions in alluvial riverscapes before and after levee construction. Pristine floodplain with meandering channel in the upper illustration and inter-levee floodplain with straightened channel in the lower illustration (Nagayama et al. 2015)

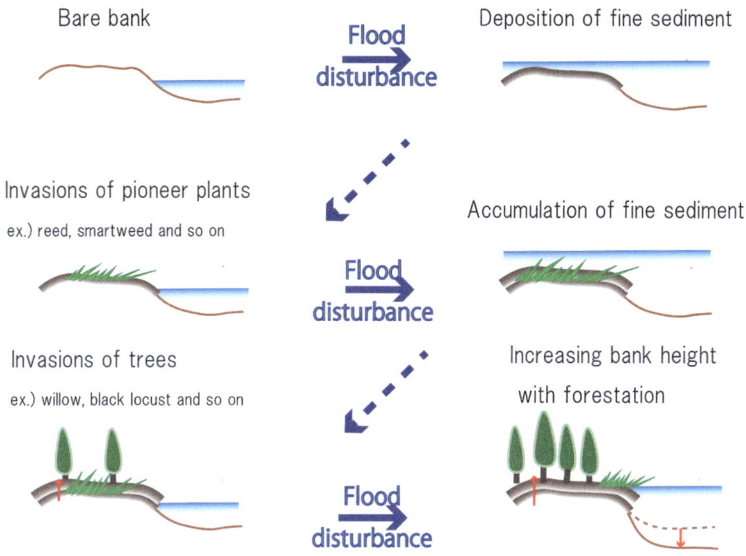

Fig. 22.2 Schematic image of typical patterns and processes of riparian vegetation development in the inter-levee floodplain

isolated in the outer-levee area with the main channel (e.g. Nakamura et al. 2006). Some of the approaches to sustainable management of the floodplain environment require cyclic cutting of plants and excavation of accumulated sediment in the inter-levee floodplain, (Sanuki et al. 2010; Nagayama et al. 2015). The Japan National Census on River Environments, especially Sanuki et al. (2010) reviewed Class A rivers, with the data showing distribution patterns of plant species causing river forestation in the inter-levee floodplain. Willow, black locust, and bamboo were dominant along these rivers, with their habitation zones varying in relation to slope, sediment size, and height from the surface of normal flow. The impacts that cutting and uprooting these trees had on river forestation could be measured. The results suggest that uprooting was the most effective practices, especially for bamboo management. Moreover their compositional patterns were different in relation to the local conditions in each of the regions of Japan. Makishima et al. (2013) estimated the unit costs and effect durations of plant loggings by surveying the combinations between plant community types, characterized by willow, black locust, and bamboo, with concrete logging operations such as cutting and uprooting trees. They were also able to suggest scenarios for preventing riparian forestation from the viewpoint of cost-effective management, as shown in Fig. 22.4. Conversely, the temporary water areas designed for flood retention are reformed in the outer-levee areas such as the "Azame-no-Se" project along the Matsuura River basin in Saga prefecture (Nakamura et al. 2006). This concept was established for managing the alluvial riverscape with its outer-levee area by promoting flood retention in the floodplain, which can be matched to the social environment

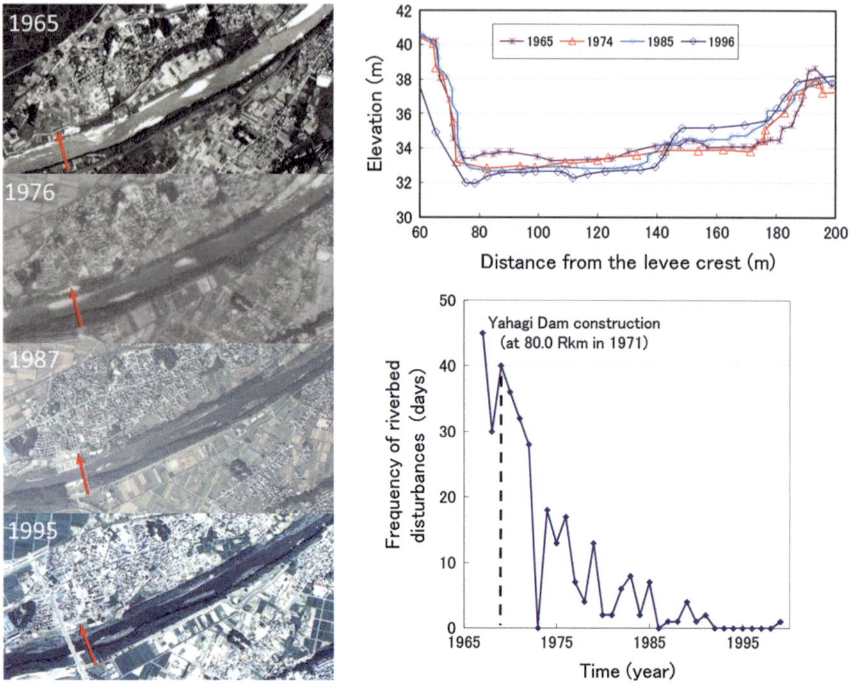

Fig. 22.3 Temporal landscape changes in the alluvial reach of the Yahagi River at 42 river km (Rkm) from the mouth (cross-sections marked on the *left*) (altered from Tashiro 2004). *Left* aerial photographs; *Upper right corner* horizontal riverbed profiles; *Lower right corner* frequency of riverbed disturbances

regarding population decreases in the near future, which may be similar to the original alluvial riverscapes with pristine floodplains (Fig. 22.1). Therefore, the land-use zoning in outer-levee areas should be introduced by being based on the screening of values and risks in whole lowland areas along the river in alluvial plains.

In medium alluvial rivers (Class B), because there is less space for fluvial processes between their levees, sediment accumulated in the riverbed needs to be removed and bank protection maintenance is required. These activities intensively in demand often conducted by local prefectural governments operating as river managers using heavy machinery to prevent flooding. Nowadays, fences and settling basins are installed to prevent turbid water runoff from work sites in order to mitigate the impacts on aquatic ecosystems. However, these kinds of management strategies are unsustainable due to financial constrictions of local governments. Moreover, riparian forestation is often observed even in the small spaces in river inter-levees due to upstream dam and weir constructions, similar to large alluvial rivers. On the other hand, local communities maintain small streams utilized for irrigation and/or drainage of agricultural lands every few years. There are many

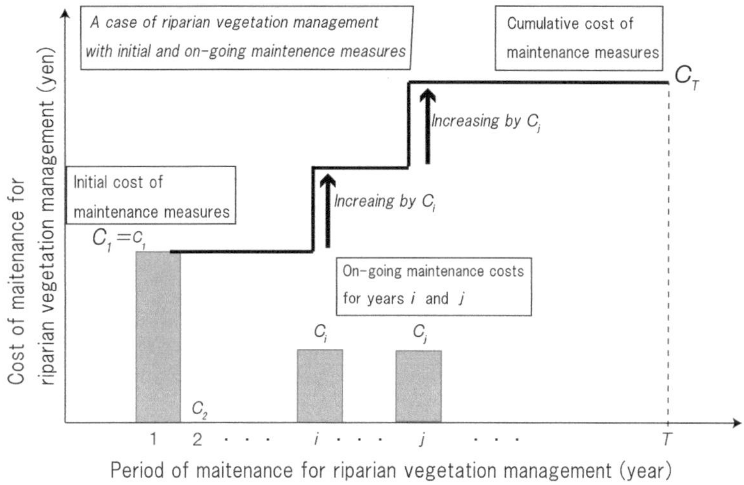

Fig. 22.4 Schematic image of maintenance cost evaluation for riparian vegetation management (Makishima et al. 2013)

small units, known as "*deai*" in some regions, which are organized by neighboring farmers for cooperating in the management of local agricultural lands and canals. Although sediment removal using heavy machinery usually affects several aquatic organisms, particularly benthic species, these impacts can be lessened if removal is conducted through manual operations. Currently, maintenance is becoming less frequent in smaller communities with its decline in accordance with that of farming populations, the increase in deserted cultivated land, and government financial restrictions. However, there is noteworthy river management in the Harai River (Fig. 22.5a), a medium river with natural banks and riparian forests flowing through Matsusaka City and Meiwa Town in Mie Prefecture, Japan. Local residents living within the Harai Watershed organized the Promotion Council of Environmental Clean-up in Harai River ("*Harai-Gawa-Kankyo-Bika-Suishin-Kyougikai*") in order to conserve the river's environment with stakeholder collaboration (Fig. 22.5b). Every winter they voluntarily remove woody debris from river channels (Fig. 22.5c), and educate local children about the river's ecology (Fig. 22.5d). These programs are generally implemented with financial support from central and local government. These kinds of cooperative administrations can provide new viewpoints for the management of medium rivers and small streams in the near future.

In the mountainous streams (mostly Class B), where environmental conditions have been considerably altered, there are many examples demonstrating the effects of small and large dam constructions and channel modification as introduced in Chap. 7. We hereby introduce some operations and measures to mitigate the dam effects of stagnation or regulation of sediment, as shown in Fig. 22.6. Sediment replenishment (Fig. 22.6a) was experimentally designed for restoring habitat

Fig. 22.5 Activities related to integrated management by local residents of the Harai river. **a** Harai river, **b** meeting of the promotion council of environmental clean-up in Harai river, **c** removing woody debris, **d** integrated education of river ecology

conditions of riverbeds below dams. A series of field surveys have partly verified its effects on the aquatic ecosystem (e.g. Katano et al. 2009; Tashiro et al. 2014; Yamada et al. 2014) according to the "RCI (reference-control-impact) design" (Nakamura 2003). There are several examples of construction of flood retention dams (basins) (Fig. 22.6b) designed only for flood control, with outlets at the original riverbed level for draining inflow from the upstream (Sumi 2008). Currently, the Masudagawa and Tatsumi dams are in operation with construction of the other two dams planned in Japan (Sumi 2008). They are expected to mitigate not only flood flow, but also transport water and materials such as sediment and nutrients. Regarding the slit and open check dams (Fig. 22.6c, d), they are also designed to mitigate debris flow hazard with less impacts on sediment regimes, based in comparable principles as conventional solid body dams (e.g. Piton and Recking 2016). Because Japan has the fourth highest dam density (Gleick et al. 2002), updating existing dams could be an effective measure for mitigating the effects of sediment regimes on aquatic ecosystems.

Although fishermen and anglers are familiarized with rivers, nowadays rivers are generally considered hazardous, with adults discouraging children from playing in them. As a result, residents living within watersheds have become increasingly disinterested in river health. In recent years, however, various activities are

Fig. 22.6 Practical measures to mitigate dam effects in mountainous streams. **a** Sediment replenishment on streambank below dam (in Agigawa stream, Gifu pref.), **b** flood retention dam (basin) (Tatsumi dam, Ishikawa pref.), **c** slit check dam (Ishikawa pref.), **d** open check dam (Ishikawa pref.)

increasingly conducted in cooperation with river managers and residents, while being supported by researchers, to enhance rivers and sustain their management. For example, communities cooperate in conducting workshops and extermination of exotic species. These cooperative works appear to promote daily observation of river environments and increase residential stewardship in river management. Thus, a new approach to managing is being established in some regions in which government and residents cooperate.

22.3 Paddies and Canals

Paddy fields are generally inundated from spring to summer (irrigation period). Water depths are shallow, and water temperatures often are high (Fig. 22.7a). However, during the non-irrigation period from autumn to winter, the fields are in a state of drought (Fig. 22.7b). Paddy fields and agricultural canals, function as seasonal wetlands during irrigation periods as if river floodplain become temporarily inundated with shallow water during early summer (Fig. 22.1). Many fish species, such as carp and catfish, migrate paddy fields and canals from river channels for spawning (Fig. 22.7c). In paddy fields and canals, there are few aquatic predators due to shallow water and both phytoplankton and zooplankton are richly produced under warm water conditions, providing food for juvenile fish born there. Hence, paddy fields and canals highly function as nursery habitats for juvenile fishes, being essential for their life history cycles. However, there are some risks for juvenile fish: predation from fish-eating bird species and drought in the non-irrigation seasons.

Considering the historical changes with agricultural developments in floodplain landscapes (see Chap. 7), it could be surmised that paddies and canals had substituted original habitats for floodplain-dependent fish until the early Showa era (early 20th century). At that time, paddies and canals maintained by farmers probably contributed to the regulation of aquatic habitats in agricultural lands and sustained characteristic fish communities and agroecosystems (Fig. 22.8a). Similar to the management of small rivers (mentioned above), farmers organized small cooperative units to share the maintenance. For example, they removed mud and debris accumulated on the canal beds once a year and repaired collapsed canal banks by hand as needed (Fig. 22.8a). If this maintenance was abandoned, the quality of the canal's habitat would degrade, resulting in a decline of the aquatic community, as shown in Fig. 22.8b.

Fig. 22.7 Seasonal changes in typical paddy landscapes with irrigation canals. **a** Irrigation period in spring-summer season, **b** non-irrigation period in autumn-winter season, **c** catfish in paddy for spawning in early summer

The modern development of agricultural lands since the Showa 20s (1945) increased crop productivity and decreased farmer labor accounts for maintaining paddy fields and canals. Paddies reformed from wet to dry type remarkably enhanced efficiency of farm works. Furthermore, straightened and concrete-lined canals are durable and reduce the need of bank repair (Fig. 22.9a, b). However, those reformations to agricultural lands deteriorate biological communities due to qualitative and quantitative loss of aquatic habitat (see Chap. 7). In addition, relationship between farm work and aquatic organisms gradually went out of people's mind after the reformation.

Currently, an interest in agroecosystems, including aquatic organisms, is steadily increasing among farmers. Cooperative approaches for improving habitat in agricultural lands provide a good opportunity of reconstructing local communities associated with management of agricultural lands. In some regions, farmers voluntarily made uneven concrete banks for ecological rehabilitation with gently sloped ladders for small animals to escape the deepened canals, and with a fishway to enable fish migration from the canal to paddy during spawning (Fig. 22.10). Farmers have also begun to maintain perennial water in paddies and canals in order to provide aquatic habitat throughout all seasons. Moreover, pools were also created inside canals to provide overwintering habitat. Labor accounts, which once decreased with modern reformation of agricultural land, will be predicted to increase with increases in work for recovering habitats. If farmer labor accounts of each management work,

Fig. 22.8 Importance of maintenance work for sustainable use of farmland. **a** Farmers working cooperatively to maintain canals, **b** idle farmland without any maintenance work

Fig. 22.9 Diverse landscapes of irrigation canals before and after modern development. **a** Canal with muddy banks, **b** straightened and concrete-lined canals with durable banks in non-irrigation period

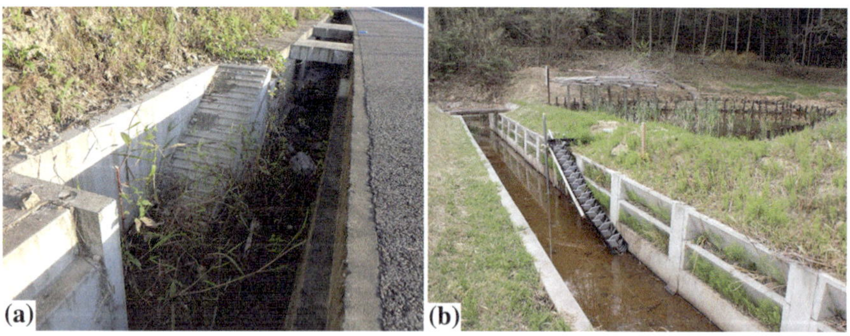

Fig. 22.10 Uneven concrete banks along canals help with ecological rehabilitation. **a** Ladder for small animals to escape deep canals, **b** fishway for fish to migrate from the canal to paddy

and biological responses to the management works are numerically evaluated, cost-effectiveness between them could be calculated. The cost-effectiveness can be used for establishing management strategies of agricultural lands with biological conservation. On the other hand, farmer populations are decreasing due to aging farmers and falling crop prices, resulting in increases in idle and deserted farmlands

(Fig. 22.8b). Canals that are not used and managed, often become desolate, and hence their biological functions decline. Deserted farmland particularly increases in valleys, where farming efficiency is lower due to steep and small cultivation areas. Considering decreases in farmer populations and government management budgets, sources may need to be appropriately allocated to yieldable and work-efficient farmland. In addition, biological values, such as biodiversity and rarity, should also be considered in the benchmark for cost-effectiveness.

22.4 Reservoirs

Reservoirs are artificial water bodies designed to stabilize water resources for beneficial uses, such as flood management, hydroelectric power generation, agricultural crop cultivation, drinking water, and recreation. Some were constructed for agricultural irrigation, in which isolated water areas, called "*Tame-ike*" in Japanese, were established (Fig. 22.11a, Uchida 2003, 2008). "*Tame-ike*" reservoirs have similar functions of lentic habitat to isolated water bodies found within the floodplain of large rivers (Fig. 22.1). Accumulation of sediment and leaf litter is a common problem, which causes deterioration of habitat quality in the reservoir and the floodplain (Nagayama et al. 2016). In particular, those accumulations progress environmental succession and impair the function of irrigation due to the deterioration of water quality and the reduction of pondage (Fig. 22.11b, c). In floodplain water bodies, habitat quality appears to be maintained by frequent flooding, which can flush out accumulated materials (Negishi et al. 2012). In reservoirs, which are usually well managed, accumulated materials are artificially removed in the manipulation of "reservoir drying", referred to as "*Ike-boshi*" in Japanese (Fig. 22.11d), which can substitute for flooding. Hence, reservoirs can function as if they are secondary nature, similar to a floodplain water body, when reservoir drying is periodically implemented.

On the other hand, reservoirs are easily invaded by exotic fish species, such as bluegills and the carnivorous largemouth bass. In these reservoirs, native fish populations decrease and are often threatened with extinction due to invasive species' predation (e.g. Katano et al. 2015). As the result, endemic fish biodiversity, including some rare and endangered species, is lost and exotic species predominate in the reservoir.

In according to the above discussion, the following practical measures for environmental conservation of reservoirs could be suggested:

1. Improve the environment, for example, perform reservoir drying every few years (Fig. 22.11d).
2. Conserve endemic species by prohibiting the introduction of exotic species and instead encourage their extermination.

Fig. 22.11 Diverse features of reservoir landscapes. **a** "*Tame-ike*" reservoir, **b** reservoir without artificial management, **c** buried reservoir, **d** reservoir drying

3. When needed, rare and endangered species should be reintroduced to reservoirs based upon reintroduction guidelines (The Ichthyological Society of Japan 2005).
4. After drying, monitor environmental conditions and succession and the illegal introduction of exotic fish.

There have been some cases where reservoir drying improved environmental conditions, exterminated exotic fish species, and reintroduced native fish species (Kanoh et al. 2005; Takahashi 2006; Kitamura 2008). In addition, it has been reported that the reproduction rate of the freshwater unionid mussels and the growth and reproduction rates of the Japanese rosy bitterling, a fish that spawns its eggs into the gills of live unionid mussels, increase after reservoir drying (Matsuba et al. 2009). As discussed in the above section ("paddies and canals"), the cost-effectiveness between farmer labor accounts and resulting biological effects is important for strategic management of reservoirs when future farming populations and management budgets decrease. Financial costs for reintroduction of native fish species should be also considered in the evaluation of the cost-effectiveness. It is imperative we find the reservoirs with high cost-effectiveness and allocate management sources to them.

22.5 Conclusion

Inland waters have been artificially altered in order to control water security and flood risk, which has caused degradation to many kinds of ecosystems in each of the water landscapes, such as rivers, streams, ponds, and floodplains with paddies, canals, and reservoirs. While the natural waterscape was originally composed of rivers, streams, and ponds, the paddies, canals and reservoirs that have been mostly formed in floodplains have come to function as secondary natures, compensating the floodplain-dependent habitats. Although efficiency of flood control and water utilization has been increasingly improved through heavy machinery, original natural environments of inland waters have degenerated with the disappearance of some migratory species and the extra-growth of marginal vegetation resulting in riparian forestation. In this chapter, we used examples found in conventional literature to review the processes by which these ecosystems deteriorate and measures for sustainably managing each landscape, in order to mitigate further damage.

According to these findings, it could be surmised that most of erosion to the ecosystems is irreversible and, therefore, requires maintenance (costs or labors) for rehabilitation. Consequently, both natural and social contexts should be taken into account for ecosystem rehabilitation in each of the waterscapes. Toward their sustainable management, we furthermore suggest that a cooperative stewardship approach involving regional stakeholders could be key solving these problems. Scientific methodology shows us that shared community work will provide the greatest effectiveness in maintenance costs or labors.

References

Gleick PH, Burns WCG, Chalecki EL et al (2002) Number of dams, by country. In: Gleick PH (ed) The world's water 2002–2003: the biennial report on freshwater resources. Island Press, Washington (DC), pp 296–299

Infrastructure Development Institute, Japan (1999) The River Law with commentary by article: legal framework for river and water management in Japan supervised by River Bureau, Ministry of Construction Japan, IDI Water Series No. 4

Kanoh Y, Harada Y, Kawamura, K (2005) Japanese rosy bitterling—influence by exotic species and isolation. In: Katano O, Mori S (eds) The current and future of the rare freshwater fishes—scenario of aggressive conservation, Shinzansha, Tokyo, pp 122–132 (in Japanese)

Katano I, Negishi JN, Minagawa T et al (2009) Longitudinal macroinvertebrate organization over contrasting discontinuities: effects of a dam and a tributary. J N Am Benthol Soc 28:331–351

Katano O, Ogawa H, Nakamura T et al (2015) Long-term effects of bluegill on the population sizes of two freshwater fishes in artificial ponds. Jpn J Ichthyol 62:169–175

Kitamura J (2008) Bitterling fishes (Cyprinidae: Acheilognathinae): current threats and conservation. Jpn J Ichthyol 55:139–144 (in Japanese)

Kitamura T, Tashiro T, Tsujimoto T (2001) Frequency of disturbance on river-bed as a habitat evaluation index. Adv River Eng Jpn Soc Civil Eng 7:282–297 (in Japanese with English Abstract)

Makishima M, Akamatsu F, Taya U et al (2013) Cost-effectiveness of different additional treatments in logging operations for controlling reproduction of riparian trees. Adv River Eng Jpn Soc Civil Eng 19:429–434 (in Japanese with English Abstract)

Matsuba H, Kimura S, Tsujii Y et al (2009) The mystery of *Ike-Boshi*: draining and drying a pond —new insights into ancient wisdom of water management. Bull Reg Res Inst Osaka Univ Econ Law 1:17–34

Miyamoto H, Akamatsu Y, Toda Y (2013) Current status and future prospects for research on riverine forest issues. Adv River Eng Jpn Soc Civil Eng 19:441–446 (in Japanese with English Abstract)

Nagayama S, Harada M, Kayaba Y (2015) Can floodplains be recovered by flood-channel excavation?—an example from Japanese lowland rivers. Ecol Civil Eng 17:67–77 (in Japanese with English abstract)

Nagayama S, Harada M, Kayaba Y (2016) Distribution and microhabitats of freshwater mussels in waterbodies in the terrestrialized floodplains of a lowland river. Limnology 17:263–272

Nakamura K, Tockner L, Amano K (2006) River and wetland restoration: lessons from Japan. Bioscience 56:419–429

Nakamura F (2003) Restoration strategies for rivers, floodplains and wetlands in Kushiro Mire and Shibetsu River, northern Japan. Ecol Civil Eng 5:217–232 (in Japanese with English abstract)

Negishi JN, Sagawa S, Kayaba Y et al (2012) Mussel responses to flood pulse frequency: the importance of local habitat. Freshw Biol 57:1500–1511

Piton G, Recking A (2016) Design of sediment traps with open check dams I: hydraulic and deposition processes. J Hydraul Eng 142. doi:10.1061/(ASCE)HY.1943-7900.0001048

Sanuki S, Ohishi T, Miwa J (2010) The present state of development of thick growth of trees and vegetation management in class A rivers. Adv River Eng Jpn Soc Civil Eng 16:241–246 (in Japanese with English Abstract)

Sumi T (2008) Designing and operating of flood retention dry dams in Japan and USA. In: Proceedings of ICHE conference on hydro-science and engineering, Nagoya, Japan

Takahashi K (2006) The influence by which a large mouth bass gives it to a community of freshwater fishes. In: Hosoya K, Takahashi K (eds) Get rid of black bass, message from the Shinaimotugo no Sato no Kai. Kouseisha Kouseikaku CO. Ltd., Tokyo, pp 29–36 (in Japanese)

Tashiro T (2004) Study on aquatic ecosystem evaluation in alluvial reach below dam with references to cobble bed with less disturbances. Doctoral dissertation, Nagoya University (in Japanese)

Tashiro T, Okuda C, Tsujimoto T (2014) Impacts of riverbed armoring below dam and their mitigation due to gravel addition on habitat of benthic fish community. J Jpn Soc Civil Eng Ser B1, 70:I_1321–I_1326 (in Japanese with English Abstract)

The Ichthyological Society of Japan (2005) The guideline in discharging fish for conservation of biodiversity. Jpn J Ichthyol 52:80–82 (in Japanese)

Uchida K (2003) Reservoir of Japan. Disaster prevention and environmental conservation. Kaiseisha Press, Otsu (in Japanese)

Uchida K (2008) Reservoir—multilateral function and utilization. Foundation Statistical Society of Agriculture and Forestry, Tokyo (in Japanese)

Yamada K, Tashiro T, Tsujimoto T (2014) Effects of sediment replenishment on stream ecosystem below dam and reservoir: their verification with particulate organic matter, aquatic organisms and their stable isotope rations of carbon and nitrogen. Limnol Tokai Reg Jpn 64:11–21 (in Japanese)

Yoshimura C, Omura T, Furumai H et al (2005) Present state of rivers and streams in Japan. River Res Appl 21:93–112

Chapter 23
Coastal Planning: Biodiversity Restoration and Ownership

Hiromi Yamashita and Taito Yasufuku

Abstract This section discusses the characteristics of labor accounts in coastal areas, coastal economy and biodiversity restoration, and ownership; and regional management of linked ecosystems. Suggestions for future labor accounts in coastal areas in a shrinking society include: a reduction in general labor accounts and funding for them, in preference of more fundamental labor accounts and projects dealing with core causes of coastal environmental degradation; investing time and effort in initial labor costs for restoration activities; determining which ecosystems should be prioritized for protection; coastal decision making taking regional and watershed zone environments into account; funding coastal ecosystem labor forces through taxes and donations from cities and upstream areas; and maintaining and designing environmental management activities that are fun and interesting for ordinary citizens.

Keywords Reducing general labor accounts · Restoration · Perceived landscape complex

23.1 Introduction

As we have discussed in the previous chapter, the coastal environment has been under immense pressure from economic, environmental, and social changes over decades. Many of the coastal areas originally flourished through the fishing industries, followed by tourism, heavy metals, or port industries. The same coastal

H. Yamashita (✉)
Ritsumeikan Asia Pacific University (APU),
1-1 Jyumonji-baru, Beppu, Oita 874-8577, Japan
e-mail: hiromiya@apu.ac.jp

T. Yasufuku
Graduate School of Environmental Studies, Nagoya University,
Furo-cho, Chikusa-ku, Nagoya, Aichi 464-8601, Japan
e-mail: taitoyasufuku@gmail.com

H. Shimizu et al. (eds.), *Labor Forces and Landscape Management*,
DOI 10.1007/978-981-10-2278-4_23

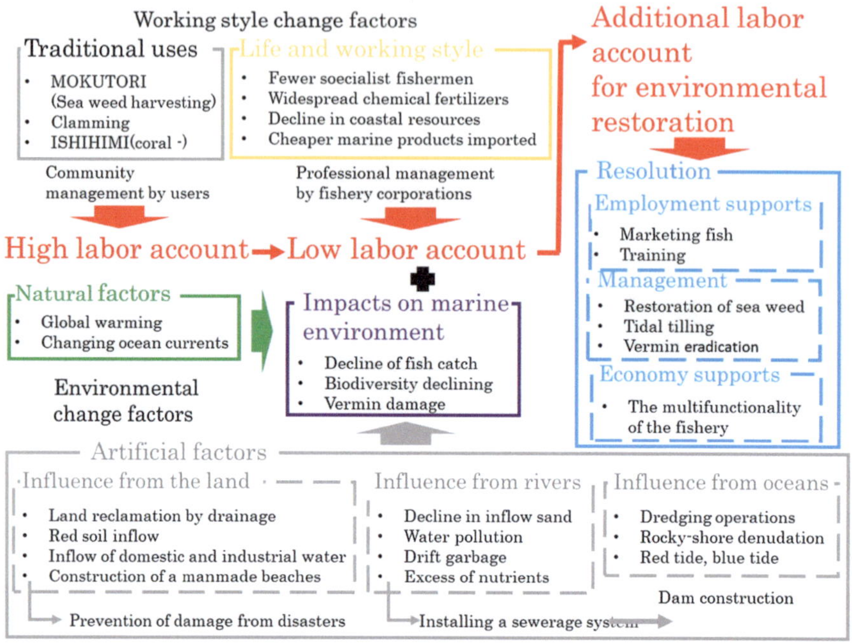

Fig. 23.1 Changes of labor account types in coastal areas

area can be utilized in various ways, often competing with one another for space and resources along the coast.

The labor force types in coastal areas have also changed over the years (see Fig. 23.1). Traditionally, coastal areas had high labor forces, directly linked to local fishing and resource foraging activities. Fishing activities were conducted by not only fishermen but also many families in the communities, and this brought a sense of cohesion in looking after the area, with residents voluntarily maintaining the environment.

The coastal labor force then declined, especially after the 1960s, due to labor-saving modern machines and artificial chemicals, diminishing the communities' involvement. However, it was during this era of heavy and multiple usages of coastal resources that coastal ecosystems were significantly damaged. Currently, many of the coastal labor costs relate to repairing the environment lost over the past 40–50 years. The types and amount of restoration work needed are increasing in Japan due to internal and global environmental change factors. This is within the context of the country's declining and aging population, which brings significantly less tax income than at the time the environment was being destroyed.

Given this circumstance, it is important for us to grasp the characteristics of coastal labor accounts clearly and consider what kinds of labor types might need to be abandoned, maintained, or created for future environmental management work.

23.2 Characteristics of Labor Accounts in Coastal Areas

When considering labor accounts in a coastal area, there are various issues related to defining boundaries, as coastal areas cannot be considered without examining the relationships within a "neighborhood landscape complex". Under the various categories of a "landscape complex", coastal areas fit into the brief of a "geographical complex" where positive or negative impacts caused by management activities in one area travel and affect another area. They also easily travel internationally. The ocean accumulates influence from mismanagement of not only coastal ecosystems but also other types of ecosystems. For example, unmanaged artificial forests, which easily collapse with heavy rains, can cause a sudden flush of red soil from the mountains into the ocean, possibly overwhelming the natural coastal water purification biological systems. The labor accounts needed in the forests to prevent this have often not been covered by coastal management funding, as they are physically removed from the ocean.

Coastal areas also fit into the character of a "perceived landscape complex", where people are not necessarily eager to understand the current status and changes of ecosystems, both mentally (since not all the people might be particularly interested in the health of the ocean) and physically (since most of the coastal environment is underwater and therefore invisible). During our research work, one female diver in Shima City in Japan said, "those who do not dive into the sea would never understand the state of the ocean". She has conducted various environmental management activities for coastal ecosystems, such as collecting litter, planting seaweed grasses, and removing invasive species, all underwater. As the population of such divers and others familiar with the underwater environment decline, who will make this "perceived landscape complex" visible to others, including those who might allocate funding for necessary labor accounts?

One can also observe that coastal areas are a "time-latent complex" landscape. In coastal biodiversity management, some of the negative and positive influences come from actions taken many years ago. Therefore, positive or negative impacts made by current management actions might not be felt or visible for many years to come. Management plans and actions for coastal areas often need long-term planning and investment, yet, in current budgetary allocation systems, it is difficult to obtain concrete, long-term plans for continuous funding until such effects become more apparent.

23.3 Coastal Economy and Biodiversity Conservation

Historically, the fishing industries were partly responsible for damaging coastal ecosystems by polluting water and harvesting too much of its resources. At the same time, over the years the fishing industries have cared for the coastal environment by reinforcing local rules and management practices. The decline of the fishing industries, whose members have been custodians of oceanic biodiversities, will influence how we think of labor activities that need to be conducted in the future. At the same time, due to degradation of the ocean environment, the amount and types of labor are increasing, not only due to local forces but also to global forces such as global warming, plastic waste, and invasive species.

The discussion of coastal management cannot ignore economic activities and the impacts on coastal ecosystems and biodiversities. We could simply say that we should let the numbers of fishermen fall, so that the number of economic activities that damaged the environment in the past would also fall. But, to be able to go down this route, Integrated Coastal Management strategies, including time and space goals, need to be proposed and agreed among various stakeholders.

23.4 Ownership and Management of Coastal Areas

Coastal areas also have an issue related to an "administrative complex" landscape. Having different managers for various parts of the coast makes Japanese coastal management complicated. One public environmental management official joked during the interview that when they see invasive species floating along their coastline, if the boundary is close to another ministry's constituency, they are tempted to push them to their shore, especially when the other has more funding for this type of labor. Behind this joke lies the serious difficulty of sharing funding and cooperating administratively to deal with common issues (here invasive species) even when their constituencies are next to one another.

Figure 23.2 shows a part of the map, which is published by the Port and Harbor Division of Aichi Prefecture (also see Table 23.1 for the types of coasts defined by the Coast Act 1956, and the management bodies). One can see how the shoreline is divided into various colors, which represent who are responsible for each section. Each management body also has different roles regarding their shores; some ministries are only focused on protecting the land behind the shore, but not conserving the sea or coastal biodiversities.

Among these complex shoreline management puzzles, slightly off shore there are fishermen who have been utilizing and looking after the coastal environment. So far, the mandate of those coastal areas used for ports, industries, or agriculture has not included aspects of coastal environmental protection and conservation activities as one of the main mandates. However, clearly active environmental conservation mandates would need to be included in their activities.

When this is not possible, separation of "ownership" and "management" might be needed for all coastal areas. Coastal areas have been the places where various governmental organizations and industries have tried to appropriate the rights to "manage". Historically, "managers" of coastal lines, often government offices, also acted as "owners" of the coast (e.g. selling a part of the shore to private companies) and as "users" of the coast (e.g. conducting a development project).

Fig. 23.2 A coastal area map in Aichi Prefecture (2012, p.1): This map shows the sections of shoreline under different management bodies [Water management and land protection (*pink lines*), port authority (*purple lines*), rural development bureau (*orange lines*), fisheries agency (*green lines*)]

Table 23.1 Types of coasts defined by the coast act 1956 (Yamashita 2014, p.120)

Types of coasts	Management responsibility	Role
Shores for agriculture [Nochi kaigan]	Ministry of agriculture, forestry, and fisheries	To protect agricultural land and activities behind the shore from erosion and natural disasters
Shores for fisheries [Gyoko kaigan]	Ministry of agriculture, forestry, and fisheries	To protect fishing ports and fishing activities
Shores for ports [Kowan kaigan]	Ministry of land, infrastructure, transport, and tourism (the areas previously looked after by the ministry of transport)	To protect port infrastructures and related business infrastructures from erosion and natural disasters
Other protected shores [Sonota kaigan hogo chiku]	Ministry of land, infrastructure, transport, and tourism (other areas previously looked after by the ministry of construction, ministry of agriculture, forestry, and fisheries)	To protect people's lives and possessions behind the shoreline

However, coastal lines and ecosystems are "common", shared by all. In this sense, there needs to be a clearer separation between the right to own, the right to use, and the right to manage in the future. This means that those people who do not "own" (who cannot sell) can still utilize and/or cooperate to manage certain parts of the coast under the negotiated local rules.

23.5 Labor Accounts that Encourage Natural Restoration Processes

When we look at the existing types of labor needed to manage coastal ecosystems, many provide only temporary solutions for a degrading environment, without touching the fundamental causes. For example, tidal flats are being refilled with soil, with special structures put in place to stop the soil eroding, and their surfaces dug out to increase the level of oxygen (this role used to be carried out by the active benthos creatures in the tidal flats). These small-scale actions might be helpful in the short term, but they are required to be repeated regularly to maintain such slight improvements.

These short-term labor inputs act like a "band-aid", a temporary solution to the much deeper problem of how to restore a damaged environment. This approach of taking small steps targeting only one aspect of the coastal ecosystem issues is not cost effective, either financially or from the labor account perspective.

The two major reasons for Japanese coastal environment decline are losing its shallow areas of water, including tidal flats, and having numerous dams in river basins (e.g. Unoki 2005). Land reclamation and defining the ocean/land borders with concrete reduced the size of the shore dramatically as discussed in Chap. 8. The creation of dams reduced the flow of natural soil into the sea, and this causes coastal squeeze and thinning of coastal wetlands, which help the functions of coastal and oceanic ecosystems. Yet, restoration of tidal flats and coastal environments and removal of dams have not been practiced as much as they could have been. Removing old dams in the US is becoming a major environmental management project. They have proved to be more effective in terms of solving multiple environmental issues than expected. Once the initial work has been conducted, nature tries to restore itself, and annual operational costs and environmental conservation labor forces for the basin decreases year on year. In Japan, the removal of Arase dam in Kyushu has been conducted as the first of its kind, and researchers are eagerly anticipating the results (e.g. Kawasaki et al. 2015).

Most concrete structures built after the 1950s, including dams, are now due to be rebuilt or destroyed as they reach the end of their lifespan. Already it is becoming clear that Japan will not have enough funding to replace all the concrete structures of coastal areas. This could be an opportunity to conduct some fundamental activities. Not only could they restore the environment but also reduce the number of small labor activities that solve issues only in the short term at a time when the numbers of people to conduct them are declining.

There are some small coastal wetland restoration projects happening in Shimacity in Mie Prefecture, which do not need future labor forces or much of the current funding (Matsuda et al. 2009; Yamashita 2014). Tidal flat restoration projects here only involve opening the water gates of existing sea dykes on farmland that used to be tidal flats. Opening gates in this way floods some parts of unused farmland to create natural tidal flats, and a flow of nutrients from the river to the sea (Kokubu and Yamada 2011). These activities require initial labor forces of negotiating various arrangements: checking available unused farmland; talking to affected landowners and surrounding landowners; and making local arrangements within different sections of the municipal and prefectural governments. The general labor accounts are almost down to zero, except if community members want to conduct educational activities. Larger scale projects of this kind can be seen in the UK, including the Steart Peninsula.

Some might argue that citizens would find it difficult to get involved in relatively large restoration projects rather than small general labor activities in their local areas. However, the various restoration projects, such as the Arase dams in Japan and Steart Peninsula in the UK, proved that is not the case. A big project can also provide many roles for citizens through educational activities, monitoring, and foraging activities. For coastal area management, local knowledge (experiential knowledge and tacit knowledge) would also be crucial in understanding specific changes in biodiversities.

23.6 Watershed Management: The Regional Management of Linked Ecosystems

It would also be important to introduce a "land complex" concept, making links between the land use policies of coastal areas to take into account wider basin zones as a whole. There have been various attempts in recent years, at the policy, practical, and funding levels.

One example is a research project on the importance of conserving networks of tidal flats and coastal ecosystems by showing how Asari clams move around in Ise Bay, Japan. Yahagi River in Aichi Prefecture has basin zone planning in its management, as well as the Akasuka area and the Ibi River (e.g. Suzuki et al. 2002; Furukawa 2007).

One practice widely used in Japan to help people understand the links between forests, rivers, and the ocean is the "Forests are the lovers of the sea [mori wa umi no koibito]" initiative (Hatakeyama 2012). Hatakeyama explained the importance of nutrients and irons from the forests naturally flowing into the sea for various creatures to thrive in its ecosystems. This tree-planting exercise has now been practiced in various parts of the coast, and is now being funded by both private organizations and local/national governments.

Another issue to be resolved with wider cooperation is that of accumulating garbage. One island in Mikawa Bay (Toshijima) is situated in the mouth of this big bay and all the waste from forests, rivers congregates around this small island

(Yasufuku et al. 2014). There are clearing operations by locals, with fishermen's boards. However, often the amount and size of the waste means they face huge difficulties in dealing with it. These labor work would not be possible with a declining population, especially of young people on the islands and further afield.

When many of the causes of coastal environmental damage are land based, the funding mechanisms for coastal labor costs need to reflect this. Discussions on setting up a funding mechanism are needed so that people upstream of the rivers (both rural and urban) contribute to some of the activities happening downstream (Manaki 2011). Other countries have a funding system for NGOs or local organizations to obtain funding simply by looking after rare or important ecosystems such as tidal flats or seaweed beds. For example, in the UK, the Environment Agency under the Ministry of the Environment has a list of price calculations for the maintenance cost of various types of coastal environments, and the amount of funding each area can receive.

We also need to consider how to combine labor types that cannot be done without funding and those that can. Another consideration is how the funding can be calculated for those with long-term budget planning. Some activities need to have long-term investment, especially those activities where it is difficult for people to appreciate the positive changes for a long time. Who uses or can tap into long-term funding to conduct environmental conservation activities (e.g. civil servants, ordinary local citizens, citizens from outside) would also need to be clearly identified and checked against all types of labor accounts in coastal areas.

23.7 Summary: Creating New "Common" and Labor Accounts

This section discussed the characteristics of labor accounts in coastal areas, and biodiversity restoration, and ownership; and regional management of linked ecosystems. The suggestions for the future labor accounts in coastal areas in a shrinking society could include the following:

1. Reducing general labor accounts and funding for them, using those resources instead for more fundamental labor accounts and projects, which deal with core causes of coastal environmental degradation
2. Investing time and effort in the initial labor costs of restoration activities (especially at the beginning of projects) to encourage the subsequent natural cycle of restoration in coastal ecosystems. Reducing overall general labor accounts, but still maintaining enjoyable labor account elements for citizens to participate in (e.g. monitoring and observing natural restoration processes)
3. Determining the important ecosystems to be protected and setting up funding for national and local governments to pay local people, NGOs, or external organizations who look after such important ecosystems

4. Coastal decision making taking into account regional and basin zone environments in policy and funding strategies
5. Funding coastal ecosystem labor forces through not only taxes in coastal areas, but also taxes and donations from cities and upstream areas
6. Maintaining and designing environmental management activities that are fun and interesting for ordinary citizens to participate in and appreciate

The lines of responsibilities for looking after coastal ecosystems are vague. Yet this ambiguity has the potential to create an environment where various stakeholders can enthusiastically become involved as labor forces, if existing operational systems also allow a similar level of flexibility to encourage a variety of activities on the coast.

References

Furukawa, K (2007) Civil engineering approach: Example of Tokyo bay [Doboku kogakuteki approach—Tokyo wan o reinishite]. In: Yamamoto T, Furuya K (eds.) Environmental restoration of enclosed coastal sea [Heisasei kaiikino kankyo saisei]. Kouseisha Koseikaku, Tokyo

Hatakeyama S (2012) Healthy forest means healthy ocean [Morio yutaka nisureba umimo yutakani]. Daisan Bunmei 625:71–73

Kawasaki T, Ohtsuki K, Nihei Y, Hirose Y, Yoshimori Y, Hanasaki T, Iyooka H, Onikura N (2015) Monitoring and evaluation of sedimentology on reservoir to estuarine tidal flat for assessment of dam removal effects [Arase damu tekkyo hyoka nimuketa damu tansuiiki, karyukaasen, higata no doshadotai no cyosa to kaiseki]. J Jpn Soc Civil Eng Ser B1 (Hydraul Eng) 71(4):1_979–1_984

Kokubu H, Yamada H (2011) Evaluation of the tidal flat restoration effect in the coastal unused reclaimed area by promoting tidal exchange in ago bay [Agowan ni okeru engan yukyuchi no higata saisei]. In: Proceedings of the Japanese conference on coastal engineering 67(2):956–960

Manaki S (2011) Economics of biodiversity and ecosystem services [Seibutsu Tayosei to Seitaikei Service no Keizai Bunseki]. Wetland Research [Shicchi Kenkyu], No 161

Matsuda O, Kokubu H, Uranaka H (2009) Summation and future perspectives of the ago bay restoration project [Agowan saisei project no sokatsu to sonogono tenkai]. Nippon Suisan Gakkaishi 75(4):737–742

Suzuki T, Ichikawa T, Momoi M (2002) The approach to predict sources of pelagic bivalve larvae supplied to tidal flat areas by receptor mode model: a modeling study conducted in mikawa bay [riseputa mode model o riyoshita higataiki ni kanyu suru nimaigai fuyu yosei no kyokyugen yosokuni kansuru kokoromi]. Suisan Kaiyo Kenkyu 66(2):88–101

Unoki S (2005) How river management projects changed the sea [Kasenjigyowa umio dokaetaka]. Seibutsu kenkyusha, Tokyo

Yamashita H (2014) Planning invisible landscapes: making invisible tidal flat landscapes visible for future sustainability. In: Shimizu H, Murayama A (eds) Basic and clinical environmental approaches in landscape planning: urban and landscape perspectives, vol 17. Springer, Japan, pp 113–131

Yasufuku T, Shimizu H, Murayama (2014) Study on estimation of maintenance scheme and calculation of management cost in coastal area: focusing on Toshi Island in Toba City, Mie Prefecture [Enganiki niokeru kanrikatudo no haakuto sagyokosutono sansyutu]. City Plan Tokai Branch Res Rep 52:681–684

Chapter 24
Connection to the Ecosystem Service Studies

Makoto Ooba and Kiichiro Hayashi

Abstract This section explains the relationship between landscape management (LM) and ecosystem services (ESs). A proper LM would have two aspects, namely pros and cons, from the perspective of biodiversity and ESs. Improper LM may cause negative impacts on the ESs and ecosystems. On the other once an LM is done properly, several positive benefits can be derived from it. Then the trade-off between the impacts on wildlife habitat and the ES provisions made by the LM also needs to be considered. Currently, LM has been reviewed for many fields. However, information about the priority of the increase of ESs and the decrease of ecosystem disservices in this area is not readily available. This section also describes a simple spatial ES assessment for an urban area, namely Nagoya City, which was conducted using the method for estimating ES potential supply and land use priority, by taking LM into consideration.

Keywords Ecosystem service · GIS · Landscape management

24.1 Overview of the Landscape Management and Ecosystem Services

The linkage between nature and the human society has been widely studied as an emerging topic in the field of interdisciplinary biodiversity science. In recent decades, benefits provided from nature have been recognized as ecosystem services (ESs) and their degradation has become one of the biggest environmental issues in

M. Ooba (✉)
National Institute for Environment Studies, Fukushima Branch,
10-2 Fukasaku, Miharu, Fukushima 963-7700, Japan
e-mail: ooba.makoto@nies.go.jp

K. Hayashi
Institute of Materials and Systems for Sustainability, Nagoya University,
Furo-cho, Chikusa-ku, Nagoya, Aichi 464-8603, Japan
e-mail: maruhaya@esi.nagoya-u.ac.jp

H. Shimizu et al. (eds.), *Labor Forces and Landscape Management*,
DOI 10.1007/978-981-10-2278-4_24

the world. The Millennium Ecosystem Assessment was a remarkable report on this topic (Millennium Ecosystem Assessment (MA) 2005). According to the MA (2005), the ESs include provisioning, regulating, cultural and supporting services. One important finding of this report was that the degradation of ESs has been happening and will continue to happen in the future.

As for the Japanese situation, based on the Japan Biodiversity Outlook (MOE 2010), under use and under management are some of the most serious biodiversity issues in Japan, although the certain level of management activities and/or intervention to protect nature have been undertaken to ensure the sustainable use of natural resources in Japan. Nowadays, aging and depopulation have made it difficult to ensure appropriate interactions between the human society and nature (MOE 2010). Sufficient labor force and funding for the management activities and/or intervention are needed to ensure appropriate ES provisioning potential. Once that is ensured, the labor account estimation for landscape management (LM) is calculated to better understand the future needs of the LM work.

24.2 The Pros and Cons of LM from the Perspective of ES Assessment

This section explains the relationship between LM and ESs. A proper LM would have two aspects, namely pros and cons, from the perspective of biodiversity and ESs. For example, an urban forest provides several positive ESs such as, the provision of a recreational opportunity for citizens, air pollution regulation, micro climate regulation, and absorption and stock of carbon (MA 2005). However, it also has some negative impacts on visitors and/or residents directly and/or indirectly, such as, fallen leaves and branches, disamenity of insects and animals, etc. These are called ecosystem disservices (EDS), and have been discussed mainly as agricultural pests and vectors of diseases elsewhere (Lyytimäki and Sipila 2009; Escobedo et al. 2011). Improper LM may cause EDSs in some cases. It may also lead to a negative impact on the forest ecosystem itself, for example, by allowing the intrusion of bamboo into the secondary forest, which is one of the issues concerning forest management in Japan (MOE 2010). So, improper LM may cause negative impacts on the ESs and also ecosystems.

On the other hand, once an LM is done properly, several positive benefits can be derived from it. For example, an appropriate LM may lead to better access to a forest for the citizens. Also, if the citizens are actively involved in a community based LM activity, the community activity in that area will be robust. Regarding the impacts of management, Ooba et al. (2015a) evaluated the effects of forest management on selected ESs under several forest management scenarios. Also Dhakal et al. (2014) made an assessment of forest management on the habitats of several animals.

However, the following relation between the impacts on wildlife habitat and the ES provisioning potential and LM needs to be considered. For example, most of the planted forest in Japan is too dense. So an appropriate management strategy, such as thinning, would be recognized as one of the ways to overcome this issue (MOE 2010). However, after conducting strength thinning, one option is to leave the forest to nature (Toyota City 2007). This means that no more LM work is needed for that particular forest, including forest production, so that it will remain as a natural forest in the future.

24.3 Priority of LM for Urban Forest Park

LM has been reviewed for many fields including urban forest park, plantation forest, paddy field, residential garden, etc. For example, an urban forest park in Nagoya City was studied as a case in Chap. 12. The case study focused on questions like: What is the purpose of conducting the LM for this urban forest park? How are the benefits generated from the LM? Are they based on both, the expected benefits to citizens and the cost and amount of labor input for this LM work? One part of the benefits is linked to the increase of ESs and the decrease of EDSs by the LM activity. However, information about the priority of the increase of ESs and the decrease of EDSs in this area is not readily available, although there is some data related to the priority of ESs. Figure 24.1 shows the priority of ESs provided for the Higashiyama forest in the eastern part of Nagoya City based on a questionnaire

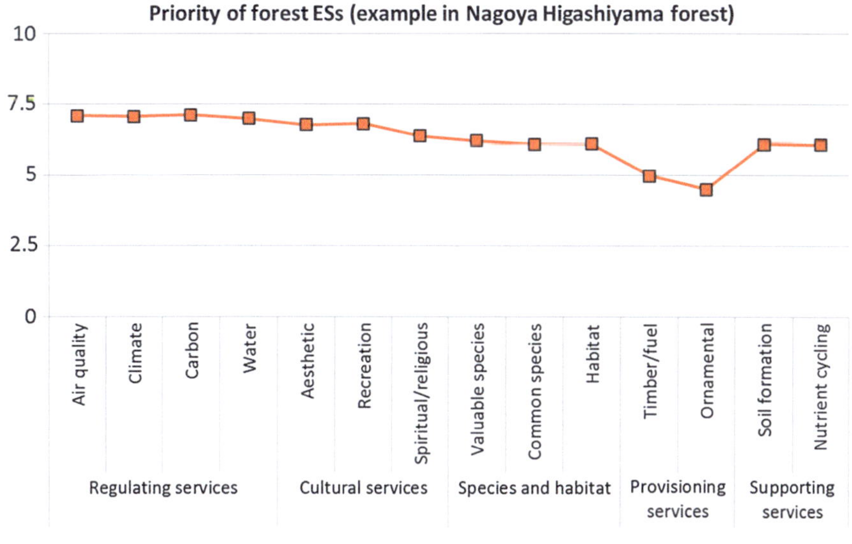

Fig. 24.1 Importance of ES for Higashiyama forest in Nagoya City, Japan (0 to 10 scale) (*Source* Hayashi et al. (2014) revised)

survey for citizens (Hayashi et al. 2014 revised). Regulating and cultural services have a higher priority compared to provisioning services.

For an appropriate LM, the amount of labor force available and their accessibility to the relevant forest are also important factors that need to be taken into account.

The next section describes a simple spatial ES assessment for an urban area, namely Nagoya City, which was conducted using the method for estimating ES potential supply and land use priority, by taking LM into consideration.

24.4 Spatial Evaluation of Ecosystem Services

Geographic Information Systems (GIS), in particular, are often used to spatially evaluate ESs. However, integrated quantitative evaluation tools which encompass all ES concepts are still at the research and development stage. A number of conservation planning support softwares that use GIS to conserve certain specified species have been developed and used, such as C-Plan (GDARD 2014), Marxan (Seebacher and Franklin 2012), and Zonation (Moilanen et al. 2011). Moilanen et al. (2011) have used Zonation, for instance, to show the biodiversity, carbon storage services, agricultural supply services and competitive land use analysis in some cities in Great Britain. In this section, we will depict an assessment scheme using a relatively simple method for providing ES in Nogaya City (Ooba et al. 2015b). This method is easily replicable in other cities too, as the methodology is clear and the input data mainly depends on the land use map.

The provision of ES from an ecosystem can be estimated by the representative geographical, biophysical, or social variables (proxy variables). It is assumed that the variables depend on the type of ecosystem ordu the land use pattern. Based on the land use map of Nagoya City, our previous study (Ooba et al. 2015b) indicated that some variables have a high correlation coefficient with other variables. Five variables that had relatively low correlation with the other variables and were selected as the proxy variables are the four MA categories and biodiversity conservation (Table 24.1).

Digital land use maps of Nagoya from 1997, at 10 m mesh level, were used to categorize five types of land use: water surface, residential and industrial areas, roads (urban areas), farmlands, and forests. Green space was defined as farmlands and forests. Digital elevation maps (10 m grid, Geospatial Information Authority of Japan) were also used for the estimation of soil erosion (regulation service) based on the Revised Universal Soil Loss Equation (RUSLE, Renard et al. 1997). The variables of ES were calculated from the land use map at the Japanese standard mesh resolution, which translates to approximately 1 km. Carbon sequestration and food supply, that represent support- and provisioning services respectively, were estimated from Ogawa et al. (2002) and Aich prefecture (2012). The proxy for biodiversity conservation was estimated from the continuity of green space, using the focal statistics (ArcGIS) and setting the proximity radius to 2 km (Li 2014).

Table 24.1 Methods used to estimate ES

Service category	Proxy variable	Basic units, method details[a]	Unit	Weight[b]
Support	Carbon sequestration	3.09 (F)	t/y ha	0.200
Provision	Food production	2.98(A)[c]	t/y ha	0.180
Regulation	Reduction of the soil erosion coefficient compared to bare soil	$S = 65.41 \sin^2\theta + 4.56 \sin\theta + 0.065$ C = 1(U), 0.33(A), 0.0085(F)	–	0.207
Culture	Economic value of the green space	Value per unit area as green area A(ha) $V = 3.0184\ A^{-0.437}$	10^6 JPY/y	0.209
Habitat	Continuity of the green space	ArgGIS tool (Focal statistics) proximity = 2 km radius	–	0.205

[a]Land-use codes—*U* Roads and urban areas, *F* Forestland, *A* Farmlands
[b]The values used for the integration of ES are an average of the values that denote their importance to ES
[c]The average value of crop production for rice, 5.068 t/y ha, and soybean, 0.887 t/y ha (Aich prefecture 2012)

The economic value from recreation was estimated by the generalized round-trip costs per capita, which was calculated based on the necessary costs such as transportation and transfer facility usage costs (Ooba et al. 2016). Four major green areas (area: 6.18–84.9 ha) were selected and the generalized round-trip costs to the green areas were gathered from a questionnaire survey in 2014 (2,800 samples from Aichi Prefecture). The results were used to estimate the economic value of cultural services. The economic value V (10^6 JPY ha^{-1} y^{-1}) per unit area was estimated as follows:

$$V = aA^b \tag{24.1}$$

where A is area (ha) and a and b are the parameters obtained by a regression calculation ($b < 0$). For a green space of 1 ha or greater, the corresponding services were evaluated.

The estimated values of services were represented by different unit systems. So, for the purpose of comparison, the values were converted into a [0, 1] scale based on the maximum and minimum value. Soil erosion variables have a negative effect. Then, high variables mean high negative ecosystem services. Based on this, the variable was then converted into a [0, 1] scale. Although all the ES variables have exponential distributions, they were converted to normal distribution by logarithmic conversion before applying the [0, 1] scale conversion. 0-value of a variable replaces a minimum value of the corresponding ES distribution obtained by the logarithmic conversion.

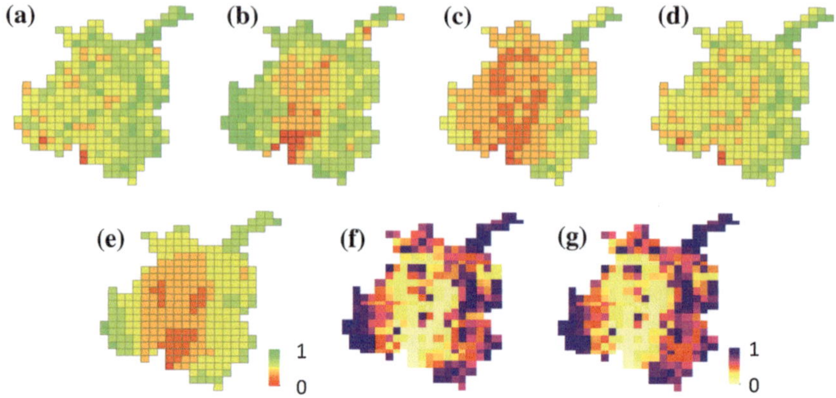

Fig. 24.2 Distribution of ecosystem services and conservation priority. **a** Support. **b** Provision. **c** Regulation. **d** Culture. **e** Habitat. **f** Conservation priority. **g** Priority of scenario (2)

Using the proxy variables of ES and biodiversity thus calculated, the integrated evaluation of the green space in Nagoya is carried out by the Zonation software, which calculates conservation priorities (Moilanen et al. 2012). The software identifies the mesh with the greatest overall loss, when one mesh is removed. By repeatedly removing the mesh that was identified and searching again among the remaining meshes for the cell with the maximum loss, conservation priority is determined. The Core Area Zonation, that minimizes loss by assigning a high rank to spatially continuous and heavy weight ecosystem services, was used. For the evaluation of species biodiversity, the number of species was calculated to evaluate the loss of diversity of species. For ES (in this study), the value of ES is calculated to evaluate the loss.

To integrate each ES value, the weight of the ES that was obtained through the web questionnaire was estimated using a five-point scale (Ooba et al. 2016) in this study (Table 24.1). The costs and other Zonation parameters were set to their default values.

The ES map of Nagoya is shown in Fig. 24.2. Cultural services were estimated based on Eq. (24.1), where $a = 3.0184$ and $b = -0.437$ ($R^2 = 0.63$). If the area increases, the value per unit area usually decreases. However, in this case, as area-based value V is multiplied by the area A, the value of the cultural services increases gradually with the increase in area. Carbon sequestration was found to be high in eastern Nagoya City and Moriyama, where there are many wooded areas. Food production was found to be high in western Nagoya, where there are relatively large areas of paddy and farmland. Cultural services and habitat quality tended to be low in the central area but higher in the surrounding areas.

The conservation priority calculated by the Zonation software is shown in Fig. 24.2. Priority was high in the areas surrounding the city, such as Higashiyama, Moriyama, and western Nagoya City. Several green spaces such as the Atsuta Shrine, which is isolated inside the city, also earned high priority.

24.5 ES Evaluation Connected to LM

Compared to the current state of ES in Nagoya, two future scenarios were considered, (1) Increase in the percentage of green space in all the meshes of the city by LM, and (2) Increase in the percentage of the continuous green space in the eastern region of the city by LM. In both the scenarios, the net rate of increase of green space is assumed to be about 5 %. In scenario (2), it is assumed that ES in the waterside areas would be enhanced in the future by afforestation or greening over a relatively large area (larger than 1 ha), to design a semi-natural ecosystem and develop a natural corridor for biodiversity conservation inside the city. In scenario (1), land use for forest increases in every mesh (but continuity was not recalculated), while land use for water-surface that is located in the selected mesh (Fig. 24.2), is replaced by forests. Recreation value and continuity are recalculated in scenario (2). It should be noted that the labor cost for the management of green spaces in both the scenarios is the same.

The priority in scenario (1) is almost the same as the current map, because the relatively continuous forests and the isolated forests have the same value. However, in scenario (2), the isolated forests were relatively less important than in scenario (1), and the green space in the eastern region of Nagoya was assigned a higher priority as it covers a wider area here than in scenario (1).

24.6 Conclusion

The following issue should be considered while planning a green space in the future, for an urban area that has shrunk. At present, it is difficult to reserve continuous green spaces inside an urban area. If the population decreases inside an urban area in the future, sparse green spaces may appear in places that had been residences, business areas, or industrial areas. Based on the result of scenario (1), increasing the sparse green space inside a city does not change the state of its ES significantly. However, if a large scale plan like scenario (2) is applied, the state of ES inside the city may change.

References

Aich prefecture (2012) Nougyou no ugoki: Shiryou hen (Trends of agriculture in Aichi prefecture: statistics 2012). http://www.pref.aichi.jp/nourin/seisaku/ugoki/ugoki2012/nougyoushiryou.pdf (in Japanese)

Dhakal A, Ooba M, Hayashi K (2014) Assessing impacts of forest conversion on terrestrial vertebrates combining forestry cost with HS and InVEST: case of Toyota city, Japan. Int J Biodivers Sci Ecosystem Serv Manag 10(3). http://dx.doi.org/10.1080/21513732.2014.920420

Escobedo F, Kroeger T, Wagner JE (2011) Urban forests and pollution mitigation: analyzing ecosystem services and disservices. Environ Pollut 159:2078–2087

GDARD (2014) Technical report for the gauteng conservation plan (Gauteng C-Plan v3.3). Gauteng Department of Agriculture and Rural Development. Nature conservation directorate. 60 p. http://bgis.sanbi.org/gauteng/Gauteng%20C-Plan3%203%20Technical%20Report%202014.pdf

Hayashi K, Sumi H, Ooba M, Hasegawa Y, Yonekura Y (2014) Preliminary study on comprehensive assessment of biodiversity offset and banking. In: Proceedings of the annual conference 2014, pp 103–106. The Japan Society for Impact Assessment (in Japanese)

Li R (2014) The potential distributions of mammalian by maxent model—case of Nagoya city. Master thesis, Graduate School of Environmental Studies, Nagoya University, Japan

Lyytimäki J, Sipila M (2009) Hopping on one leg—the challenge of ecosystem disservices for urban green management. Urban For Urban Greening 8:309–315

Millennium Ecosystem Assessment (2005) Ecosystems and human well-being: synthesis. Island Press, Washington, DC, USA

Ministry of the Environment (MOE) (2010) Japan biodiversity outlook—report of comprehensive assessment of biodiversity in Japan. Global biodiversity strategy office, biodiversity policy division, nature conservation bureau, the ministry of the environment

Moilanen A, Meller L, Leppanen J, Pouzols FM, Arponen A, Kujala A (2012) Zonation version 3.1—user manual. Biodiversity Conservation Informatics Group, Department of Biosciences, University of Helsinki, Finland. http://cbig.it.helsinki.fi/software/zonation/

Moilanen A, Anderson BJ, Eigenbrod F, Heinemeyer A, Roy DB, Gillings S, Armsworth PR, Gaston KJ, Thomas CD (2011) Balancing alternative land uses in conservation prioritization. Ecol Appl 21:1419–1426

Ogawa K, Miwa M, Shimada T, Ogawa S (2002) Nihon ni okeru ryokuchi no taikijyoukakinou to sono keizaitekihyouka (Air purification function and economic value of green space in Japan). Ann Rep Center Environ Sci Saitama 1:106–117 (in Japanese)

Ooba M, Hayashi K, Fujii M, Fujita T, Machimura T, Matsui T (2015a) A long-term assessment of ecological-economic sustainability of woody biomass production in Japan. J Clean Prod 88:318–325

Ooba M, Hayashi K, Suzuki T, Li R (2015b). Analysis of urban ecosystem services considering conservation priority. Int J Environ Rural Dev (in press)

Ooba M, Ito H, Hayashi K (2016) Socio-economic and spatial evaluation of ecosystem services in Nagoya, Japan. Int J Environ Rural Dev (under review)

Renard KG, Foster GR, Weesies GA, McCool DK, Yoder DC, coordinators. (1997) Predicting soil erosion by water: a guide to conservation planning with the revised universal soil loss equation (RUSLE) (agricultural handbook 703). U.S. Government Printing Office, Washington, DC

Seebacher F, Franklin C (2012) Determining environmental causes of biological effects: the need for a mechanistic physiological dimension in conservation biology. Phil Trans Royal Soc B Biol Sci 367(1596):1607–1614

Toyota City (2007) Toyotashi 100nen no moridukurikousou (100 years strategy of forest management in Toyota city), Toyota City (in Japanese)

Chapter 25
Conclusion: Towards Sustainable Labor Force Management of Landscapes

Hiroyuki Shimizu, Chika Takatori and Nobuko Kawaguchi

Abstract This chapter summarizes the key concepts of this book. In the shrinking and aging societies throughout Japan, the issues of landscape management are serious. In this book, the concept of landscape management labor accounts is proposed toward a comprehensive and integrated understanding of labor force management of the ecological elements and units in different landscapes from a small to large scales. The framework of landscape management labor accounts will provide a good viewpoint for analyzing current problems and determining future solutions that involve the cooperation of various stakeholders. In the future, collaborative development of functional studies on ecological services, structural studies on green infrastructure, and affordability studies on landscape management labor accounts might provide an integrated planning and design process with not only the aspects of ecological values considered but also economic and social values in a sustainable and resilient manner as well.

Keywords Green infrastructure · Ecosystem services · Future perspectives

H. Shimizu · C. Takatori (✉) · N. Kawaguchi
Graduate School of Environmental Studies, Nagoya University,
Furo-Cho, Chikusa-Ku, Nagoya, Aichi 464-8601, Japan
e-mail: takatori@corot.nuac.nagoya-u.ac.jp

H. Shimizu
e-mail: shimkankyou@msn.com

N. Kawaguchi
e-mail: nobqou@gmail.com

H. Shimizu et al. (eds.), *Labor Forces and Landscape Management*,
DOI 10.1007/978-981-10-2278-4_25

25.1 What Has Been Clarified Though the Consideration of Landscape Labor Accounts?

In this book, the concept of landscape management labor accounts are proposed for the comprehensive and integrated understanding of labor force management of ecological elements and units in different landscapes ranging from small-scale to a large-scale.

25.2 Issues and History of Policies of Various Landscape Types

In Part 2, current issues and a history of policies and institutional frameworks of various landscape types were explored.

Chapter 3 discussed the issues of urban landscape types and policies and institutional frameworks in Japan, demonstrating the need to examine regional spatial calculations of conservation costs and how these play an important role. This chapter also explored how the need for green infrastructure development in urban areas, including urban centers, inner-cities, and suburbs will require intervention of private lands, due to scarce public financial resources.

Chapter 4 showed the issues of urban periphery landscapes in Japan, which is often threatened by urban sprawl, land abandonment, or unmanaged land uses resulting from loose land use regulations. Farmland is protected, on some level, through the easement of land taxes, but mostly it has succumbed to urban sprawl. The remaining forests, former coppiced woodlands, are attracting public attention as Satoyama, but applying appropriate management practices is a struggle. Urban peripheries face a host of challenges, but need to be re-designed in order to effectively respond to depopulation and city shrinkage.

Chapter 5 described the historical shift of agricultural landscape types in Japan by reviewing causes of agricultural land abandoned through the two perspectives of "Socio-economic causes" and "Biophysical causes". The chapter explored Japanese land improvement projects and the borrowing and lending of farmland since WWII and refers to recent trends in agricultural land abandonment and extensification in Japan. In addition, by comparative research on the direct payment system between Japan and the EU, we reviewed the history and research of direct payment schemes in agricultural and rural development policies in Japan.

Chapter 6 presented current issues revolving around different forest landscape types. Because of high demand for wood materials during and after WWII, expansive afforestation had been promoted during the 1950s and 60s in Japan. With economic growth in Japan, higher wood demands continued, however, imported wood had been used given its lower price and the fact that forestry was declining. The accumulation of forest volume has been increased three times during the past 50 years. Recently, forest management has been accelerated with higher inputs of subsidy. Forest soils support ecosystem service for carbon storage, nutrient cycling, biodiversity and controlling water quality, but can be degraded without sustainable forest managements.

Chapter 7 introduced structural and ecological functional changes of riverscapes, including water bodies in agricultural landscapes, with socioeconomic backgrounds in Japan. Riverscapes, which are characterized by hydrogeomorphic processes of rivers in each geomorphic division, dramatically change along the longitudinal axis of rivers from mountain to sea. The longitudinal diversity of riverscapes, coupled with an old agricultural system, provides essential habitat for riverine and floodplain organisms and supports freshwater biodiversity. However, socio-economic growth led to intensive alterations of freshwater habitat started during the Meiji era for rivers and the early Showa era for agricultural lands, in accordance with innovations in civil engineering technology. This chapter also provides a general understanding of the present state of freshwater habitat in Japan.

The previous chapters in Part 2, clarify that each landscape type faces specific challenges, however, they share common problems in terms of landscape management in addition to being deeply interdependent. Thus, it is important to find the integrated solutions to the problems, rather than solve each problem individually.

25.3 Method of Analysis and Possibility of the Usage of Landscape Management Labor Accounts

The case studies of various landscape types detailed in Part 3, including the method of the analysis and the possibility of the usage of landscape management labor accounts at a landscape management complex unit level. From the diversity of annual personal labor accounts and densities, several basic working style types were extracted. A purified pursuit is not always the best approach when it comes to sound and sustainable management of landscapes. It is important to note that some essential working styles are not centered on efficiency; however, they are integral as they maintain the robustness and resiliency of the local community or the happiness and health of the local people.

Annual labor accounts also vary with the environmental conditions of the lands. In the hilly Satoyama area, landscape labor accounts tend to be higher than those of the flatlands, as steep land gradients require greater laborious efforts. Another example is that small divided paddy fields with channels covered by grasses, such as in the old Jori-sei type, need elaborative work compared with modern improved paddies; but the tradeoff is the former can provide great biodiversity. The government and public alike, however, generally ignore the need for extra labor in order to maintain these types of functions. The calculation of landscape labor accounts can clarify the additional labor amounts for the kinds of economic externalized costs required in sustaining nature.

In areas where populations are both shrinking and aging, the capable labor force for managing landscapes might also decrease. There are some landscapes in which biodiversity or ecosystem services can remain or increase sustainably without the intervention of human beings or management. Such landscapes can be abandoned positively. However, there also exist many landscapes by which biodiversity or ecological services must be human-maintained. Both Satoyama and paddy fields or urban secondary forests are cases requiring the latter. For the sustainable maintenance of such landscapes in this current shrinking period, new ways of allocating extra labor forces must be considered. For such a consideration, landscape management labor accounts can provide important base data. In Part 3, several future scenarios are studied. These types of scenario studies provide insight into options for appropriate future landscape management.

25.4 Neighborhood Landscape Complex Units as a Core of Local Landscape Management

Additionally, the concept of the neighborhood landscape complex units is an important proposal of this book. There are many kinds of landscape elements, such as forests, gardens, farmlands and so on, in each area identified as a local unit by local communities or governments. Usually these different landscape elements are not associated with each other and not recognized as a whole, despite the fact that they do act in a mutually complementary manner. However, for the sustainable and resilient management of landscapes, ecological environments in a targeted district must be treated as one integrated unit and be preserved under an integrated policy. This way of thinking of landscape management labor accounts for a neighborhood landscape complex unit can also provide an integral vision for such treatments on a local scale.

In this book, the importance of not only land management but also of sea management is proposed. Nowadays, ecological environments under the sea surface are being degraded due to excessive harvesting or global warming. These environments, however, in order to survive, must be maintained by local communities, fishermen, or volunteers. These activities are normally economically externalized and ignored by the public or elected officials. The consideration of landscape management labor accounts for a coastal neighborhood landscape complex unit, however, can show the weight this kind of labor provides and the clear need for developing a new environmental management system that includes seas.

As an evaluation of landscape management within a neighborhood landscape complex unit, four important indexes are proposed, i.e., total (annual) labor account (TLD), gross labor density (GLD), net labor density (NLD) and social labor intensity (SLI). Total labor accounts include the annual amount of all labor needed for the maintenance of all (ecological) landscape elements within a targeted landscape unit or neighborhood landscape complex unit. Gross labor density is the total (annual) labor account per one are of the whole area. Net labor density is the total (annual) labor account per one are of the total area of the ecological landscape elements, i.e. green covered spaces, in the unit. It is important to note that both labor densities, GLD and NLD, are reduced according to the management area ratio, if there are some parts, which are not managed, in an element, such as the forest examples shown in Chaps. 14 and 15. If the value of NLD is high, even though the value of GLD is low, it means that there are many uncovered areas with some green elements, but that these isolated greenspaces, require high levels of management. If NLD and GLD are nearly equal, then the area is covered mostly by greenspace requiring management to some extent. SLI shows the social burden of the landscape management of the unit or the complex unit, because SLI is the TLA per population. In Satoyama and agricultural landscapes, the amounts of SLI tend to present high values. In these areas, the burden will rapidly increase if the population shrinks. It means that these areas are highly vulnerable when it comes to ecological landscape management. On the contrary, in urban areas, SLI is low and has strong resiliency within this context. If there is an area with high SLI near one displaying low SLI, there might be the possibility of sending extra management labor from the low SLI area to the high SLI area through volunteers or a professional labor force. Such a combination might be considered by the landscape labor accounts analysis.

25.5 Integration on Larger Scales

The results shown in Chap. 17 are not correct enough because of the uncertainty of the identification of labor densities in each landscape element. More sophisticated research are needed.

But the landscape management labor accounts analysis can provide the seamless connection of the landscape management thinking from a small scale to a large scale. For the efficient and resilient management of landscapes, wider consideration, beyond municipalities or regions, is needed. The landscape management labor accounts can act as an information base of the collaboration of the municipalities or regions. More elaborate research are needed for such kind of decision making among municipalities and regions.

25.6 Future Prospect of Planning Each Landscape Type by Using the Concept of Landscape Management Labor Accounts

Part 4 discussed the characteristics of labor accounts, including economic and biodiversity conservation and ownership and management of areas in each landscape type. The findings showed future prospects of spatial planning for each landscape type. In the case of shrinking and hyper-aging communities, landscape management issues are grave. The framework of landscape management labor accounts provides a good viewpoint for analyzing the problems and providing appropriate solutions through cooperation among various stakeholders.

As for the urban landscape, public money is consumed by medical and welfare-related projects, leaving scarce resources for landscape management, including the new development of parks as well as the maintenance of existing parks as urban facilities. Since there is a clear limitation, urban greening policies are now shifting toward increasing greenspaces along public-owned roads and on private property. Many cities are envisioning networked compact urban forms that result in the increase of population and building density of urban centers and areas surrounding public transit stations. Furthermore, these visions predict a decrease in density for areas under-served by public transit. The concept of landscape labor accounts plays an important role in both intensifying and de-intensifying urban areas. In the former, labor could be paid through urban redevelopment or by urban dwellers themselves. In the latter, labor required for maintaining the increasing quantity of greenspace should be paid through a new system since the remaining number of decreasing dwellers would not be able to afford the maintenance of vacant properties around them. Unless this system is established, the vision of the networked compact urban form is unjust.

The urban-rural mix typifying Japanese urban peripheries, while often recognized as a failure of modern urban planning, is the result of urban-rural divisions by zoning, one of the urban planning principals from the city expansion period. This chapter discusses a new planning concept for the city shrinkage period, which uses an urban-rural mix as a potential tool for enhancing landscape sustainability. The discussion's starting point examines what will happen if current trends continue without proactive policy intervention or anticipating worse case scenarios. The chapter looks at preferable future goals and measures required to obtain them. The chapter concludes that integrating urban and rural, not dichotomizing them, is key to overcoming conventional modern urban planning problems of the 20th century. As for future rural planning, the appropriate policies, social systems, and initiatives for sustainable management of agriculture and rural areas are also needed.

Regarding forest landscapes, because of lower timber prices, high costs for management to small-scale forest owners, there has been an increase in under-managed plantation forests, which could lead to ecosystem disservices such as soil degradation and landslides. Forest planning for plantations depends on government subsidies. The trends in use of the subsidy have moved from afforestation in the 1950s to management such as weeding and thinning, and to wood production to enhancing self-sufficiency. Ecosystem disservices can be caused by deforestation for use of wood biomass materials and increased populations of Sika deer in the rural areas as a result of shrinking and aging populations. To conduct sustainable forest planning in future, human resource development must include hiring experts, such as "Foresters", who hold special knowledge and skills in forest management for not only regional but also global scales.

As for the management of inland water systems, the ecosystem deterioration due to the anthropogenic alterations is cause for concern. Inland water systems were originally composed of rivers, streams, ponds, and floodplains, and included paddies, canals and reservoirs for agricultural irrigations, hydropower generation, flood mitigation, and other water uses. These paddies, canals and reservoirs have functioned as secondary natures, compensating for floodplain-dependent habitats. Deterioration processes and measures for sustainable ecosystem management are reviewed using conventional literature for each of the landscapes. According to findings, we provided overviews of management approaches for each of the inland waterscapes with particular reference to ecosystem rehabilitation. Furthermore conserving ecosystem sustainability in some of the landscapes was also discussed.

Labor accounts along coastal areas encourage natural restoration and regional management of linked ecosystems while promoting coastal economy and biodiversity conservation and ownership and management of coastal areas. Suggestions for future labor accounts in coastal areas facing shrinking populations include a number of strategies. The first approach includes reducing general labor accounts and their funding, and using those resources, instead, for more fundamental labor accounts and projects which deal with core causes of coastal environmental degradation. Second, investing time and effort into initial labor costs for restoration activities. Third, determining which ecosystems are important and therefore require protection. Next, taking coastal decision-making into account for regional and watershed zone environments. Funding coastal ecosystem labor through taxes and donations from cities and upstream areas. Lastly, maintaining and designing environmental management activities which ordinary citizens consider enjoyable and interesting.

Chapter 24 compared the relationship between Landscape management (LM) and ecosystem services (ESs). A proper LM should have two aspects, namely pros and cons, from the perspective of biodiversity and ESs. Improper LM may cause negative impacts on the ESs and ecosystems. On the other hand, once an LM is done properly, several positive benefits can be derived from it. The trade-off between the impacts on wildlife habitat and the ES provisions made by the LM also needs to be considered. Currently, LM has been reviewed for many fields. However, information about the order of priority of the increase of ESs and the decrease of ecosystem disservices in this area is not readily available. This section also describes a simple spatial ES assessment for an urban area, namely Nagoya City, which was conducted using the method for estimating ES potential supply and land use priority, by taking LM into consideration.

25.7 Green Infrastructure, Ecosystem Services, and Landscape Management Labor Accounts

Chapter 17 discussed the relation between ecosystem services and landscape management labor accounts. The combined evaluation of both indexes might be for the identification of areas qualifying for the ecological conservation through human intervention. This chapter further extended consideration between ecosystem services, green infrastructure, and landscape management labor accounts.

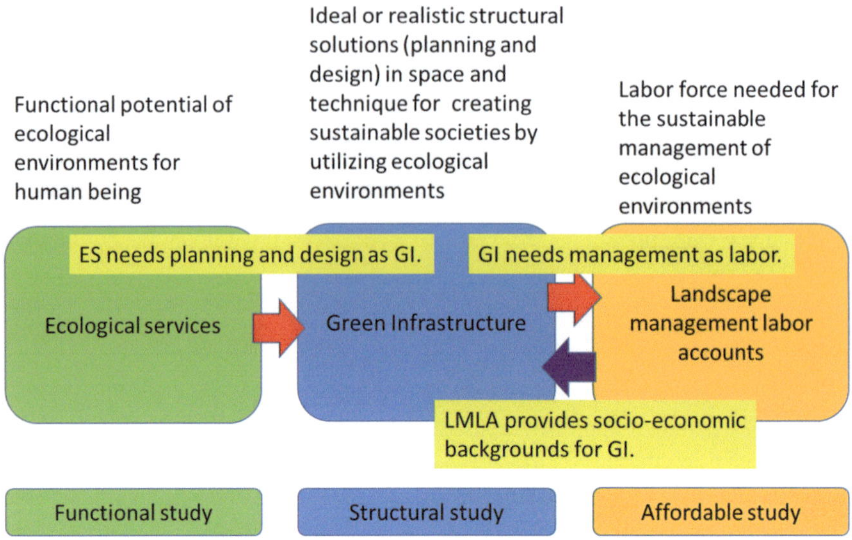

Fig. 25.1 Relation between ecosystem services and landscape management labor accounts

Figure 25.1 shows a diagram comparing studies on ecosystem services, green infrastructure, and landscape management labor accounts. Benedict and McMahon (2006) proposed the concept of green infrastructure. It is an integrated planning tool based on structural analysis, such as allocation and networks of greenspace. On the other hand, studies on ecological services are good at the functional evaluation of greenspace. Our proposal of landscape management labor accounts is good at the evaluation of sustainability or resiliency of greenspace from economic and social viewpoints. How much labor can be afforded in the targeted district or region for the appropriate management of landscapes? What kind of association or cooperative organization is needed in a community for the appropriate management? How much burden for labor can be afforded in a community for the appropriate management? What kind of the political or economic systems or measures should be arranged for the appropriate management? The concept of landscape management labor accounts can provide the base for these types of considerations.

In other word, landscape management labor accounts provide a means to study affordability of managing greenspace or landscapes. Collaborative development of functional studies on ecological services, structural studies on green infrastructure, and affordable studies on landscape management labor accounts might provide integrated planning and design approaches with not only the aspects of ecological values but also economic and social values in sustainable and resilient manners.

Reference

Benedict MA, McMahon ET (2006) Green infrastructure. Island, Washington

Erratum to: Labor Forces and Landscape Management

Hiroyuki Shimizu, Chika Takatori and Nobuko Kawaguchi

Erratum to: H. Shimizu et al. (eds.), *Labor Forces and Landscape Management*, DOI 10.1007/978-981-10-2278-4

In Chap. 2, Fig. 2.49 caption: 1975 is changed to 2005; Chap. 5, Fig. 5.2 caption: rations is changed to ratios; Fig. 5.5 caption: (Some parts of this figure were altered) text added.

The updated original online version for book can be found at DOI 10.1007/978-981-10-2278-4_2 and DOI 10.1007/978-981-10-2278-4_5

H. Shimizu (✉) · C. Takatori · N. Kawaguchi
Nagoya University, Nagoya, Aichi, Japan
e-mail: shimkankyou@msn.com

C. Takatori
e-mail: takatori@corot.nuac.nagoya-u.ac.jp

N. Kawaguchi
e-mail: nobgou@gmail.com

H. Shimizu et al. (eds.), *Labor Forces and Landscape Management*,
DOI 10.1007/978-981-10-2278-4_26

MIX
Papier aus verantwortungsvollen Quellen
Paper from responsible sources
FSC® C105338

If you have any concerns about our products,
you can contact us on
ProductSafety@springernature.com

In case Publisher is established outside the EU,
the EU authorized representative is:
Springer Nature Customer Service Center GmbH
Europaplatz 3, 69115 Heidelberg, Germany

Printed by Libri Plureos GmbH
in Hamburg, Germany